Lecture Notes in Mathematics 1565

Editors:
A. Dold, Heidelberg
B. Eckmann, Zürich
F. Takens, Groningen

Subseries: Fondazione C.I.M.E., Firenze
Adviser: Roberto Conti

T0236578

L. Boutet de Monvel C. de Concini
C. Procesi P. Schapira M. Vergne

D-modules, Representation Theory, and Quantum Groups

Lectures given at the 2nd Session of the Centro Internazionale Matematico Estivo (C.I.M.E.) held in Venezia, Italy, June 12-20, 1992

Editors: G. Zampieri, A. D'Agnolo

Springer-Verlag
Berlin Heidelberg New York
London Paris Tokyo
Hong Kong Barcelona
Budapest

Authors

Louis Boutet de Monvel
Département de Mathématiques, Université Paris VI
2, Place Jussieu, F-75252 Paris, France

Corrado De Concini
Scuola Normale Superiore
Piazza dei Cavalieri, 7, I-56126 Pisa, Italy

Claudio Procesi
Dipartimento di Matematica, Università "La Sapienza"
Piazzale Aldo Moro, 5, I-00185 Roma, Italy

Pierre Schapira
Université de Paris XIII, CSP Mathématique
F-93430 Villetaneuse, France

Michèle Vergne
Ecole Normale Supérieure
45, rue d'Ulm, F-75005 Paris, France

Editors

Giuseppe Zampieri
Andrea D'Agnolo
Dipartimento di Matematica, Università di Padova
Via Belzoni, 7, I-35131 Padova, Italy

Mathematics Subject Classification (1991): Primary: 58, 17 Secondary: 22

ISBN 3-540-57498-0 Springer-Verlag Berlin Heidelberg New York
ISBN 0-387-57498-0 Springer-Verlag New York Berlin Heidelberg

Printed in Germany

Typesetting: Camera-ready by author/editor
46/3140-543210 - Printed on acid-free paper

PREFACE

In this CIME session we aimed at proposing to a wide audience the language and the techniques of various fields (such as D-modules, K-theory, equivariant cohomology, microlocal study of sheaves and quantization of groups) converging to recent common results on index theorems and representation theory.

The contents of the lectures can be roughly divided into two groups:
D-modules and microlocal study of sheaves (P. Schapira and J.-P. Schneiders), K-theory and D-modules (L. Boutet de Monvel), G-equivariant cohomology (N. Berline and M. Vergne)
The theory of quantum groups, with particular emphasis to the quantization of some remarkable Poisson groups (C. De Concini and C. Procesi).

The last ingredient of the school was the magic of Venice.

A. D'Agnolo
G. Zampieri

Paris, March 19th, 1993

This CIME Session was held at Ca Dolfin under the sponsorship of the University of Venice. The organizers wish to heartily acknowledge the President of the University of Venice, Prof. G. Castellani, for the excellency of his hospitality.

TABLE OF CONTENTS

INDICE DES SYSTÈMES DIFFÉRENTIELS

L. Boutet de Monvel

Dans ces notes nous nous proposons de décrire la formule d'indice d'Atiyah et Singer pour les systèmes d'opérateurs différentiels et son extension au cas relatif due à Malgrange et l'auteur. Cette formule décrit le nombre de solutions d'un système d'équations différentielles — ou plus exactement, parce qu'il s'agira pratiquement toujours de complexes elliptiques d'opérateurs différentiels, la caractéristique d'Euler, somme alternée de nombres de Betti de la cohomologie du système — en termes d'invariants topologiques plus simples attachés à ce système. Il y a une version absolue du théorème d'indice, où l'indice est un nombre entier, et des versions avec paramètres, où l'indice est un fibré virtuel sur l'espace des paramètres, ou avec action de groupe compact où l'indice est une représentation du groupe, ou relative où l'indice est un nouveau système différentiel dont on décrit des invariants topologiques en fonction de ceux de l'ancien. L' énoncé que nous décrivons ici tient compte en outre plus explicitement des supports, en s'inspirant de la définition du micro-support des faisceaux constructibles et de la description de la formule de l'indice qu'en déduit P.Schapira dans ses conférences.

§1. Introduction, description du problème

Formellement l'indice d'un complexe d'opérateurs linéaires est défini comme suit: on se donne un complexe, c'est à dire une suite

$$a : \cdots \longrightarrow E_k \xrightarrow{a_k} E_{k+1} \longrightarrow \cdots$$

d'espaces vectoriels complexes E_k et d'applications linéaires $a_k \in L(E_k; E_{k+1})$, telle que $a_{k+1} \circ a_k = 0$ pour tout k (une application linéaire $a: E_0 \to E_1$ s'identifie à un complexe de longueur 2, concentré en degrés 0, 1). La cohomologie de a est le complexe de différentielle nulle $H(a) = \ker a / \operatorname{Im} a$ ($H^k(a) = \ker a_k / \operatorname{Im} a_{k-1}$). Si $H(a)$ est de dimension finie, i.e. si les $H^k(a)$ sont de dimension finie, et nuls sauf pour un nombre fini d'indices $k \in \mathbf{Z}$, l'indice (ou caractéristique d'Euler) de a est la somme alternée:

$$\operatorname{Ind} a = \sum (-1)^k \dim H^k(a)$$

(on a $\operatorname{Ind} a = \sum (-1)^k \dim E^k$ si le complexe a est lui même de dimension finie)

Les complexes dont on s'occupera ici ne sont pas abstraits mais ont une origine ou une signification géométrique: ce sont des complexes d'opérateurs différentiels sur une variété et la formule d'indice d'Atiyah et Singer relie leur indice à des invariants géométriques de a.

La formule d'indice pour un système elliptique d'équations aux dérivées partielles a son origine dans le théorème de Riemann-Roch de la géométrie complexe, plus précisément sous la forme que lui ont donné Hirzebruch, et Grothendieck dans le cas relatif. Voici une description de cette formule, en termes de K-théorie, résumant celle de Baum, Fulton, Mac Pherson: soient X un espace analytique projectif complexe, et $Z \subset X$ une partie compacte. A tout faisceau cohérent M de $\mathcal{O}_X$-modules à support dans Z est associé un élément $[M]$ du groupe de Grothendieck $K_Z^{\mathrm{an}}(X)$, qui est le groupe engendré par les classes d'isomorphismes $[M]$ de $\mathcal{O}_X$-modules cohérents à support dans Z et les relations $[M] = [M'] + [M'']$ s'il existe une suite exacte $0 \to M' \to M \to M'' \to 0$. Il existe un homomorphisme canonique $K_Z^{\mathrm{an}}(X) \to K_Z^{\mathrm{top}}(X)$, où $K_Z^{\mathrm{top}}(X)$ est le groupe d'Atiyah des "fibrés vectoriels virtuels à support dans Z", qui décrit les propriétés additives et invariantes par déformation des complexes de fibrés vectoriels topologiques exacts en dehors de Z (ce groupe sera défini plus en détail au §2). Suivant Baum, Fulton, Mac Pherson, le théorème de Riemann-Roch relatif exprime que cet homomorphisme commute aux images directes propres (il commute aussi aux images inverses). Ce théorème doit être complété par la description de l'image directe K-théorique, qui est définie au moyen du théorème de périodicité de Bott. On exprime aussi le théorème d'indice en termes de cohomologie au moyen de l'isomorphisme de Chern:

$$ch : K_Z^{\mathrm{top}}(X) \otimes \mathbf{Q} \longrightarrow H^{\mathrm{pair}}(X, \mathbf{Q})$$

qui permet de traduire en termes de cohomologie ce qui ne dépend pas de la torsion. Le théorème de Riemann-Roch "classique" de Hirzebruch est la traduction en termes cohomologiques (via le caractère de Chern) du cas particulier de l'énoncé ci-dessus lorsque le but est un point ($K(\text{point}) = Z$).

Les fonctions holomorphes sont les solutions du système elliptique des équations de Cauchy-Riemann et on sait que pour celles-ci, les théorèmes de finitude sur une variété compacte sont un cas particulier d'un théorème de finitude analogue pour les solutions des systèmes elliptiques d'équations aux dérivées partielles linéaires (la définition sera rappelée plus loin); il était alors naturel de chercher à généraliser aux systèmes elliptiques la formule de Riemann-Roch de Hirzebruch. Les objets qui nous intéresseront d'abord ici sont les systèmes elliptiques d'équations différentielles linéaires sur une variété compacte, ou un peu plus généralement les systèmes dont l'espace des solutions est de dimension finie.

Un système, ou complexe, d'équations différentielles linéaires peut être décrit comme suit: on se donne une variété X, complexe, ou réelle de classe C^∞, une suite de fibrés vectoriels complexes $E = (E_j)$, $j \in \mathbf{Z}$ ($E_j = 0$ sauf pour un nombre fini d'indices j), et une suite $P = (P_j)$ d'opérateurs de type $E_j \to E_{j+1}$ telle que $P_{j+1} \circ P_j = 0$ pour tout j (nous considérerons $(E, P) = ((E_j), (P_j))$ comme un objet gradué et écrirons simplement $P^2 = 0$). Un opérateur P de type $E_0 \to E_1$ peut toujours être considéré comme un complexe de longueur 2. On considère alors le faisceau gradué de cohomologie de P : $H(P) = \ker(P)/\operatorname{Im}(P)$ ($H^j(P) = \ker P_j / \operatorname{Im} P_{j-1}$) et leurs espaces de sections globales. Par exemple si P est un opérateur de type $E_0 \to E_1$, assimilé à un complexe de longueur 2, $H^0(P) = \ker P$ est l'espace des solutions de l'équation $Pf = 0$, et l'espace $H^1(P)$ des sections de $E_1 \mod \operatorname{Im} P$ mesure l'obstruction à la résolution de l'équation $Pf = g$ (il est aussi étroitement lié à l'espace des solutions de l'équation adjointe $P^*g = 0$).

Les complexes d'opérateurs qui se présentent naturellement sont souvent localement exacts, et ont été construits pour calculer la cohomologie, au sens géométrique, de X à coefficients dans le faisceau des solutions d'un système d'équations différentielles (parce que le faisceau des fonctions C^∞ ou analytiques sur une variété C^∞ ou analytique réelle est cohomologiquement trivial). Ainsi sur une variété X réelle, le faisceau constant $\mathbf{C}_X$ (dont les sections sont les fonctions localement constantes) est le faisceau des solutions de l'équation $df = 0$, et la cohomologie $H^*(X, \mathbf{C})$ est celle du complexe de De Rham de la différentielle extérieure opérant sur les formes différentielles; si X est une variété complexe, les fonctions holomorphes sont les solutions des équations de Cauchy-Riemann: $\overline{\partial} f = 0$, et la cohomologie $H^*(X, \mathcal{O}_X)$ à coefficients dans le faisceau $\mathcal{O}_X$ des fonctions holomorphes est celle du complexe de Dolbeault, de la différentielle extérieure antiholomorphe $\overline{\partial}$. Le Laplacien, ou plus généralement l'opérateur de Laplace-Beltrami opérant sur les formes différentielles d'une variété Riemannienne, est un exemple plus ancien d'opérateur elliptique; il est subordonné à l'exemple précédent: $\Delta = -d\,d^* + d^* d$.

On définit le symbole d'un opérateur, ou d'un complexe d'opérateurs différentiels: si P est d'ordre m son symbole est le polynôme σ_P homogène de degré m qu'on obtient en ne gardant que les termes de plus haut degré m lorsqu'on écrit P comme polynôme des dérivations $\partial/\partial x_j$, dans n'importe quel système de coordonnées. On a, pour toute fonction C^∞ φ et toute section u de E

$$e^{-t\varphi} P(e^{t\varphi} u) = t^m \sigma_P(d\varphi) u + O(t^{m-1})$$

plus généralement si (E, P) est un complexe d'opérateurs différentiels, σ_P s'interprète comme un complexe de fibrés vectoriels sur le fibré cotangent T^*X, de type $\pi^{-1}E \to \pi^{-1}E$, où π est la projection $T^*X \to X$ et $\pi^{-1}E$ désigne le relèvement de E à T^*X (dans la suite nous écrirons E au lieu de $\pi^{-1}E$). Si X est une variété réelle, on dit alors que P est elliptique si pour tout covecteur imaginaire pur $\xi \neq 0$, le symbole $\sigma_P(\xi)$ est une suite exacte d'espaces vectoriels[1]. L'ellipticité est une propriété simple d'un système P qui, jointe à une condition de compacité du domaine sur lequel on calcule ces solutions (par exemple que la variété de base soit compacte), assure que les groupes de cohomologie sont de dimension finie. La dimension individuelle de chacun des groupes de cohomologie $H^j(P)$ d'un complexe elliptique P n'est pas stable par déformation ou par petite perturbation, et peut être difficile à calculer; mais l'indice, ou caractéristique d'Euler: $\chi(P) = \sum(-1)^j \dim H^j(P)$, est très stable, de même que la condition de finitude (ellipticité). Pour un système elliptique P il est alors naturel d'imaginer que l'indice ne dépend que de caractères topologiques plus grossiers du symbole σ_P, et est donné par une extension de formule analogue à la formule de Riemann-Roch. Les deux exemples ci-dessus (d et $\overline{\partial}$) sont elliptiques et le théorème de Riemann-Roch de la géométrie complexe apparaîtra ainsi comme cas particulier du théorème d'indice pour les systèmes elliptiques, appliqué à $\overline{\partial}$.

Comme nous l'avons mentionné la formule de l'indice pour les systèmes elliptiques est très stable par perturbation ou déformation, aussi n'est il pas essentiellement restrictif de se limiter aux systèmes elliptiques à coefficients analytiques: toute variété paracompacte X[2] possède une structure analytique réelle, et tout système elliptique sur X peut être déformé en un système à coefficients analytiques (une telle isotopie préserve l'indice, et, pour l'essentiel, les objets topologiques évoqués ci-dessus). Dans toutes ces notes nous supposerons donc que les variétés considérées sont analytiques, et nous considérerons toujours une variété analytique réelle comme germe d'une variété complexe au voisinage d'une sous-variété U totalement réelle. Un inconvénient est que ce point de vue est plus rigide, et certaines constructions ou déformations topologiques sont moins faciles que dans le cadre C^∞, mais elles ont le plus souvent un analogue grâce au fait que le faisceau des fonctions

---

[1] La définition complète qu'on reverra plus loin est la suivante: P est elliptique si la variété caractéristique $\operatorname{car} P$ ne contient pas de covecteur imaginaire pur (ou réel) non nul; c'est le cas si $\sigma_P(\xi)$ est exact pour ξ imaginaire pur non nul.

[2] Dans la suite les variétés que nous considérerons seront toujours paracompactes et ce mot sera omis.

analytiques réelles est cohomologiquement trivial. Cet inconvénient est largement compensé par le fait que les systèmes d'équations différentielles à coefficients analytiques sont commodément décrits par la théorie des $\mathcal{D}$-modules, qui est décrite au §3. On dispose pour les $\mathcal{D}$-modules analytiques de notions raisonnables de finitude ou de cohérence, et d'image directe ou inverse par une application analytique, qui n'existent pas pour les systèmes à coefficients C^∞. La notion d'image directe, que nous rappelons au §3, est essentielle dans notre description du théorème d'indice relatif. En particulier l'image directe par immersion fermée (plongement) permettra de plonger n'importe quel système dans une variété plus simple telle que $\mathbf{R}^n$ ou un espace projectif, de façon plus naturelle que dans l'article original d'Atiyah et Singer.

L'ellipticité n'est pas préservée par plongement, et on est amené à en généraliser la définition, en introduisant, suivant l'idée de B. Malgrange, la notion de presqu'ellipticité qui est décrite au §4. La presqu'ellipticité est stable par image directe et donne lieu aux mêmes théorèmes de finitude, et au même théorème d'indice que l'ellipticité. Mais elle s'applique dans plus de cas; en particulier un $\mathcal{D}$-module holonôme est toujours prequ'elliptique, et le théorème d'indice pour les modules holonômes est ainsi démontré dans le même cadre que celui des systèmes elliptiques.

Jusqu'à présent nous n'avons parlé que de solutions C^∞ ou analytiques réelle; on peut aussi étudier les solutions distributions, ou plus généralement à coefficients dans un faisceau F convenable. Suivant Schapira et Schneiders il sera commode d'incorporer au faisceau des coefficients la description du domaine dans lequel on calcule les solution, qui ne sera pas toujours X tout entier, ce qui sera particulièrement utile sur une variété X complexe où on est encore plus amené à varier le domaine. Dans les notes de Schapira-Schneiders, $F = \mathcal{O}_X \otimes \phi$ est associé à un faisceau réel-constructible ϕ. Ici en nous inspirant du travail de M. Ohana, nous nous limiterons au cas plus simple où F est le faisceau des germes de fonctions holomorphes au voisinage d'une sous-variété à bord (ou à coins) de X.

La formule de l'indice comportera ainsi plusieurs éléments:

1) un objet topologique décrivant des propriétés additives et invariantes par déformation d'un système différentiel elliptique P. Dans ces notes cet objet sera un fibré virtuel (à support) $[P]$, élément de $K_{\operatorname{car} P}(T^*X)$, où $\operatorname{car} P$ est la variété caractéristique de P; on peut le remplacer par une classe de cohomologie à support, grâce au caractère de Chern comme mentionné plus haut.
2) un objet topologique $[F]$ décrivant de façon analogue les propriétés du faisceau de fonctions ou distributions F des coefficients dans lesquels on calcule les solutions. Ici, $[F]$ sera un élément de $K_{\mathrm{SS}\,\phi}(T^*X)$ où $\mathrm{SS}\,\phi$ est le micro-support de ϕ défini par Kashiwara et Schapira.
3) une formule universelle produisant à partir de ces données, et moyennant une condition supplémentaire de compacité, un nombre (ou plus généralement un fibré virtuel ou une classe de cohomologie sur l'espace des paramètres, dans le cas relatif). La condition de compacité est que $\operatorname{car} P \cap \mathrm{SS}\,\phi$ soit compact, ce qui complète la condition d'ellipticité: $\operatorname{car} P \cap \mathrm{SS}\,\phi \subset \{0\}$ (on a noté $\{0\}$ la section nulle). Le produit $[P][F]$ est alors bien défini, à support compact et le théorème d'indice affirme que l'indice de P (à coefficients dans F) en est l'image K-théorique (définie au §2).

Le premier ingrédient de notre démonstration est la K-théorie, qui est décrite au §2. Le deuxième ingrédient est la théorie des $\mathcal{D}$-modules. La presqu'ellipticité et le théorème d'indice dans la cas absolu sont décrits au §4, et le cas relatif au §5.

§2. K-THÉORIE ET OPÉRATEURS DE TOEPLITZ

2.1 Rappels de K-théorie.

a. Définitions.

Soient X un espace topologique paracompact, $Z \subset X$ une partie fermée (ou plus généralement une famille de supports[3]). Nous renvoyons à Atiyah ou Steenrod pour la définition des fibrés vectoriels sur X. Nous nous intéressons aux morphismes $a : E \to F$ de fibrés vectoriels complexes sur X tels que a soit un isomorphisme en dehors de Z, et plus généralement aux complexes de fibrés vectoriels exacts en dehors de Z. On note $K_Z(X)$ le groupe des classes d'équivalence $[a]$ de tels complexes a pour la relation d'équivalence engendrée par les relations:

(i) $[a] + [b] = [a \oplus b]$

(ii) $[a] = 0$ s'il existe une déformation exacte en dehors de Z de a sur un complexe exact.

On note encore $K(X) = K_X(X)$. Si a et b sont des complexes de fibrés on a $[a] = [b]$ si et seulement s'il existe un complexe exact c et une déformation exacte en dehors de Z de $a \oplus c$ sur $b \oplus c$; en particulier on a $[a] = [b]$ si a et b sont quasi-isomorphes, en particulier s'il existe un morphisme $a \to b$ qui induise un isomorphisme en cohomologie. Muni de la loi ci-dessus $K_Z(X)$ est un groupe; on a $-[a] = [a^*] = [a(1)]$ où $a(1)$ désigne le complexe décalé (a_{k-1}) si $a = (a_k)$ et a^* désigne l'adjoint de a (pour n'importe quelles métriques hermitiennes).

Le produit tensoriel de deux complexes a et b ($(a \otimes b)_k = \oplus_p (a_p \otimes 1 + (-1)^p 1 \otimes b_{k-p})$) est exact en tout point où un des deux facteurs l'est; par passage au quotient on définit ainsi une loi de produit bilinéaire $K_Z(X) \times K_{Z'}(X) \to K_{Z \cap Z'}(X)$ pour laquelle $K_X(X) = K(X)$ est un anneau et les $K_Z(X)$ sont des $K(X)$-modules.

Soit H un espace de Hilbert. On note $\mathrm{Fred}(H) \subset L(H)$ l'ensemble des opérateurs de Fredholm. Si $X \supset Z$ sont comme plus haut on note $F_Z(X)$ le groupe des classes d'homotopie de fonction continues $A : X \to \mathrm{Fred}(H)$ telles que A soit inversible en dehors de Z. Parce que $GL(H)$ est contractile (th. de N. Kuiper) $F_Z(X)$ s'identifie aussi au groupe des classes d'homotopie de familles continues d'opérateurs de Fredholm sur un fibré hilbertien (un tel fibré est toujours trivial). Une telle famille peut toujours être considérée comme un complexe de longueur 2 et de cette façon $F_Z(X)$ s'identifie aussi au groupe des classes d'homotopie de complexes de Fredholm hilbertiens exacts en dehors de Z (c'est à dire de complexes de fibrés hilbertiens dont la cohomologie est de dimension finie en tout point et nulle en dehors de Z; deux tels complexes sont équivalents s'il existe une déformation du premier sur le second, exacte en dehors de Z). De fait si

$$D : \cdots \longrightarrow H_k \longrightarrow H_{k+1} \longrightarrow \cdots \tag{2.1}$$

est un tel complexe, et qu'on note $[D]_Z$ (ou simplement $[D]$) sa classe dans F_Z, on montre élémentairement, par récurrence sur la longueur de D, qu'on a $[D] = [\partial(D)]$ si $\partial(D)$ est la famille d'opérateurs de Fredholm:

$$\partial(D) = D - D^* : \oplus H_{2k} \longrightarrow \oplus H_{2k+1}. \tag{2.2}$$

[3] On notera K_c la K-théorie à support compact. Atiyah suppose toujours que le support Z est compact (ou contenu dans la famille des parties compactes). Nous avons omis cette condition, inutile au niveau des définitions, et qui ne sera jamais réalisée dans ces notes où X est un fibré cotangent, et Z partie conique de X.

Un complexe de fibrés vectoriels de rang fini sur X est un cas particulier de complexe de Fredholm (on peut toujours rajouter un complexe hilbertien exact de dimension infinie). On a donc une application canonique du groupe $K_Z(X)$ des fibrés virtuels à support dans Z dans $F_Z(X)$. On montre élémentairement (cf. Jänisch) que si Z est compact, ou si X est métrique de dimension finie[4], cette application est bijective; c'est en particulier le cas si (X, Z) est une paire d'espaces analytiques ou sous-analytiques, ou une paire de CW-complexes, ce qui suffit largement pour les formules d'indice. Dans ce cas on note Ind_Z l'isomorphisme inverse $\mathrm{Ind}_Z(A) \to K_Z(X)$ l'élément correspondant à un complexe de Fredholm A exact hors de Z: $\mathrm{Ind}_Z(A) = [a]$ signifie qu'il existe une déformation B de A et un quasi-isomorphisme $a \to B$.

b. Produits.

La somme et le produit tensoriel de complexes munissent $F(X) = F_X(X)$ d'une structure d'anneau et $F_Z(X)$ d'une structure de $F(X)$-module qui prolongent celles de $K(X)$ et $K_Z(X)$ (on a $A \oplus B \sim A \circ B$, et $A \otimes B$ est un complexe de Fredholm si A et B le sont, et est exact en tout point où l'un des deux complexes A ou B l'est).

Pour $Z \subset Y \subset X$ (fermés) on a encore un produit $F_Z(Y) \otimes F_Y(X) \to F_Z(X)$ défini comme suit: si A (resp. B) est une famille continue d'opérateurs de Fredholm sur Y (resp. X) inversible hors de Z (resp. Y), on choisit un prolongement continu $\widetilde{A} : X \to L(H)$ de A; comme B est exact hors de Y, le produit $\widetilde{A} \otimes B$ l'est aussi, et en particulier c'est un complexe de Fredholm hors de Y; sur $Y - Z$ il est exact comme A; enfin il est Fredholm sur Z puisque A et B le sont; l'ensemble des prolongements de A est connexe (en fait convexe) donc la classe $[\widetilde{A} \otimes B]$ ne dépend pas du choix du prolongement $\widetilde{A}$.

c. Images inverses.

Soit $f : X' \to X$ continue, et $Z' \subset X'$ fermé, $f(X' - Z') \subset X - Z$. L'image inverse $F_Z(X) \to F_{Z'}(X')$ est définie par $f^{-1}[A] = [A \circ f]$. Si f est l'application identique d'un ouvert X' de X, et $Z = X$, f^{-1} est un isomorphisme (excision), parce que $GL(H)$ est contractile donc toute famille continue B d'opérateurs de Fredholm sur X' inversible hors de Z se déforme en une famille égale à 1 hors d'un voisinage de Z, qui se prolonge.

Si $U \subset X$ est un voisinage de Z, l'application de restriction $F_Z(X) \to F_Z(U)$ (image inverse par l'inclusion canonique $U \to Z$) est un isomorphisme, parce qu'une application $a : U \to L(H)$ égale à 1 en dehors d'un voisinage de Z se prolonge continûment par 1 sur $X - U$, (théorème d'excision).

2.2 Indice d'une famille d'opérateurs de Toeplitz. Isomorphisme de Bott (Thom-Gysin).

a. homomorphisme de Bott.

X désigne un espace topologique comme ci-dessus. Soit $p : N \to X$ un fibré vectoriel complexe sur X. On note k_N le complexe de Koszul de N[5]:

(2.3) $k_N : \cdots p^{-1} \wedge^{-k} N^* \to p^{-1} \wedge^{-k+1} N^* \to \cdots N^* \to \mathbf{C} \to 0$ (k_N est en degrés ≤ 0)

[4] $\dim X \leq d$ signifie ici que tout recouvrement ouvert peut être raffiné en un recouvrement pour lequel l'intersection de $d+2$ ouverts distincts est vide. Pour le deuxième isomorphisme il faut accepter pour la définition de K_Z des fibrés localement mais non globalement de rang borné.

[5] Le complexe de Koszul est aussi défini pour un fibré réel, mais ne donne plus lieu dans le cas réel à un "isomorphisme de Bott" (la période des groupes d'homotopie stables du groupe orthogonal réel est 8).

dont la différentielle au point $n \in N$ est le produit intérieur $\omega \to n \llcorner \omega$. Ce complexe est exact en dehors de la section nulle de N, que nous identifierons encore à X, et la multiplication par $[k_N] \in K_X(N)$ définit pour tout $Z \subset X$ fermé un homomorphisme

$$b_{N/X} : K_Z(X) \longrightarrow K_Z(N) \tag{2.4}$$

(comme on verra ci-dessous b_N ne dépend en fait que de la structure spin^c associée à la structure complexe de N, qui est bien définie à homotopie près).

Le théorème de périodicité de Bott (voir encore ci-dessous) implique que b_N est un isomorphisme lorsque $X = Z$ et X est compact. En fait c'en est toujours un lorsque X est de dimension finie; nous admettrons ce résultat, qui se démontre de façon élémentaire — la présence du support Z complique un peu la situation parce que $X - Z$ n'est pas compact même si X et Z le sont. L'inverse de l'homomorphisme de Bott est donné par l'indice des familles d'opérateurs de Toeplitz, que nous décrivons maintenant, et qui est un ingrédient essentiel de la formule de l'indice.

b. Indice d'une famille d'opérateurs de Toeplitz.

Dans ce qui suit nous supposons le fibré N trivial: $N = X \times \mathbf{C}^k$ (le cas général où N est un fibré vectoriel de rang fini quelconque se ramène au cas où N est trivial, du moment que N est complémenté, i.e. qu'il existe un fibré N' tel que $N \oplus N'$ soit trivial).

Soit B un ellipsoïde de $\mathbf{C}^k$, par exemple la boule unité, par exemple l'ellipsoïde B_ε défini par l'inégalité $|\operatorname{Re} z|^2 + \varepsilon^{-2} |\operatorname{Im} z|^2 \leq 1 + \varepsilon$ ($\varepsilon > 0$). Un opérateur de Toeplitz de degré 0 sur B est un opérateur linéaire continu sur l'espace $\mathcal{O}^0(B)$[6] des fonctions holomorphes de carré sommable sur B, de la forme $A = T_a+$ opérateur compact, où a est une fonction continue sur ∂B, T_a est l'opérateur $u \to B(au)$, et B le projecteur de Bergman $L^2(B) \to \mathcal{O}^0(B)$ (cf. Boutet de Monvel 1). Modulo les opérateurs compacts A ne dépend que de son symbole $a_{|\partial B}$ et on a : $T_a T_b \sim T_{ab}$.

On a une notion de famille continue d'opérateurs de Toeplitz paramétrée par X, dont le symbole est une fonction continue sur $X \times \partial B$, et plus généralement d'opérateur de Toeplitz de type $E \to F$ où E et F sont deux fibrés vectoriels sur X, ou de complexe d'opérateurs de Toeplitz paramétré par X, dont le symbole est un complexe de fibrés sur $X \times \partial B$. Rappelons que A est elliptique si son symbole est exact; c'est alors un complexe de Fredholm.

Nous dirons encore que A est *basique* s'il est de la forme T_a où a ne dépend que de la variable $x \in X$, autrement dit si pour chaque $x \in X$ A_x est l'opérateur $f \to a(x)f$. Pour tenir compte des supports on introduit la définition suivante:

Définition 2.1. *On appelle famille admissible (relativement à Z) d'opérateurs de Toeplitz elliptiques un couple (A, A^t) $(0 \leq t \leq 1)$, où A est une famille elliptique d'opérateurs de Toeplitz d'ordre 0 (de type $E \to F$) sur B, paramétrée par X, inversible hors de Z, et A^t est une homotopie d'opérateurs de Toeplitz inversibles paramétrés par $x \in X - Z$ reliant $A^1 = A_{|X-Z}$ à une famille basique A^0.*

Une famille d'opérateurs de Toeplitz inversible en dehors de Z, et en particulier une famille admissible, a un indice $\operatorname{Ind}_Z(A) \in K_Z(X)$.

[6] Il y a beaucoup d'autres notations pour l'espace de Hardy; nous avons repris le cas $s = 0$ de la notation $\mathcal{O}^s$ utilisée dans Boutet de Monvel 3, inspirée de celle des espaces de Sobolev pour la théorie des opérateurs pseudodifférentiels tout en évitant la lettre H (qui représente trop d'autres choses par ailleurs).

Une famille admissible définit d'autre part un élément $[A]_Z \in K_Z(X \times B)$ comme suit: si (a, a^t) est le symbole de (A, A^t) on pose $a(x, tz) = a(x, z)$ si $t = 1$, $z \in \partial B$, resp. $a^t(x, z)$ si $t \leq 1$, $x \in X - Z$, $z \in \partial B$; a est ainsi bien défini sur $X \times \partial B \cup (X - Z) \times B$ (parce que a^0 est basique). Il définit donc un élément de la K-théorie de $X \times B$ à support dans $Z \times \overset{\circ}{B}$, qui est égale à $K_Z(X \times B)$ (par déformation).

Si a est un complexe de fibrés vectoriels de rang fini sur X exact en dehors de N, son image $a \otimes k_N$ par l'homomorphisme de Bott définit un complexe d'opérateurs de Toeplitz $T_{a \otimes k_N}$ exact en dehors de X, qui se prolonge canoniquement en une famille admissible: $A_t = T_{a \otimes k_N(tz)}$.

Théorème 2.2. *On suppose X métrique de dimension finie. Alors*

(i) *Tout élément $\xi \in K_{Z \times \{0\}}(X \times B)$ est de la forme $[A]_Z^{\mathrm{top}}$ pour une famille admissible $A = (A_X, A^t)$ convenable.*[7]

(ii) *Si $A = (A_X, A_X^t)$ est une famille admissible d'opérateurs de Toeplitz $\mathrm{Ind}_Z(A)$ ne dépend que de $[A]_Z^{\mathrm{top}}$.*

(iii) *L'application $[A]_Z^{\mathrm{top}} \to \mathrm{Ind}_Z(A)$ de $K_{Z \times \{0\}}(X \times B)$ dans $K_Z(X)$ ainsi définie est l'inverse de l'homomorphisme de Bott.*

Les résultats i) et ii), ainsi que le fait que l'application canonique $K_Z \to F_Z$ soit un isomorphisme, se démontrent à partir de variantes du résultat suivant: si Y est de dimension finie et si A est une famille continue d'opérateurs de Fredholm sur un fibré hilbertien H, E un sous fibré de rang assez grand de H, il existe une déformation $A_t = A + q_t$, avec q_t compact, $q_0 = 0$ telle que A_1 soit injectif sur $E^\perp$ (la condition est $\mathrm{rg}(E) \geq \mathrm{Ind}(A) + \dim X$ en tout point); on peut dans une telle déformation ne rien bouger sur un ensemble fermé où A est déjà bon, ou forcer A_t à rester inversible sur un ensemble ouvert où A l'est.

Preuve de iii): si a est un complexe de fibrés sur X exact hors de Z, k le complexe de Koszul de $X \times B$, $k^t = k(x, tz)$, $A = T_{a \otimes k}$, $A^t = T_{a \otimes k^t}$, on a par construction $[A]_Z = [k][a] = b([a])$ et il est élémentaire de vérifier $\mathrm{Ind}_Z(A) = [a]$ (l'injection canonique $a \to A$ est un quasi-isomorphisme). Ainsi l'application Ind_Z définie par i) ii) est inverse à gauche de l'homomorphisme de Bott b. Un argument de G. Segal montre alors simplement que c'est aussi un inverse à droite.

2.3 Image directe

a. Structures spin^c.

Soit X comme ci-dessus, et soit N un fibré vectoriel réel de dimension paire sur X. Une structure spin^c[8] sur N est la donnée

(i) d'une métrique euclidienne (continue) sur N

On note C_N l'algèbre de Clifford négative de N: fibré en $\mathbf{C}$-algèbres $\mathbf{Z}_2$-graduées sur X, engendré par le sous fibré vectoriel réel N et les relations $n \cdot n = - \| n \|^2$ pour $n \in N$.

(ii) d'un C_N-module $\mathbf{Z}_2$-gradué $\mathcal{S}_N = \mathcal{S}_N^+ + \mathcal{S}_N^-$ (fibré), simple en tout point de X.

Si N est muni d'une structure spin^c, il est canoniquement orienté: une base orthonormale $e_1, e_2, \ldots e_{2n}$ de N_X ($\dim_{\mathbf{R}} N_X = 2n$) est orientée si $e_1 e_2 \cdots e_{2n}\ \xi = i^n \xi$ pour tout $\xi \in \mathcal{S}_{N_X}^+$.

---

[7] Nous avons noté $\{0\}$ la section nulle de T^*X — aussi notée T_X^*X dans les notes de Schapira-Schneiders.

[8] sur les structure spin^c on pourra aussi consulter l'article de Atiyah-Bott-Shapiro.

Les structures spin^c s'ajoutent: si N et N' sont munis de structures spin^c, $N \oplus N'$ est muni de la métrique somme, et de $\mathcal{S}_{N\oplus N'} = \mathcal{S}_N \otimes^g \mathcal{S}_{N'}$, produit tensoriel gauche, qui est un $C_{N\oplus N'}$-module $\mathbf{Z}_2$-gradué ($C_{N\oplus N'} \sim C_N \otimes^g C_{N'}$). Elles sont complémentées: si N est muni d'une structure spin^c, ainsi que $N' \subset N$ (N' muni de la métrique induite), l'orthogonal $N'' = N'^{\perp}$ est muni de la métrique induite et de $\mathcal{S}_{N''} = \mathrm{Hom}^{\mathrm{gr}}_{C_{N''}}(\mathcal{S}_{N'}, \mathcal{S}_N)$, qui en fait un $C_{N''}$-module simple $\mathbf{Z}_2$-gradué. C'est l'unique structure spin^c sur N'' pour laquelle N, avec sa structure spin^c, soit isomorphe à $N' \oplus N''$.

On peut alors parler de fibré spin^c virtuel et former le groupe K^{spin^c} analogue au groupe $K^{\mathrm{réel}}$ des fibrés vectoriels virtuels; on a un homomorphisme $K^{\mathrm{spin}^c}(X) \to K^{\mathrm{réel}}(X)$.

Exemple. soit E un fibré vectoriel complexe de rang fini, muni d'une métrique hermitienne, et soit N le fibré réel sous-jacent. Alors N est muni d'une structure spin^c canonique, pour laquelle la métrique euclidienne est celle déduite de la métrique hermitienne de E, et

$$\mathcal{S}_N = \wedge_{\mathbf{C}} E^* \quad (= \wedge^{\mathrm{pair}}_{\mathbf{C}} E^* \oplus \wedge^{\mathrm{impair}}_{\mathbf{C}} E^*$$

où E^* désigne le dual de E, l'indice $(\cdot)_{\mathbf{C}}$ signifie qu'il s'agit de l'algèbre extérieure complexe du fibré vectoriel complexe E; la multiplication de Clifford est donnée par

$$n \cdot \omega = n \llcorner \omega - n^* \wedge \omega$$

où $n \in N = E$, $\omega \in \mathcal{S}_N = \wedge E^*$, et $n^* \in E^*$ est l'élément correspondant à n par la métrique hermitienne ($\omega \to n * \wedge \omega$ est adjoint de $\omega \to n \llcorner \omega$).

Comme les métriques hermitiennes de E sont toutes homotopes la structure spin^c virtuelle définie par la structure ci-dessus ne dépend que de la structure complexe de E.

Le théorème de complémentation ci-dessus montre que si E est un fibré vectoriel réel sur X (pas virtuel), une structure spin^c virtuelle sur E (c'est à dire un élément de $K^{\mathrm{spin}^c}(X)$ dont l'image dans $K^{\mathrm{réel}}(X)$ soit $[E]$) provient d'une vraie structure spin^c sur E, unique à isomorphisme près. Il n'y a pas de résultat semblable pour les structures complexes.

La définition de l'isomorphisme de Bott: $K_Z(X) \to K_Z(N)$, que nous avons donnée lorsque N est un fibré complexe, se prolonge canoniquement au cas où N est un fibré réel muni d'une structure spin^c: c'est la multiplication par $[c_N]$ où $c_N : p^{-1}S_N^+ \to p^{-1}S_N^-$ est le morphisme de fibrés sur N défini par le produit de Clifford ($c_N = \delta(k_N)$) lorsque N est un fibré vectoriel complexe muni d'une métrique hermitienne); on voit que l'homomorphisme de Bott $b_{N/X} : [a] \to [c_N][a]$ de $K_Z(X)$ dans $K_Z(N)$ est un isomorphisme lorsque X est compact ou métrique de dimension finie en choisissant un supplémentaire de N c'est à dire un fibré spin^c N' tel que $N \oplus N'$ soit trivial — donc en particulier complexe — et en composant avec $b_{N\oplus N'/N}$.

b. Image directe K-théorique.

On définit alors l'image directe K-théorique $f_* : K_Z(Y) \to K_{Z'}(X)$ lorsque X et Y sont deux variétés, $f : Y \to X$ une application de classe C^1 propre sur Z, $Z' \supset f(Z)$, et f est muni d'une structure spin^c, i.e. on s'est donné une structure spin^c (virtuelle) sur le fibré normal (virtuel) $f^{-1}(TX) - TY$, par exemple une structure complexe virtuelle. Elle est caractérisée axiomatiquement par les conditions suivantes:

(I1) Elle est covariante: $(fg)_* = f_* g_*$

(I2) Elle commute aux changements de base.

(I3) Si X est un fibré vectoriel spin^c sur Y, f la section nulle, munie de la structure spin^c déduite de celle des fibres de X, on a $f_* = b_{X/Y}$ (isomorphisme de Bott).

La signification de (I2) est la suivante: soit un diagramme

$$\begin{array}{ccc} Y' & \xrightarrow{G} & Y \\ {\scriptstyle F}\downarrow & & \downarrow{\scriptstyle f} \\ X' & \xrightarrow{g} & X \end{array}$$

où X, Y, X', Y' sont des variétés, f, g, F, G sont de classe C^1, f et g sont transverses et Y' s'identifie au produit fibré $Y \times_X X'$, f est muni d'une structure spin^c et F de la structure spin^c qui s'en déduit. On se donne des fermés $Z \subset Y$, $Z' \subset Y'$, $T \subset X$, $T' \subset X'$ tels que $Z' \supset G^{-1}(Z)$, $T' \supset g^{-1}(T)$, $T \supset f(Z)$, $T' \supset F(Z')$. Alors on a $G^{-1}f_* = F_* g^{-1}$: $K_Z(Y) \to K_{T'}(X')$. En particulier l'image directe commute aux restrictions aux ouverts (cas où g et G sont des plongements ouverts); ainsi si $Y \subset X$ est ouvert $f : Y \to X$ est l'inclusion canonique munie de la structure spin^c triviale, $Z = T$ est fermé dans X, $f_* : K_Z(Y) \to K_Z(X)$ est l'isomorphisme d'excision.

Ceci étant l'image directe f_* est bien définie si f est une projection $Y = X \times \mathbf{C}^n \to X$, comme composé de l'homomorphisme de restriction $K_Z(Y) \to K_B(Y)$ où $B \supset Z$ est un fibré en boules de base T, de rayon assez grand, et de l'inverse de l'isomorphisme de Bott $K_B(Y) \to K_T(X)$. Dans le cas général f spin^c de classe C^1 est composée d'une telle application, d'un plongement ouvert, et de la section nulle d'un voisinage tubulaire spin^c, de sorte que f_* est bien défini par ces axiomes.

Remarque. dans Boutet de Monvel 1 nous avons aussi défini les opérateurs de Toeplitz de degré $\neq 0$, auxquels le théorème d'indice ci-dessus se généralise. Ainsi soit $E^k = \oplus_l E^k_l$ une famille de fibrés gradués sur X. Soit

$$D : \cdots E^k \otimes \mathcal{O} \longrightarrow E^{k+1} \otimes \mathcal{O} \longrightarrow \cdots$$

une famille de complexes d'opérateurs différentiels à coefficients holomorphes sur B. On suppose D de degré 0, i.e. D_{ij} est d'ordre $\leq i - j$ si (D_{ij}) est la matrice de D dans la décomposition $E^k = \oplus_l E^k_l$; en particulier $D_{ij} = 0$ si $i < j$, et dans ce cas D est exact comme complexe d'opérateurs différentiels ssi sa partie diagonale d'ordre 0 l'est en tout point de $Y = B \times X$.

Supposons D elliptique, exact hors de Z. Alors à D est canoniquement associée une famille admissible, qu'on construit en deux crans: 1) pour $1/2 \leq t \leq 1$ on pose $D^t = ((2t-1)^{i-j} D_{ij})$; c'est bien défini parce que $D_{ij} = 0$ si $i - j < 0$; $D^{1/2}$ est "diagonal", d'ordre 0. 2) pour $0 \leq t \leq 1/2$ on pose $D^t = D^{1/2}(2tz)$; D^0 est basique.

Le symbole $[D]^{\text{top}}_Z$ associé à cette famille peut encore être décrit ainsi: c'est l'élément de K-théorie associé à $\sigma_D(\eta)$, où η est une section de $X \times T^*B$ de direction voisine de celle de du au voisinage de $X \times \partial B$, où $u = 0$ est l'équation du bord ∂B (une description équivalente est donnée au n°3). Le théorème 2 est donc encore vrai dans ce cas: $\text{Ind}_Z(D) = p_*[D]^{\text{top}}_Z$.

§3. $\mathcal{D}$-Modules

Dans l'introduction nous avons rappelé ce qu'est un système différentiel sur une variété. En fait il est plus commode de décrire les systèmes à coefficients analytiques en termes de $\mathcal{D}$-modules. Ce paragraphe contient un rappel des définitions concernant les $\mathcal{D}$-modules qui forment un cadre naturel pour décrire les systèmes différentiels, et à peu près indispensable pour décrire les opérations fondamentales (images directe ou inverse, etc.) au moyen desquelles s'énonce et se démontre le théorème d'indice dans le cas relatif. Le lecteur pourra aussi se reporter au séminaire Borel et al où ces objets sont décrits avec précision (dans un cadre algébrique), ainsi qu'aux livres de Sato-Kawai-Kashiwara, de Kashiwara 5, Schapira, et de Björk, ou au séminaire de Grenoble 1965 (Boutet de Monvel-Lejeune-Malgrange). Plusieurs opérations sur les $\mathcal{D}$-modules représentent une superposition de nombreuse manipulations et sont commodément décrites en termes de complexes et de catégorie dérivée; nous renvoyons aux mêmes références pour cela, ainsi qu'à Verdier, Borel et al 2, et Kashiwara-Schapira 2.

3.1 $\mathcal{D}$-Modules.

Soit X une variété analytique complexe, ou une variété analytique réelle, que nous considérerons toujours comme le germe d'un variété complexe au voisinage des points réels. On note $\mathcal{O}_X$, ou simplement $\mathcal{O}$, le faisceau des fonctions analytiques sur X et Ω_X, ou Ω, le faisceau des formes différentielles de degré maximum. On note $\mathcal{D}_X$ (ou $\mathcal{D}$) le faisceau des opérateurs différentiels sur X: $\mathcal{D}$ opère à gauche sur $\mathcal{O}$ et à droite Ω. $\mathcal{D}$ est muni d'une filtration croissante canonique, pour laquelle $\mathcal{D}_m$ est le sous-faisceau des opérateurs d'ordre $\leq m$, et pour laquelle le faisceau gradué $\operatorname{gr}\mathcal{D}$ s'identifie au faisceau des sections de l'algèbre symétrique STX, ou au faisceau des fonctions polynomiales à coefficients dans $\mathcal{O}_X$ sur le fibré cotangent T^*X.

Le faisceau $\mathcal{D}$ est cohérent, autrement dit, pour tout homomorphisme de $\mathcal{D}$-modules (à droite ou à gauche) $a: \mathcal{D}^p \to \mathcal{D}^q$, le faisceau $\ker a$ est localement de type fini sur $\mathcal{D}$. Rappelons qu'un faisceau M de $\mathcal{D}$-modules[9] (à droite ou à gauche) est cohérent s'il est localement engendré sur $\mathcal{D}$ par un nombre fini d'éléments soumis à un nombre fini de relations. Rappelons la définition des bonnes filtrations:

Définition 3.1. *Une bonne filtration sur un $\mathcal{D}_X$-module M à droite (ou à gauche) est une filtration croissante $M = \cup M_k = \cup F_k M$ telle que*

(F1) *M_k est un $\mathcal{O}_X$-module cohérent, nul pour $k << 0$.*
(F2) *$M_k \mathcal{D}_p \subset M_{k+p}$, avec égalité si $k >> 0$.*

La filtration canonique $\mathcal{D} = \cup\mathcal{D}_k$ est bonne. Si un $\mathcal{D}$-module M possède une bonne filtration il est cohérent car, si M est muni d'une bonne filtration, le gradué $\operatorname{gr} M$ est un module gradué cohérent sur le faisceau $\operatorname{gr}\mathcal{D}$ des fonctions polynomiales à coefficients analytiques sur T^*X. Inversement si M est un $\mathcal{D}$-module cohérent, il possède localement de bonnes filtrations[10]. Dans la suite nous supposerons le plus souvent que les $\mathcal{D}$-modules que nous considérerons possèdent une bonne filtration globale, sans nécessairement la préciser (modules globalement bien filtrables).

Définition 3.2. *On appelle variété caractéristique de M et on note* $\operatorname{car} M$ *le sous-ensemble analytique conique de T^*X support de $\operatorname{gr} M$ pour une bonne filtration.*

La variété caractéristique $\operatorname{car} M$ ne dépend pas du choix d'une bonne filtration sur M, et est donc défini de façon globale (on peut aussi le définir comme support du faisceau $\mathcal{E}\otimes_{\mathcal{D}}M$ sur T^*X, où $\mathcal{E}$ est le faisceau des opérateurs microdifférentiels analytiques, pour lequel nous renvoyons à Sato, Kawai, Kashiwara; cette deuxième définition s'étend à n'importe quel

---

[9] Dans la suite nous parlerons de $\mathcal{D}$-module, en précisant au besoin s'il est à gauche ou à droite, et en omettant le mot faisceau

[10] Il en possède aussi globalement s'il est "algébrique" ou holonôme régulier; mais on ne sait toujours pas si ce résultat d'existence globale est vrai en général.

faisceau de $\mathcal{D}$-modules). Si P est un complexe de $\mathcal{D}$-modules, on note car P la réunion des variétés caractéristiques de ses faisceaux d'homologie.

Sato, Kawai et Kashiwara ont montré que car M est toujours **involutif** lorsque M est cohérent, autrement dit si f et g sont différentiables et s'annulent sur car M, le crochet de Poisson $\{f,g\}$ s'annule aussi sur car M[11].

3.2 Lien avec les systèmes différentiels; solutions.

Dans la suite il sera commode de noter de la même manière les fibrés vectoriels et leurs faisceaux de sections. A un complexe d'opérateurs différentiels

(3.1) $$p: \cdots E^k \xrightarrow{p_k} E^{k+1} \longrightarrow \cdots$$

où les E^k sont des fibrés vectoriels analytiques sur X on associe une complexe de $\mathcal{D}$-modules à droite localement libres

(3.2) $$P^d: \cdots \longrightarrow \mathcal{E}^k \longrightarrow \mathcal{E}^{k+1} \longrightarrow \cdots$$

où $\mathcal{E}^k = \mathrm{Diff}(\mathcal{O}, E^k)$ est le faisceau des opérateurs différentiels de type $\mathcal{O} \to E^k$, la différentielle étant donnée en degré k par $Q \to p_k \circ Q$.

De même on associe à P un complexe de $\mathcal{D}$-modules à gauche localement libres:

(3.3) $$P^g: \cdots \mathcal{E}'^k \to \mathcal{E}'^{k+1} \cdots$$

où $\mathcal{E}'^k = \mathrm{Diff}(E^{-k}, \mathcal{O})$ est le faisceau des opérateurs différentiels de type $E^{-k} \to \mathcal{O}$, et la différentielle en degré k est donnée par $Q \to Q \circ p_{-k-1}$.

Exemples 1. Soit $f: X \to Y$ une submersion. Notons T^*X/Y le fibré cotangent relatif. Le complexe de De Rham relatif $d_{X/Y}$ est le complexe

(3.4) $$d_{X/Y}: 0 \longrightarrow \mathcal{O}_Y \longrightarrow T^*X/Y \longrightarrow \cdots \wedge^k T^*X/Y \longrightarrow \wedge^{k+1} T^*X/Y \longrightarrow \cdots$$

défini par la différentiation extérieure verticale. Il lui correspond un complexe de $\mathcal{D}_X$-modules à droite (en degrés positifs) $\mathrm{DR}^d_{X/Y}$, et un complexe de $\mathcal{D}_X$-modules à gauche (en degrés négatifs) $\mathrm{DR}^g_{X/Y}$. Le symbole de $d_{X/Y}$ est le complexe de l'algèbre extérieure

(3.5) $$0 \longrightarrow \mathbf{C} \longrightarrow T^*X/Y \longrightarrow \cdots \wedge^k T^*X/Y \longrightarrow \wedge^{k+1} T^*X/Y \longrightarrow \cdots$$

dont la différentielle au point $\eta \in T^*X$ est le produit extérieur $\omega \to \eta' \wedge \omega$, où $\eta' \in T^*X/Y$ est la partie verticale de $\eta \in T^*X$.

2. Si $\mathcal{M}$ est un complexe de $\mathcal{O}_X$-modules, on lui associe le $\mathcal{D}_X$-modules à droite

(3.6) $$M = \mathcal{M} \otimes_{\mathcal{O}_X} \mathcal{D}_X$$

---

[11] On rappelle que le crochet de Poisson est le crochet de Lie défini par $\{f,g\} = \sum \partial f/\partial\xi_j \partial g/\partial x_j - \partial f/\partial x_j \partial g/\partial \xi_j$ pour f, g fonctions différentiables sur T^*X (pour tout choix de coordonnées locales $x_1, \ldots, x_n$ sur X, $\xi_1, \ldots, \xi_n$ désignant les coordonnées duales dans les fibres de T^*X). On a $\sigma_{m+n-1}([P,Q]) = \{\sigma_m(P), \sigma_n(Q)\}$ si P et Q sont deux opérateurs différentiels d'ordres m, n et σ_* désigne le symbole (partie principale).

3. Si $E \xrightarrow{p} Y$ est un fibré vectoriel complexe, X la variété analytique sous-jacente à E, le complexe de Koszul de E est le complexe de $\mathcal{O}_X$-modules (en degrés négatifs)

$$(3.7) \qquad k_E : \cdots p^{-1} \wedge^{-k} E' \longrightarrow p^{-1} \wedge^{-k+1} E' \longrightarrow \cdots p^{-1} E' \longrightarrow \mathcal{O}_X \longrightarrow 0$$

dont la différentielle au point $x \in X = E$ est le produit intérieur $\omega \to x \llcorner \omega$ où E' désigne le fibré dual de E. Il lui correspond un complexe de $\mathcal{D}_X$-modules à droite $K_E^d = k_E \otimes_{\mathcal{O}_X} \mathcal{D}_X$ (il y a de même un $\mathcal{D}_X$-module à gauche K_E^g).

Pour la suite il sera plus commode de travailler dans la catégorie dérivée de la catégorie des $\mathcal{D}$-modules: nous noterons $\mathrm{D}^b_d(\mathcal{D})$ (resp. $\mathrm{D}^b_g(\mathcal{D})$) la catégorie dérivée de la catégorie de $\mathcal{D}$-modules à droite (resp. à gauche), déduite de la catégorie des complexes bornés de $\mathcal{D}$-modules en inversant les quasi-isomorphismes (c'est à dire les morphismes qui induisent un isomorphismes en homologie). Un objet (nous dirons encore par abus "complexe") de $\mathrm{D}^b_d(\mathcal{D})$ est dit cohérent s'il est à cohomologie cohérente.

Faisceau des solutions.

Les faisceaux de solutions du système différentiel (3.1) sont les faisceaux de cohomologie du complexe P de (3.1). Si M est le complexe de $\mathcal{D}$-modules correspondant à P, on notera $\mathrm{Sol}\, P$ l'objet de la catégorie dérivée de la catégorie des faisceaux d'espaces vectoriels complexes défini par le complexe P lui-même; il s'identifie de façon évidente à $P^d \otimes_{\mathcal{D}} \mathcal{O}$ (resp. à $Hom_{\mathcal{D}}(P^g, \mathcal{O})$) où P^d (resp. P^g) est le complexe de $\mathcal{D}$-modules à droite (resp. à gauche) associé à P comme ci-dessus (3.2) resp. (3.3). En général si M est un complexe de $\mathcal{D}$-modules à droite, (resp. à gauche), non nécessairement localement libres, on note $\mathrm{Sol}\, M$ l'objet dérivé $M \otimes^L_{\mathcal{D}} \mathcal{O}$ (resp. $\mathrm{R}\mathcal{H}om_{\mathcal{D}}(M, \mathcal{O})$). Dans ces notes nous travaillerons plutôt avec les $\mathcal{D}$-modules à droite pour lesquels l'image directe est un peu plus simple à décrire; en fait tant qu'on se limite aux complexes de $\mathcal{D}$-modules à cohomologie cohérente, il revient au même de travailler avec les $\mathcal{D}$-modules à droite ou à gauche pour beaucoup des questions qui nous intéressent; en effet si M est un objet cohérent de $\mathrm{D}^b_d(\mathcal{D})$ (resp. $\mathrm{D}^b_g(\mathcal{D})$ à gauche), le dual bête $M' = \mathrm{R}\mathcal{H}om_{\mathcal{D}}(M, \mathcal{D})$[12] est un objet cohérent de $\mathrm{D}^b_g(\mathcal{D})$ (resp. $\mathrm{D}^b_d(\mathcal{D})$); M est canoniquement isomorphe à son bidual et $\mathrm{Sol}\, M = M \otimes^L_{\mathcal{D}} \mathcal{O}$ est canoniquement isomorphe à $\mathrm{Sol}\, M' = \mathrm{R}\mathcal{H}om_{\mathcal{D}}(M', \mathcal{O})$. Dans l'exemple (3.1-3) P^g est le dual bête de P^d.

Il est encore utile d'introduire des faisceaux de solutions dans un cadre un peu élargi: si M est un complexe de faisceaux de $\mathcal{D}$-modules à droite et si N est un complexe de faisceaux de $\mathcal{D}$-modules à gauche, on appelle faisceau des solutions de M à coefficients dans N le complexe $M \otimes^L_{\mathcal{D}} N$, considéré comme objet de la catégorie dérivée de la catégorie des faisceaux d'espaces vectoriels complexes. En pratique dans ce qui suit M sera cohérent et la définition de N reflétera le domaine d'existence des solutions que l'on veut étudier.

3.3 Modules de transfert.

Soient X, Y des variétés analytiques et $f : X \to Y$ une application analytique.

Définition 3.3. *On note $\mathcal{D}_{X\to Y}$ le faisceau sur Y des opérateurs différentiels de type $f^{-1}\mathcal{O}_Y \to \mathcal{O}_X$ (f^{-1} désigne l'image inverse faisceautique); localement un tel opérateur s'écrit sous la forme $u \in \mathcal{O}_Y \to P_X((Q_Y u) \circ f)$ avec $P_X \in \mathcal{D}_X$, $Q_Y \in \mathcal{D}_Y$.*

[12] Il y a d'autres façons d'associer un $\mathcal{D}$-module à droite à un $\mathcal{D}$-module à gauche ou inversement: une usuelle est d'associer à un $\mathcal{D}$-module à gauche M le $\mathcal{D}$-module à droite $M \otimes_{\mathcal{O}} \Omega$

De même on note $\mathcal{D}_{Y\leftarrow X}$ le faisceau sur Y des opérateurs différentiels de type $f^{-1}\Omega_Y \to \Omega_X$, i.e. localement de la forme $\omega \in \Omega_Y \to P_X(f^{-1}(\omega Q_Y))$, avec $Q_Y \in \mathcal{D}_Y$, P_X opérateur différentiel de type $f^*(\Omega_Y) \to \Omega_X$.

$\mathcal{D}_{X\to Y}$ est un $(\mathcal{D}_X, f^{-1}\mathcal{D}_Y)$-bimodule monogène, engendré comme bimodule par l'opérateur $\varepsilon : u \to u \circ f$. De même $\mathcal{D}_{Y\leftarrow X}$ est un $(f^{-1}\mathcal{D}_Y, \mathcal{D}_X)$-bimodule localement monogène (il n'y a pas de générateur canonique).

Exemples 4. Supposons f submersive. Alors $\mathcal{D}_{X\to Y}$ est monogène, de générateur ε, comme $\mathcal{D}_X$-module à gauche; il est plat sur $f^{-1}\mathcal{D}_Y$, de même que $\mathcal{O}_X$ est plat sur $f^{-1}\mathcal{O}_Y$. Le complexe $\mathrm{DR}^g_{X/Y}$ en est une résolution localement libre, d'augmentation $P \in \mathcal{D}_X \to P \circ \varepsilon$.

De même $\mathcal{D}_{Y\leftarrow X}$ est localement monogène comme $\mathcal{D}_X$-module à droite, plat sur $f^{-1}\mathcal{D}_Y$, et $\mathrm{DR}^d_{X/Y}$ est une résolution localement libre de $\mathcal{D}_{Y\leftarrow X}[-\dim X/Y]$.

5. Soit $p : E \to Y$ un fibré vectoriel, et soient X la variété sous-jacente à E, $f : Y \to X$ la section nulle. Alors $f_*(\mathcal{D}_{X\to Y})$ est un $\mathcal{D}_Y$-module à droite de support X, monogène engendré par ε. Il admet pour résolution localement libre le complexe K^d_E déduit du complexe de Koszul (augmentation: $P \in \mathcal{D}_Y \to \varepsilon \circ P$, en degré 0).

3.4 Images directe et inverse.

Si Z est une variété analytique, on notera $\mathrm{D}^b(\mathcal{D}_Z)$ la catégorie dérivée (fabriquée avec les complexes à cohomologie bornée) de la catégorie des $\mathcal{D}_Z$-modules à droite. Soit $f : X \to Y$ un morphisme de variétés analytiques. Si $M \in \mathrm{ob}(\mathrm{D}^b(\mathcal{D}_X))$ l'image directe $f_+(M)$ est l'élément

$$f_+M = \mathrm{R}f_*(M \otimes^L_{\mathcal{D}_X} \mathcal{D}_{X\to Y}) \in \mathrm{ob}(\mathrm{D}^b(\mathcal{D}_Y)) \tag{3.8}$$

Pour $M \in \mathrm{ob}(\mathrm{D}^b(\mathcal{D}_Y))$ on définit l'image inverse $f^+(M)$ par

$$f^+M = f^{-1}(M) \otimes^L_{\mathcal{D}_Y} \mathcal{D}_{Y\leftarrow X}[-d] \in \mathrm{ob}(\mathrm{D}^b(\mathcal{D}_Y)) \quad (d = \dim X - \dim Y) \tag{3.9}$$

($\mathrm{R}f_*$ désigne le foncteur dérivé du foncteur image directe faisceautique, f^{-1} le foncteur image inverse (il est exact), $\otimes^L$ le produit tensoriel dérivé). Il y a des constructions similaires pour les $\mathcal{D}$-modules à gauche.

Exemples 6. Supposons que f soit une immersion fermée. Alors l'image directe f_* est exacte, et $\mathcal{D}_{X\to Y}$ est localement libre sur $\mathcal{D}_X$. Donc si M est un $\mathcal{D}_X$-module à droite, $f_+(M)$ est le $\mathcal{D}_Y$-module (= complexe pur de degré 0)

$$f_+(M) = f_*(M \otimes_{\mathcal{D}_X} \mathcal{D}_{X\to Y}) \tag{3.10}$$

D'après M. Kashiwara, $M \to f_+(M)$ est dans ce cas une équivalence de la catégorie des $\mathcal{D}_X$-modules à droite sur la catégorie des $\mathcal{D}_Y$-modules à droite portés par $f(X)$ (i.e. dont toute section est tuée par une puissance de l'idéal de définition de $f(X)$); elle admet pour équivalence inverse (à homotopie près) le foncteur $N \to f^+N[2d]$. En outre f préserve la cohérence: $f_+(M)$ est $\mathcal{D}_Y$-cohérent si et seulement si M est $\mathcal{D}_X$-cohérent (tout ceci marche aussi bien pour les $\mathcal{D}$-modules à droite ou à gauche).

7. Image inverse submersive (cf. Boutet de Monvel-Malgrange; aussi Honda, J. Fac. Sci. Tokyo 38, 351-358 (1991)). Supposons que f soit une submersion de dimension relative $d = \dim X/Y$. Notons $\mathcal{D}_{X/Y}$ le faisceau des opérateurs différentiels relatifs, et $T_{X/Y}$ le fibré tangent relatif, dont le faisceau des sections s'identifie au sous faisceau des champs de vecteurs verticaux dans $\mathcal{D}_{X/Y}$. On notera $H \subset T^*X$ le fibré des covecteurs horizontaux: c'est l'orthogonal de $T_{X/Y}$ et l'application cotangente ${}^tf'$ est un isomorphisme de $X \times_Y T^*Y$ sur H.

Les modules de transfert $\mathcal{D}_{X\to Y}$ et $\mathcal{D}_{Y\leftarrow X}$ sont plats sur $f^{-1}\mathcal{D}_Y$, et $\mathrm{DR}^g_{X/Y}$ est une résolution de $\mathcal{D}_{X\to Y}$ (resp. $\mathrm{DR}^d_{X/Y}$ est une résolution du module décalé $\mathcal{D}_{Y\leftarrow X}[-d]$, cf. exemple 4). Le produit tensoriel total $\mathcal{D}_{Y\leftarrow X} \otimes^L_{\mathcal{D}_X} \mathcal{D}_{X\to Y}[-d]$ est quasi-isomorphe à $\mathrm{DR}^d_{X/Y} \otimes_{\mathcal{D}_X} \mathcal{D}_{X\to Y}$ donc à $f^{-1}\mathcal{D}_Y$ (localement X est un produit $Z \times Y$ et le complexe ci-contre s'identifie au complexe des f-opérateurs différentiels de type $\mathcal{O}_Y \to \Omega_{X/Y}$, i.e. de la forme $P(z, y, \dfrac{\partial}{\partial y})$, à coefficients formes relatives, muni de la différentielle $P \to d_{X/Y}P$; la cohomologie se réduit aux opérateurs dont les coefficients ne dépendent pas de z en degré 0, et est nulle en degré > 0).

Si M est un $\mathcal{D}_Y$-module à gauche on a $f^+M \approx \mathcal{D}_{X\to Y} \otimes_{f^{-1}\mathcal{D}_Y} f^{-1}M$, parce que f^{-1} est exact et $\mathcal{D}_{X\to Y}$ plat sur $f^{-1}\mathcal{D}_Y$; en particulier f^+M est pur de degré 0. Or on a $\operatorname{car} f^+M = F\overline{f}^{-1}(\operatorname{car} M) \subset \operatorname{car} d_{X/Y} = H$ où $\overline{f}$ désigne la projection $X \times_Y T^*Y \to T^*Y$ et F l'application cotangente $X \times_Y T^*Y \to H \subset T^*X$); enfin f^+M est "régulier le long de H", i.e. il possède une bonne filtration $f^+M = \cup N_k$ telle que $T_{X/Y}N_k \subset N_k$ ($\Leftrightarrow \mathcal{D}_{X/Y}N_k \subset N_k$), par exemple celle déduite d'une bonne filtration de M: $N_k = \sum_{p+q=k} \mathcal{D}_{X\to Y,p} \otimes f^{-1}M_q$. Enfin le sous faisceau des sections de f^+M annulées par $T_{X/Y}$ s'identifie à $\mathcal{D}_{Y\leftarrow X} \otimes_{\mathcal{D}_X} f^+M \approx f^{-1}M$, de sorte que le morphisme canonique $M \to f_+f^+M$ est un isomorphisme si Y est un germe de variété analytique au voisinage d'une section continue de f (ou plus généralement si f est relativement contractile).

On vérifie qu'inversement, un $\mathcal{D}_Y$-module cohérent à gauche N tel que $\operatorname{car} N \subset H$, régulier le long de H, est localement, au voisinage de tout point de X, isomorphe à une image inverse f^+M où M est un $\mathcal{D}_Y$-module cohérent (M est isomorphe au $\mathcal{D}_Y$-module des sections de N annulées par $T_{X/Y}$ au voisinage de ce point; N est globalement isomorphe à une image inverse f^+M si de plus X/Y est relativement connexe et le faisceau de ses sections de N annulées par $T_{X/Y}$ n'a pas de monodromie relative, en un sens que nous laissons au lecteur le soin de préciser; ceci est en particulier le cas si X est un germe de variété analytique au voisinage d'une section continue de f, ou si f est relativement contractile, et on a alors $N \approx f^+f_+N$).

Dualement, si M est un $\mathcal{D}_Y$-module à droite cohérent, f^+M est un complexe de $\mathcal{D}_X$-modules cohérents, pur de degré $d = \dim X/Y$, à caractéristique $\subset H$ et régulier le long de H; et inversement si N est un $\mathcal{D}_X$-module à droite cohérent, à caractéristiques $\subset H$ et régulier le long de H, N est isomorphe à f^+f_+N si Y est le germe d'une variété au voisinage d'une section de f (ou si f est relativement contractile), de sorte que f^+ et f_+ sont encore dans ce cas des équivalences de catégories inverses l'une de l'autre comme plus haut. (Dans le cas général le germe N_s de N le long d'une section continue s de f est isomorphe à $f^+f_+N_s$, et on peut énoncer une condition de monodromie relative analogue à celle ci-dessus qui assure que N soit globalement une image inverse, bien que l'analogue du faisceau des sections annulées par $T_{X/Y}$ ne soit plus un sous faisceau de N comme dans le cas des modules à gauche).

8. Soit X une variété complexe. Notons $\overline{X}$ la variété complexe conjuguée, et $X_{\mathbf{R}}$ la variété réelle sous-jacente: diagonale de $X \times \overline{X}$ munie de la restriction de $\mathcal{O}_{X\times X}$ (ou ce qui revient au même, germe de la diagonale dans $X \times \overline{X}$). Soit $f : X_{\mathbf{R}} \to X$ la projection. Le complexe de De Rham relatif $d_{X_{\mathbf{R}}/X}$ est le complexe de Dolbeault ($\overline{\partial}$ ou d''). Si M est un $\mathcal{D}_X$-module, on note $M_{\mathbf{R}} = f^+(M)$ (c'est le produit tensoriel externe complété analytiquement de M et d''); si M correspond à un système d'équations différentielles, $M_{\mathbf{R}}$ correspond au même système, auquel on a adjoint le système des équations de Cauchy-Riemann, sur $X_{\mathbf{R}}$. L'exemple 7 montre que $M \to M_{\mathbf{R}}$ est une équivalence de la catégorie des $\mathcal{D}_X$-modules sur la catégorie des $\mathcal{D}_{X_{\mathbf{R}}}$-modules à caractéristiques $\subset \operatorname{car} d''$, réguliers le long de $\operatorname{car} d''$; l'image directe f_+ est une équivalence inverse: l'homomorphisme canonique $M \to f_+(M_{\mathbf{R}})$ est un isomorphisme.

Remarque. On a $(fg)_+ = f_+g_+$ si g est propre, en particulier si c'est une immersion fermée. En outre l'image directe propre f_+ commute aux changements de base submersifs: si $g : X' \to X$ est submersive, $Y' = Y \times_X Y$ et $F : Y' \to X'$, $G : Y' \to Y$ sont les projections, on a $G^+f_+ = F_+g^+$ (si f est propre).

3.5 Symbole, élément K-théorique associé à un $\mathcal{D}$-module

Soit X une variété complexe, ou analytique réelle comme ci-dessus, et M un $\mathcal{D}$-module. Si Z est une partie conique fermée de T^*X contenant $\operatorname{car} M$ on note:

$$[M]_Z^{\mathrm{an}} \in K^{\mathrm{an}}(T^*X) \tag{3.11}$$

l'élément du groupe de Grothendieck des $\operatorname{gr}\mathcal{D}_X$-modules cohérents à support dans Z défini par $\operatorname{gr} M$. Si Z est conique de base compacte, $[M]_Z^{\mathrm{an}}$ ne dépend que de M, et pas du choix d'une bonne filtration. (cf. Sato-Kawai-Kashiwara, Björk, ou le séminaire Boutet de Monvel-Lejeune-Malgrange). $[M]_Z^{\mathrm{an}}$ est encore défini si M est un complexe borné de $\mathcal{D}$-modules, à cohomologie bien filtrée, à caractéristiques dans Z: $[M]_Z^{\mathrm{an}} = \sum(-1)^j[H^jM]_Z^{\mathrm{an}}$.

Toujours sous les hypothèses ci dessus, on définit l'élément[13]

$$[M]_Z^{\mathrm{top}} \in K_Z(T^*X) \tag{3.12}$$

par les trois conditions suivantes:

(ST1) $[M]_Z^{\mathrm{top}}$ est additif: si $0 \to M' \to M \to M'' \to 0$ est une suite exacte de $\mathcal{D}_X$-modules bien filtrés, à caractéristiques $\subset Z$, on a $[M]_Z^{\mathrm{top}} = [M']_Z^{\mathrm{top}} + [M'']_Z^{\mathrm{top}}$.

(ST2) Si $M = \operatorname{Diff}(\mathcal{O}, P)$ provient d'un complexe P d'opérateurs différentiels tel que le symbole $\sigma(P) = \operatorname{gr} P$ soit exact en dehors de Z, on a

(3.13)
$$[M]_Z^{\mathrm{top}} = [\sigma(P)]_Z, \quad \text{élément de } K_Z(T^*X) \text{ défini par le complexe de fibrés } \sigma(P)^{14}.$$

[13] L'homomorphisme $K_Z^{\mathrm{an}}(X) \to K_Z^{\mathrm{top}}(X)$ a été défini par Baum-Fulton-Mac Pherson lorsque X est projectif, qui utilisent une déformation au cône normal. La définition présenté ici passe par la variété réelle sous-jacente à X, qui n'est pas supposée projective. Ce passage au réel fait perdre toute l'information modulaire contenue dans $K_Z^{\mathrm{an}}(X)$, mais on retrouve bien sûr la définition de Baum-Fulton-Mac Pherson dans le cas projectif.

Si $f: X \to Y$ est une submersion, notons $F: f^{-1}T^*Y = X \times_Y T^*Y \hookrightarrow T^*X$ l'application cotangente. Son image $H \subset T^*Y$ est l'ensemble des covecteurs horizontaux. Notons $\overline{f}: f^{-1}T^*Y \to T^*Y$ la projection. Le $\mathcal{D}_X$-module à droite $\mathcal{D}_{X\to Y}$ est bien filtré, et on a

$$\text{(3.14)} \qquad \operatorname{car}(f^+(M)) = F\overline{f}^{-1}(\operatorname{car} M) \subset H.$$

La troisième condition s'énonce alors :

(ST3) Si $f: X \to Y$ est une submersion, et si M est un $\mathcal{D}_Y$-module à droite bien filtré, on a avec les notations ci-dessus, et $Z \supset \operatorname{car} M$, $f^+Z = F\overline{f}^{-1}(Z)$

$$\text{(3.15)} \qquad [f^+M]^{\text{top}}_{f^+Z} = \overline{f}^{-1}[M]^{\text{top}}_Z \cdot (-1)^d [\mathcal{D}_{X\to Y}]^{\text{top}}_{Z'} = \widetilde{F}_* \overline{f}^{-1}[M]^{\text{top}}_Z \quad (d = \dim X/Y)$$

où $\widetilde{F}_*$ désigne le conjugué de l'image K-théorique (multiplication par l'élément de $K_{Z'}(T^*X)$ correspondant au complexe de l'algèbre extérieure, symbole de $d_{X/Y}$).

Ces axiomes permettent de construire $[M]^{\text{top}}_Z$ lorsque X est compact dans une variété de Stein, car M possède alors une bonne résolution localement libre au voisinage de X, dont le symbole fournit $[M]$. Dans le cas général, $X_{\mathbf{R}}$ est de Stein, donc $[M_{\mathbf{R}}]^{\text{top}}_{Z_{\mathbf{R}}}$ est défini et l'homomorphisme $F_* f^{-1}$ de $K_Z(T^*X)$ dans $K_{fZ}(T^*X_{\mathbf{R}})$ est bijectif dans ce cas. En se ramenant au cas où X est réelle on démontre de façon analogue:

Proposition 3.1. *Soit $f: X \to Y$ une immersion fermée, M un $\mathcal{D}_X$-module bien filtré, $Z = \operatorname{car} M$. Alors f_+M est bien filtré, on a $\operatorname{car} f_+M = \overline{f}F^{-1}(Z)$, et*

$$\text{(3.16)} \qquad [f_+M]^{\text{top}}_{Z'} = \overline{f}_* F^{-1}[M]^{\text{top}}_Z \quad \textit{(image K- théorique)}$$

(où $\overline{f}: f^{-1}(T^*Y) \to T^*Y$ est la projection, et $F: f^{-1}(T^*Y) \to T^*X$ l'application cotangente)

Remarque. si $\mathcal{M}$ est un $\mathcal{O}_Y$-module cohérent de support Z, on note $[\mathcal{M}]^{\text{top}}_Z \in K_Z(Y)$ l'élément dont l'image inverse est $[\mathcal{M} \otimes_{\mathcal{O}_Y} \mathcal{D}_Y]^{\text{top}}_{p^{-1}Z} \in K_{p^{-1}Z}(T^*Y)$. Il résulte de la proposition 1 que cette définition commute aux immersions fermées. Cette définition coïncide évidemment avec celle de Baum, Fulton, Mac Pherson lorsque Y est une variété et $\mathcal{M}$ admet une résolution localement libre finie (par exemple Y de Stein ou projective), donc dans tous les cas; dans l'article de Baum, Fulton, Mac Pherson le résultat sur l'image directe immersive est démontré par "déformation au cône normal".

§4. Cas absolu. Presqu'ellipticité.

Nous décrivons maintenant le théorème d'indice dans le cas absolu, c'est à dire pour un opérateur différentiel elliptique, ou un complexe elliptique d'opérateurs différentiels. Le théorème dans le cas absolu s'étend assez facilement au cas des familles de tels opérateurs dépendant continûment d'un paramètre (l'indice est alors un fibré virtuel sur l'espace des paramètres), et au cas des systèmes équivariants sous l'action d'un groupe compact G (l'indice est alors un élément de l'anneau $R(G)$ des représentations de dimension finie de G).

Le théorème d'indice d'Atiyah et Singer s'occupe des solutions C^∞ dans un domaine donné; pour un système elliptique ce sont les mêmes que les solutions distributions ou que les solutions analytiques s'il s'agit d'un complexe à coefficients analytiques et comme annoncé (parce que cela ne limite pas la généralité de la formule) nous nous limiterons à l'étude de l'indice pour les solutions analytiques d'opérateurs à coefficients analytiques.

Soit donc X une variété analytique complexe, ou une variété analytique réelle, que, comme précédemment, nous considérerons comme germe d'un variété complexe au voisinage des points réels. Le premier ingrédient de la formule d'indice est la donnée d'un $\mathcal{D}_X$-module (ou d'un objet de $\mathrm{D}^b_d(\mathcal{D})$), représentant un système différentiel X. Les solutions analytiques de ce système correspondent à l'élément $\mathrm{Sol}(M, \mathcal{O}_X) = M \otimes^L_{\mathcal{D}} \mathcal{O}_X$ de la catégorie dérivée de la catégorie des faisceaux de $\mathbf{C}$-espaces vectoriels sur X. Pour le théorème d'indice nous devrons en outre préciser le domaine de définition U dans lequel nous calculons les solutions:

(4.1) si U est une partie fermée de X, nous notons $\mathcal{O}_U$ le faisceau des germes de fonctions holomorphes au voisinage de U, et $\mathrm{Sol}(M, \mathcal{O}_U) = M \otimes^L_{\mathcal{D}} \mathcal{O}_U$ (c'est un objet de $\mathrm{D}^b_d(\mathcal{D})$).

Plus généralement on peut examiner le faisceau $\mathrm{Sol}(M, F) = M \otimes^L_{\mathcal{D}} F$ des solutions de M à coefficients dans un faisceau F de $\mathcal{D}$-modules à gauche et l'indice $\mathrm{Ind}(M, F)$ est la caractéristique d'Euler de $\mathrm{R}\Gamma(\mathrm{Sol}(M, F))$, somme alternée des dimensions des groupes d'hypercohomologie, lorsque ceux-ci sont de dimension finie[15]. Dans ses conférences Schapira montrera comment étudier le cas $F = \mathcal{O}_X \otimes \phi$ où ϕ est un faisceau "réel-constructible" sur X (i.e. tel qu'il existe une stratification sous-analytique réelle sur les strates desquelles ϕ est à cohomologie localement constante), et comment on peut associer à un tel faisceau une classe d'Euler qui intervient dans la formule d'indice. Nous nous contenterons ici, en nous inspirant du travail de M. Ohana, du cas où F est le faisceau $\mathcal{O}_U$ des germes de fonctions holomorphes au voisinage d'un sous-variété à bord analytique réelle U fermée dans X, ce qui correspond au cas $\phi = \mathbf{C}_U$ (image directe sur X du faisceau constant $\mathbf{C}$ sur U): le complexe qui décrit les germes de solutions de M au voisinage de U est $\mathrm{R}\Gamma\,\mathrm{Sol}(M, \mathcal{O}_U)$ $(= \mathrm{R}\Gamma(U, \mathrm{Sol}(M, \mathcal{O}_X), U))$, et l'indice $\mathrm{Ind}(M, \mathcal{O}_U)$ est la caractéristique d'Euler de ce complexe. Deux cas nous intéressent particulièrement: celui où U est une sous variété totalement réelle de X, qui correspond à l'étude des systèmes elliptiques sur une variété réelle, et celui où U est un "ouvert à bord" de X, i.e. a même dimension que X.

4.1 $\mathcal{D}$-modules elliptiques.

Soit M un complexe de $\mathcal{D}$-modules sur X et U une partie fermée de X. En nous inspirant de Kashiwara et Schapira, nous noterons $\mathrm{SS}\,U$ le micro-support de U: lorsque U est assez régulier (par exemple une variété à coins), c'est l'ensemble des covecteurs de partie réelle sortante sur U, c'est à dire les covecteurs ξ de base $x \in U$ tels que $\langle \mathrm{Re}\,\xi, f\rangle \geq 0$ pour toute fonction f réelle de classe C^1 négative sur U. Par exemple si U est un ouvert à bord de X, $\mathrm{SS}\,U$ est la réunion de la section nulle de U et du fibré normal sortant du bord. Si U est une sous-variété totalement réelle, sans bord, $\mathrm{SS}\,U$ est la partie purement imaginaire de $U \times_X T^*X$ (restriction à U du fibré cotangent de X). Dans ce qui suit nous nous limiterons au cas où U est une sous-variété à bord analytique réelle de X, ou occasionnellement un produit de telles variétés (variété à coins).

Comme indiqué plus haut nous notons $\mathrm{Sol}(M, \mathcal{O}_U) = M \otimes^L_{\mathcal{D}} \mathcal{O}_U$ le complexe des solutions de M à coefficients dans $\mathcal{O}_U$ et $\mathrm{Ind}(M, \mathcal{O}_U)$ l'indice de ce complexe, s'il existe.

---

[15] Dans la formule usuelle F est un faisceau des fonctions $\mathcal{C}^\infty$ ou analytiques réelles, et les $\mathrm{R}^j\Gamma(X, F)$ sont nuls pour $j > 0$. En général on doit tenir compte $\mathrm{R}^j\Gamma(X, F)$ dans la formule de l'indice.

Définition 4.1. *On dit que M est elliptique le long de U*[16] *si* car $M \cap \mathrm{SS}\,U$ *est contenu dans la section nulle.*

Lorsque U est une sous-variété complexe à bord de X, de même dimension que X, ceci signifie que ∂U est non caractéristique pour M. Si U est totalement réelle sans bord, $\mathrm{SS}\,U$ est l'ensemble des covecteurs imaginaires purs aux points de U, et la définition équivaut à la définition usuelle, i.e. que la variété caractéristique car M ne contienne pas de covecteur imaginaire pur (ou réel) non nul au dessus de U. Si U est totalement réelle avec bord, la condition car $M \cap \mathrm{SS}\,U \subset \{0\}$ aux points du bord de ∂U signifie qu'en outre la condition d'ellipticité au bord de Shapiro Lopatinsky y est satisfaite, sans qu'il y ait à ajouter de condition limite. Dans tous ces cas on sait que l'espace des solutions globales est de dimension finie si de plus U est compact.

4.2 $\mathcal{D}$-modules presqu'elliptiques.

Pour démontrer le théorème d'indice nous nous ramènerons au cas où U est une boule ou un ellipsoïde dans un espace numérique $\mathbf{C}^n$ en remplaçant le $\mathcal{D}$-module M par son image directe par un plongement dans une telle boule. Il est évident que l'image directe d'un $\mathcal{D}$-module elliptique par une immersion de codimension > 0 n'est pas elliptique en général, car la variété caractéristique d'un $\mathcal{D}$-module porté par une sous variété stricte contient en général les points du fibré conormal de cette sous-variété; mais on peut remédier de façon raisonnable à ceci en généralisant la notion d'ellipticité.

Définition 4.2. *Si U est une partie compacte de X on appelle famille adaptée de voisinages de U une famille fondamentale décroissante U_ε $(\varepsilon > 0)$ de voisinages de U telle que $\mathrm{SS}\,U_\varepsilon$ dépende continûment de ε et tende vers $\mathrm{SS}\,U$ pour $\varepsilon \to 0$.*

Dans le cas auquel nous nous limitons, U est une sous variété à bord analytique réelle compacte de X, et il possède toujours une telle famille de voisinages; par exemple

(4.2) $U_\varepsilon =$ l'ensemble des $x \in X$ tels que $\delta^2 + 2\varepsilon u \leq \varepsilon^2$,

où δ est la distance pour une métrique hermitienne analytique[17] sur X, et U une fonction analytique réelle négative $(u < 0)$ dans l'intérieur de U, et qui s'annule transversalement sur le bord ∂U ($u = 0$ et $\mathrm{Re}\, d\,u$ non orthogonal à TU le long de ∂U).

En fait, dans ce choix ci dessus $\mathrm{SS}\,U_\varepsilon$ est une déformation de topologie constante de $\mathrm{SS}\,U$. Comme $\mathrm{SS}\,U_\varepsilon$ se déforme aussi sur l'intérieur de U_ε, il possède un voisinage structure spin^c (en fait presque complexe) canonique (à homotopie près) ainsi à la limite que $\mathrm{SS}\,U$. Suivant Ohana, nous définissons l'élément canonique $[\mathcal{O}_U] \in K_{\mathrm{SS}\,U}(T^*X)$, image du fibré trivial de $\mathrm{SS}\,U$ par l'application de Bott $K(\mathrm{SS}\,U) \to K_{\mathrm{SS}\,U}(T^*X)$.

Si M est un $\mathcal{D}$-module cohérent non caractéristique le long de ∂U_ε pour ε assez petit (c'est à dire car $M \cap \mathrm{SS}\,U_\varepsilon \subset \{0\}$ pour ε assez petit), le théorème de prolongement de Bony et Schapira, qui s'appuie sur une généralisation du théorème de Cauchy-Kowalewski, montre que pour tous ε, ε' assez petits avec $\varepsilon > \varepsilon'$ l'homomorphisme $\mathrm{Sol}(M, \mathcal{O}_{U_\varepsilon}) \to \mathrm{Sol}(M, \mathcal{O}_{U_{\varepsilon'}})$, de restriction de U_ε à $U_{\varepsilon'}$ est un quasi-isomorphisme. Alors à la limite l'homomorphisme de restriction $\mathrm{Sol}(M, \mathcal{O}_{U_\varepsilon}) \to \mathrm{Sol}(M, \mathcal{O}_U)$ est un quasi-isomorphisme pour ε petit. Ceci motive la définition suivante:

---

[16] ou, comme Schapira et Schneiders, que la paire (M, U) est elliptique

[17] métrique Riemannienne conviendrait aussi bien ici car la définition de $\mathrm{SS}\,U$ est en fait purement réelle

Définition 4.3. *On dit que M est presqu'elliptique le long de U (ou simplement presqu' elliptique) s'il existe une famille adaptée U_ε de voisinages tubulaires de U telle que M soit elliptique sur U_ε ($\operatorname{car} M \cap \operatorname{SS} U_\varepsilon \subset \{0\}$) pour ε assez petit.*

Exemples 1. Un module elliptique est évidemment presqu'elliptique.

2. Le produit direct de deux modules presqu'elliptiques est presqu'elliptique. En effet soient X, X' des variétés complexes, U et U' des parties de X et X', M et M' des $\mathcal{D}$-modules cohérents sur X et X' presqu'elliptiques le long de U resp. U', U_ε U'_ε des systèmes adaptés de voisinages sur lesquels M et M' sont elliptiques. Alors $U_\varepsilon \times U'_\varepsilon$ est un système adapté de voisinages de $U \times U'$ sur lesquels le produit direct de M et M' est elliptique: on a en effet $\operatorname{SS}(M \otimes M') = \operatorname{SS} M \times \operatorname{SS} M'$, $\operatorname{SS}(U_\varepsilon \times U'_\varepsilon) = \operatorname{SS} U_\varepsilon \times \operatorname{SS} U'_\varepsilon$ et il est immédiat que $M \otimes M'$ est elliptique le long de $U_\varepsilon \times U'_\varepsilon$ (le lecteur vérifiera qu'on peut encore arrondir les angles de $U_\varepsilon \times U'_\varepsilon$ pour fabriquer une famille adaptée de voisinages tubulaires de $U \times U'$ qui soient des variétés à bord si U et U' sont des variétés à bord).

3. Soient X, U comme ci dessus, f une immersion fermée $X \to X'$. et M un $\mathcal{D}$-module cohérent sur X. Si M est presqu'elliptique relativement à U (en particulier s'il est elliptique), f_+M est presqu'elliptique relativement à $f(U)$. En effet le module $\delta = \mathcal{D}/\sum z_j\mathcal{D}$ sur $\mathbf{C}^n$ (ou $\mathbf{R}^n$) est évidemment presqu'elliptique (relativement à n'importe quel U puisqu'il est nul en dehors de 0); et localement l'image directe f_+M est isomorphe à un produit direct $M \otimes \delta$.

4. Un module holonôme est presqu'elliptique le long de toute sous variété réelle à bord U (cf. Boutet de Monvel 2). En effet choisissons $U_\varepsilon = \{\delta^2 + 2\varepsilon \operatorname{Re} u \leq \varepsilon^2\}$, de sorte que les ∂U_ε sont les surfaces de niveau de la fonction $v = -u + \sqrt{\delta^2 + u^2}$. Soit N l'espace analytique réel défini par la section ∂v, et $\Lambda = \operatorname{car} N$. Sur $N \cap \Lambda$ on a $\partial v = 0$ ($\partial v = \sum \zeta_j dz_j$ dans n'importe quel système de coordonnées, et $\sum \zeta_j dz_j$ est nulle sur Λ). Comme v est réelle on a aussi $\overline{\partial} v = 0$ sur l'ensemble des points réels de $N \cap \Lambda$ de sorte que v est localement constante sur cet ensemble. Comme $N \cap \Lambda$ est un ensemble analytique réel, 0 est une valeur isolée de v sur $N \cap \Lambda$ ce qui signifie que M est elliptique le long de U_ε pour ε assez voisin de 0.

4.3 Formule de l'indice.

Si M est presqu'elliptique et U compact, les applications de restriction $\mathcal{O}_{U_\varepsilon} \to \mathcal{O}_{U_{\varepsilon'}}$ sont compactes pour ε, ε' petits, $\varepsilon > \varepsilon'$, et il résulte alors de ce qui précède que la cohomologie des complexes $\operatorname{Sol}(M, \mathcal{O}_{U_\varepsilon})$ (ε petit) et donc aussi celle de $\operatorname{Sol}(M, \mathcal{O}_U)$ est de dimension finie, et que ces complexes ont le même indice. Ceci reste vrai plus généralement si $\operatorname{supp} M \cap U$ est compact. On peut alors énoncer le théorème de l'indice:

Théorème 4.4. *Si M est elliptique le long de U, et $\operatorname{supp} M \cap U$ compact, la cohomologie de $\operatorname{Sol}(M, \mathcal{O}_U)$ est de dimension finie et l'indice est l'image K-théorique par l'application $T^*X \to$ point du produit $[M]_{\operatorname{car} M}[\mathcal{O}_U]_{\operatorname{SS} U} \in K_{\operatorname{car} M \cap \operatorname{SS} U}(T^*X)$ dans $K(\text{point}) \approx \mathbf{Z}$.*

Notons que le support de $[M]_{\operatorname{car} M}[\mathcal{O}_U]_{\operatorname{SS} U}$ est contenu dans l'intersection compacte $\operatorname{car} M \cap \operatorname{SS} U$, et que l'application $E \to \dim E$ identifie $K(\text{point})$ à $\mathbf{Z}$.

Dans les cas auxquels nous nous sommes limités (U variété à bord analytique réelle) $[\mathcal{O}_U]_{\operatorname{SS} U}$ est l'élément de Bott du plongement $\operatorname{SS} U \to T^*X$, et le produit $[M]_{\operatorname{car} M}[\mathcal{O}_U]_{\operatorname{SS} U}$ est l'image par l'isomorphisme de Bott de la restriction ξ de $[M]_{\operatorname{car} M}$ à $\operatorname{SS} U$, qui est un élément de $K_{\operatorname{car} M \cap \operatorname{SS} U}(\operatorname{SS} U)$; l'indice est donc aussi bien l'image K-théorique de ξ dans $K(\text{point}) = \mathbf{Z}$, de $[M]_{\operatorname{car} M|\operatorname{SS} U}$, comme indiqué dans Boutet de Monvel-Malgrange.

Dans le cas d'un système presqu'elliptique il convient de remplacer le produit $[M]_{\mathrm{car}\,M}[\mathcal{O}_U]_{\mathrm{SS}\,U}$ (ou la restriction $[M]_{\mathrm{car}\,M}|_{\mathrm{SS}\,U}$) par l'élément qu'on obtient en passant à la limite par déformation à partir des U_ε.

Indication de démonstration. Réductions diverses: en utilisant les remarques ci-dessus nous pouvons de diverses façons réduire le cas général à des cas où la géométrie de f est plus simple. Tout d'abord les deux membres du théorème de l'indice ne changent pas (par définition dans le cas presqu'elliptique) si on remplace U par un voisinage tubulaire U_ε convenable. Ceci permet de ramener le cas où U (ou X) est totalement réel au cas où U est "ouvert à bord" complexe. Il est d'autre part immédiat que l'énoncé du théorème d'indice est compatible avec les images inverses submersives relativement contractiles de l'exemple 3.7; en particulier on peut remplacer X par $X_{\mathbf{R}}$ et M par $M_{\mathbf{R}}$, et se ramener ainsi au cas où X et M sont réels.

Le théorème de l'indice pour les $\mathcal{D}$-modules presqu'elliptique est aussi évidemment invariant par immersion fermée. Or il est toujours possible de plonger analytiquement une variété à bord analytique réelle $(X, \partial X)$ dans une boule euclidienne $(B, \partial B)$ (de sorte que X coupe transversalement ∂B le long de ∂X); moyennant tout ceci on se ramène ainsi au cas où X est une variété réelle, plongée dans un espace numérique $\mathbf{R}^n$ et $U = X \cap B$, où B est la boule unité de $\mathbf{R}^n$ pour la métrique euclidienne usuelle de $\mathbf{R}^n$; puis par plongement de M, au cas où $X = \mathbf{R}^n$, $U = B$.

On peut enfin remplacer B par un des ellipsoïdes complexes B_ε (définis par $\| \operatorname{Im} z \|^2 + 2\varepsilon(\| \operatorname{Re} z \|^2 - 1) \leq \varepsilon^2$) qui en forment un système fondamental adapté de voisinages tubulaires. Dans ce dernier cas le groupe K-théorique $K_{\mathrm{SS}\,B_\varepsilon}(T^*\mathbf{C}^n)$ est canoniquement isomorphe (par l'isomorphisme de Bott) à $K_B(\mathrm{SS}\,B) \approx K_{\{0\}}(B)$ et ce dernier est le groupe libre engendré par le symbole du complexe de Koszul (toujours l'isomorphisme de Bott): le théorème d'indice est vrai, par construction, pour le complexe de Koszul et est donc ainsi établi dans tous les cas.

4.4 Formule de l'indice avec paramètres.

Comme l'ont montré Atiyah, Segal et Singer, le théorème de l'indice se généralise bien au cas d'un système elliptique dépendant d'un paramètre, ou au cas d'un système équivariant sous l'action d'un groupe compact.

Dans le cas d'une famille de systèmes elliptiques dépendant d'un paramètre, l'analogue analytique se décrit comme suit: on se donne d'abord une submersion $f : X \to Y$ de variétés analytiques (réelles ou complexes), et un complexe M de $\mathcal{D}$-modules relatifs bien filtrables, i.e. M est un complexe de $\mathcal{D}_{X/Y}$-modules à cohomologie cohérente, qui représente une famille de systèmes différentiels sur les fibres de X, dépendant analytiquement du point de base. Le faisceau $\mathcal{D}_{X/Y}$ des opérateurs différentiels verticaux est, comme $\mathcal{D}$, filtré par le degré des opérateurs, et le gradué $\operatorname{gr}\mathcal{D}_{X/Y}$ s'identifie à l'algèbre des fonctions polynomiales sur les fibres du fibré cotangent relatif T^*X/Y, à coefficients analytiques. La variété caractéristique relative du complexe cohérent M, que nous noterons $Z = \operatorname{car} M/Y \subset T^*X/Y$, est le support de $\operatorname{gr} M$ (pour n'importe quelle bonne filtration). Si Z est à base compacte, on associe à M l'élément

$$[M]_Z^{\mathrm{top}} \in K_Z(T^*X/Y) \tag{4.3}$$

Pour définir l'image directe et l'indice relatif, on se donne en outre, comme plus haut, une sous-variété à bord $U \subset X$, décrivant le domaine dans lequel on calcule faisceau des solutions. Pour assurer que les solutions varient de façon cohérente par rapport au paramètre de la base Y il faut en outre que U vérifie une condition de régularité relative convenable: nous supposerons que la restriction de f à U est une fibration analytique réelle. On définit alors le micro-support relatif $\mathrm{SS}(U/Y) \subset T^*X/Y$, dont la fibre en un point $y \in Y$ est $\mathrm{SS}\, U_y \subset T^*X_y$[18].

On dit que M est relativement elliptique le long de U (ou pour f et U) si $\mathrm{car}\, M/Y \cap \mathrm{SS}\, U/Y$ est contenu dans la section nulle de T^*X/Y, i.e. si M_y est elliptique le long de U_y, pour tout $y \in Y$.

Le théorème de finitude pour les systèmes elliptiques dans le cas absolu se généralise alors comme suit: si M est relativement elliptique le long de U et si f est propre sur $U \cap \mathrm{supp}\, M$ (par exemple si la restriction de f à U est propre), l'image directe dérivée $\mathrm{R}f_*(\mathrm{Sol}(M, \mathcal{O}_U))$ est cohérente sur $\mathcal{O}_Y$. Son support est contenu dans Z', où

(4.4) Z' est la projection de $Z = \mathrm{car}\, M/Y \cap \mathrm{SS}\, U/Y$.

La généralisation "avec paramètres" de la formule d'indice est alors la suivante: si de plus Y est compact (ou germe d'une partie compacte d'une variété analytique)[19], l'indice; élément de $K_{Z'}(Y)$ défini par $\mathrm{R}f_*(\mathrm{Sol}(M, \mathcal{O}_U)$ est donné par

$$[Rf_*(M_{|U})] = \overline{f}_*([M]_{Z|\mathrm{SS}(U/Y)}) \quad \text{(image } K\text{-théorique)} \tag{4.5}$$

C'est aussi l'image K-théorique du produit $b[M]_Z[\mathcal{O}_U]_{\mathrm{SS}(U/Y)} \in K_{Z \cap \mathrm{SS}(U/Y)}(T^*X/Y)$.

On démontre ce résultat en recopiant, avec paramètres, les constructions du cas absolu ci-dessus: l'assertion de cohérence et la formule d'indice relative sont stables par immersion fermée, et en utilisant les formules du §4, on voit qu'elles sont vraies pour M si et seulement si elles le sont pour $M_{\mathbf{R}}$ (noter que même si le bord de U est vide, celui des voisinages tubulaires complexes de $U_{\mathbf{R}}$ ne le sont pas, et notre méthode ne permet pas d'éviter les variétés à bord). On se ramène ainsi au cas où X et Y sont réelles. Il existe alors un plongement analytique i de X dans un espace numérique $\mathbf{R}^n$ qui transforme $(U, \partial U)$ en $i(X) \cap (B, \partial B)$[20], où B est la boule unité. Quitte à remplacer U par $B \times Y$ et M par l'image directe (relative) $(i, f)_+ M$, on est ramené au cas où f est la projection $\mathbf{R}^n \times Y \to Y$ et $U = B \times Y$. On peut enfin épaissir un peu B et le remplacer par un ellipsoïde B_ε et se ramener au cas où U est une fibré en ellipsoïdes: $U = B_\varepsilon \times Y$. Dans ce dernier cas, M représente une famille elliptique d'opérateurs de Toeplitz, à laquelle on applique le résultat du §2.

Dans l'énoncé d'Atiyah et Singer Y peut en fait être n'importe quel espace (de dimension finie pour que la version avec supports du théorème de périodicité soit vraie); ceci sort à moitié du cadre analytique ci-dessus, mais sauf pour la cohérence de $\mathrm{R}f_*(\mathrm{Sol}(M, \mathcal{O}_U))$ qui n'a plus de sens, la démonstration est la même.

Dans le deuxième cas X est muni de l'action d'un groupe compact G, U est invariant et M est équivariant. La définition de l'ellipticité ou de la presqu'ellipticité de M le long de U est la même que dans le cas absolu, mais comme G opère sur les solutions de M, il est naturel de chercher à décomposer celles-ci en représentations irréductibles de G et, lorsque M est elliptique, de lui associer un indice équivariant $\mathrm{Ind}_G(M, U)$, élément de l'anneau

[18] Il suffirait en fait que la projection $T^*X \to T^*X/Y$ soit propre sur $\mathrm{SS}\, U$, au dessus de $U \cap \mathrm{supp}\, M$. Le micro-support $\mathrm{SS}(U/Y)$ est alors la projection de $\mathrm{SS}\, U$.

[19] La définition du symboles K-théorique $[M]$ fait intervenir une résolution finie globale de M ou de $M_{\mathbf{R}}$, et n'a de sens que si la base est compacte, aussi devons nous supposer Y compact ici, bien que la K-théorie ou le théorème d'indice d'une famille d'opérateurs de Toeplitz aient un sens sur un espace de dimension finie.

[20] Il convient de remplacer $(U, \partial U)$ par $(U \cap \mathrm{supp}\, M, \partial U \cap \mathrm{supp}\, M)$ si seul $U \cap \mathrm{supp}\, M$ est compact.

$R(G)$ engendré par les représentations finies de G (ou anneau des caractères de G). Dans la situation équivariante, Atiyah a aussi défini les groupes de K-théorie équivariante $K_Z^G(X)$ pour lesquels le théorème de périodicité de Bott est encore vrai (il se démontre comme au §2 pour la K-théorie usuelle, compte tenu du théorème de Kuiper équivariant, qui affirme que l'espace $U_G(H)$ des G-automorphismes unitaires d'un G-espace de Hilbert H est contractile si chaque représentation irréductible de G est de multiplicité infinie dans H). On peut donc définir l'image K-théorique équivariante comme dans le cas $G = 1$ du §2. Au $\mathcal{D}$-module M, muni d'une bonne filtration (qu'on peut toujours supposer équivariante en prenant des moyennes sur G) on associe un élément de K-théorie $[M] \in K^G_{\operatorname{car} M}(T^*X)$. On définit de même l'élément $[\mathcal{O}_U] \in K^G_{\operatorname{SS} U}(T^*X)$ si U (donc aussi $\operatorname{SS} U$) est invariant. Le théorème d'indice équivariant affirme encore que l'indice $\operatorname{Ind}_G(M, U)$ est l'image K-théorique du produit $[M][\mathcal{O}_U]$. Ici encore il suffit de recopier la démonstration ci-dessus, où toutes les opérations peuvent être effectuées de façon équivariante: les passages de X à $X_{\mathbf{R}}$, de U à U_ε ou B à B_ε sont fonctoriels donc compatibles à l'action de G, ainsi que les images directes équivariantes; et si X est réel il est essentiellement possible de le plonger de façon équivariante dans un espace numérique (de façon précise il existe un voisinage V du compact $U \cap \operatorname{supp} M$ et un plongement analytique réel équivariant φ de V dans un G-espace vectoriel euclidien E tel que $\varphi(V \cap U) = \varphi(V) \cap B$, où B est la boule unité de E). Il est évidemment encore possible de démontrer un théorème d'indice équivariant avec paramètres englobant les deux généralisations que nous venons d'évoquer.

Dans ses conférences M. Vergne étudie un théorème d'indice équivariant plus général que celui que nous avons décrit ici, et que nous n'aborderons pas. Il concerne l'indice des systèmes transversalement elliptiques pour l'action de G sur X: un $\mathcal{D}$-module M est transversalement elliptique le long d'un sous-ensemble U invariant si $\operatorname{car} M \cap \operatorname{SS} U \cap G^\perp \subset \{0\}$, où $G^\perp$ désigne l'ensemble des covecteurs de X qui sont orthogonaux à l'action de G; on a de même une notion de $\mathcal{D}$-module presque elliptique transversalement. Pour un tel module on montre que l'espace des solutions $\operatorname{Sol}(M, \mathcal{O}_U)$ est "tempéré" i.e. chaque représentation irréductible de G n'y figure qu'avec une multiplicité finie, "à croissance polynomiale", de sorte que le caractère "trace de l'action de G dans $\operatorname{Sol}(M, \mathcal{O}_U)$" est bien défini comme distribution sur G. Le problème est alors de calculer cette multiplicité, ou de façon analogue, de calculer la distribution trace de l'action de G dans $\operatorname{Sol}(M, \mathcal{O}_U)$.

§5. Cas relatif. Ellipticité relative. Énoncé du théorème

5.1 Préliminaires.

Dans ce paragraphe nous nous proposons de décrire le théorème d'indice relatif (cf. Boutet de Monvel-Malgrange). Voici ce dont il s'agit: on se donne deux variétés analytiques X, Y, une application analytique $f : X \to Y$, et un $\mathcal{D}_X$-module cohérent, bien filtrable, M sur X (M n'est plus un $\mathcal{D}_{X/Y}$-module comme au §4). Comme au §4 on se donne encore une sous-variété à bord U analytique réelle, pour préciser le domaine sur lequel on calcule les solutions, et on supposera que U est fibré au dessus de Y (comme on l'a mentionné, il suffirait que la projection $T^*X \to T^*X/Y$ soit propre sur $\operatorname{SS} U \cap \operatorname{car} M$). L'analogue relatif du faisceau $\mathcal{O}_X$ avec lequel nous avons calculé les solutions dans le cas absolu est le module de transfert $\mathcal{D}_{X\to Y}$, et l'analogue de $\mathcal{O}_U$ est $\mathcal{D}_{U\leftarrow X} = \mathcal{O}_U \otimes_{\mathcal{O}_X} \mathcal{D}_{X\to Y}$. L'analogue relatif du faisceau des solutions dans les fibres, tenant compte des relations différentielles sur la base entre celles-ci, est le complexe $M \otimes^L_{\mathcal{D}} \mathcal{D}_{X\to Y}$, et l'analogue relatif du complexe des solutions globales dans les fibres de U est le $\mathcal{D}_Y$-module image directe

$f_+(M_{|U}) = \mathrm{R}f_*(M \otimes^L_{\mathcal{D}} \mathcal{D}_{U\leftarrow X})$.

Fixons maintenant les notations pour la suite: nous noterons

$$(5.1)\qquad \begin{cases} H = X \times_Y T^*Y \\ \overline{f} : H \longrightarrow T^*Y & \text{la deuxième projection} \\ F : H \longrightarrow T^*X & \text{l'application cotangente} \end{cases}$$

si f est submersive, F identifie H à $\operatorname{car} \mathcal{D}_{X\to Y}$, ensemble des covecteurs horizontaux de T^*X.

Lorsque f est une immersion fermée (et $U = X$), on a vu au §3 que l'image directe des $\mathcal{D}_X$-modules se décrit agréablement (équivalence de Kashiwara). En particulier $f_+\mathcal{D}_X$ est l'image directe faisceautique du module de transfert $\mathcal{D}_{X\to Y}$; sa variété caractéristique est le pull-back Z' de X dans T^*Y, et son symbole $[f_+\mathcal{D}_X]^{\mathrm{top}}_{Z'}$ est l'élément de Bott de la sous-variété Z' dans T^*Y. En général si M est un $\mathcal{D}_X$-module bien filtré, $f_+M = f_*(M \otimes \mathcal{D}_{X\to Y})$ est bien filtré, et le gradué $\operatorname{gr} f_+M$ s'identifie à $\overline{f}_*F^{-1}(\operatorname{gr} M)$. Dans ce cas le théorème d'indice se réduit au fait que le symbole $[f_+M]^{\mathrm{top}}_{Z'}$ est l'image K-théorique $\overline{f}_*F^{-1}([M]^{\mathrm{top}}_Z)$. Aussi nous décrétons dans ce cas que tout $\mathcal{D}_X$-module est relativement elliptique.

Pour tenir compte du domaine U nous notons encore

$$(5.1)\qquad \begin{cases} U_e = \mathrm{SS}\, U \times_X H \\ F_e : U_e \longrightarrow T^*X & \text{l'application qui prolonge } F (F_e(\eta, h) = \eta + h) \\ \overline{f}_e : U_e \longrightarrow T^*Y & \text{la projection} \end{cases}$$

Lorsque f est submersive, F_e est une immersion fermée et son image $F(H) + \mathrm{SS}\, U$ joue le rôle de $\mathrm{SS}(\mathcal{D}_{U\leftarrow X})$.

Remarquons que l'image directe $f_+(M_{|U})$ ne dépend évidemment que de $\operatorname{supp} M \cap U$ ($f_+(M_{|U}) = f_+(M_{|V})$ si $\operatorname{supp} M \cap U = \operatorname{supp} M \cap V$). Pour le calcul d'indice nous nous intéressons d'abord à la cohérence de l'image directe $f_+(M_{|U})$; pour assurer celle-ci il est indispensable de faire comme au §4.4 une hypothèse de régularité relative sur U: dans la suite de ce paragraphe nous supposerons (comme au §4.4) que la restriction de $f : U \to X$ est une fibration, i.e. localement un produit au sens analytique réel (il suffira en fait que ceci soit vrai au voisinage de $\operatorname{supp} M \cap U$), lorsque f est submersive, et que U est une variété à bord (ou à coins). En général on supposera qu'on peut se ramener à cette situation fibrée après immersion fermée, et en modifiant U de façon à préserver $\operatorname{supp} M \cap U$. Ceci implique que l'application F_e ci-dessus est propre (ou que la projection $\mathrm{SS}\, U \to T^*X/Y$ est propre). La définition ci-dessous de l'ellipticité relative garde un sens sans une telle hypothèse de régularité sur U, et implique, jointe à une hypothèse de propreté, un résultat de finitude en chaque point de Y, mais dans ce cas on ne peut guère espérer mieux que la constructibilité de l'image directe $f_+(M)$, en un sens qui pour les $\mathcal{D}$-module reste encore à définir.

Nous définissons plus bas l'ellipticité et la presqu'ellipticité relatives de M pour f et U, lorsque f est une submersion. En général f est de la forme $f = p \circ i$ où p est une submersion et i une immersion fermée (on peut par exemple choisir pour $i : X \to X \times Y$ le graphe de f ($i(y) = (y, f(y)) \in X \times Y$) et pour et $p : X \times Y \to Y$ la deuxième projection). Nous dirons que M est relativement presqu'elliptique pour f et U si i_+M est presqu'elliptique pour p et $i(U)$ (ou $U \times Y$) au sens de la définition ci-dessous. On se ramène ainsi au cas où f est une submersion, qui est le cas substantiel pour l'image directe comme pour la formule d'indice relative. Nous supposerons le plus souvent dans la suite que f est submersive (en fait que c'est la deuxième projection $X = X' \times Y \to Y$).

Houzel et Schapira ont montré que si M est relativement presqu'elliptique et bien filtré, et si f est propre sur $\operatorname{supp} M \cap U$ et U fibré comme ci-dessus, l'image directe $f_+(M_{|U})$ est cohérente (bien filtrée) et qu'on a une majoration: $\operatorname{car} f_+ M \subset Z'$ où Z' se déduit simplement de la variété caractéristique Z de M:

(5.3) $\operatorname{car} f_+ M \subset Z'$ avec $Z' = \overline{f} F^{-1}(Z \cap \mathrm{SS}\, U)$, $Z = \operatorname{car} M \subset T^*X$.

La formule d'indice relative décrit alors cas l'élément K-théorique $[f_+(M_{|U})]^{\mathrm{top}}_{Z'}$: on verra (théorème 5.4) que c'est l'image K-théorique du produit $[M]^{\mathrm{top}}_Z \cdot [\mathcal{D}_{U \leftarrow X}]_{H + \mathrm{SS}\, U}$ (voir énoncé ci-dessous), ou plus correctement, $[f_+(M_{|U})]^{\mathrm{top}}_{Z'} = \overline{f}_{e*} F_e^{-1}([M]^{\mathrm{top}}_Z)$. Lorsque X est un point on retrouve la formule du cas absolu ci-dessus.

Lorsque le bord ∂U est vide (X complexe, $U = X$), la condition d'ellipticité relative est toujours satisfaite. Dans ce cas si f est propre, Malgrange a démontré qu'on a une description analogue de l'élément de K-théorie analytique $[f_+ M]^{\mathrm{an}}_{Z'}$ associé à $f_+ M$: c'est l'image K-théorique $\overline{f}_* F^{-1}[M]^{\mathrm{an}}_Z$ de l'élément associé à M (ceci implique bien sûr dans ce cas la relation pour les éléments de K-théorie topologique). Dans le cas plus général que nous examinons ici (∂U non vide), il n'y a plus de formule en K-théorie analytique, car f n'est plus propre au sens de la théorie holomorphe et l'image directe en K-théorie analytique n'est plus définie; seule subsiste la formule topologique.

5.2 Bonnes filtrations, bonnes résolutions verticales, ellipticité relative.

Afin de motiver les constructions ci-dessous, nous commençons par décrire en partie l'idée de la démonstration, qui est d'essayer d'abord de décrire le gradué associé à l'image directe (pour la filtration image). Dans leur travail sur l'image direct Houzel et Schapira utilisent pour cela de "bonnes résolutions verticales" des $\mathcal{D}_X$-modules. Rappelons en la définition dans le cas où f est submersive. On note comme plus haut $\mathcal{D}_{X/Y}$ le faisceau des opérateurs verticaux. Si M est un $\mathcal{D}_X$-module bien filtré on introduit la filtration verticale $M^v_p = M_p \mathcal{D}_{X/Y}$. Le gradué associé est un $\operatorname{gr} \mathcal{D}_X$-module, noté $\operatorname{gr}^v M$. Le faisceau $\operatorname{gr}^v \mathcal{D}_X$ s'identifie au faisceau $\mathcal{D}_{[H/T^*Y]}$ des opérateurs différentiels sur H, verticaux pour la projection $\overline{f}: H \to T^*Y$, "à coefficients polynomiaux dans les fibres de T^*Y". Le gradué $\operatorname{gr}^v M$ représente un système d'opérateurs différentiels verticaux sur $H = X \times_Y T^*Y$, paramétré par T^*Y. Une bonne résolution verticale de M est une résolution localement libre bien filtrée L telle que $\operatorname{gr}^v L$ soit une résolution de $\operatorname{gr}^v M$ et $\operatorname{gr} \operatorname{gr}^v L$ une résolution de $\operatorname{gr} \operatorname{gr}^v M$. Ici nous avons encore noté gr les gradués associés aux filtrations déduites de la filtration initiale de M ou de $\mathcal{D}$; les bigradués $\operatorname{gr} \operatorname{gr}^v \mathcal{D}$ et $\operatorname{gr}^v \operatorname{gr} \mathcal{D}$ sont isomorphes mais il n'en est en général pas de même de $\operatorname{gr} \operatorname{gr}^v M$ et $\operatorname{gr}^v \operatorname{gr} M$. Il résulte du théorème des syzygies qu'il existe de telles résolutions localement, et semi-globalement (i.e. au voisinage de tout compact) si X est de Stein. La première idée est d'approcher le gradué $\operatorname{gr} f_+ M$ par l'image directe $\mathrm{R} f_* \operatorname{gr}^v M$: les deux ne sont pas égaux, mais la théorie de la suite spectrale d'un module muni de deux filtrations montre qu'il existe une suite spectrale dont le deuxième terme est $\mathrm{R} f_* \operatorname{gr}^v M$, et qui souhaiterait converger vers $\operatorname{gr} f_+ M$. Ceci motive en tout cas la définition suivante:

Définition 5.1. *On suppose f submersive. On dit qu'un $\mathcal{D}_X$-module bien filtré M est relativement elliptique (par rapport à f et U) si le $\mathcal{D}_{H/T^*Y}$-module $\operatorname{gr}^v M$ est relativement elliptique (pour F et $F^{-1}U$), au sens du §4.4.*

Dans cette définition nous supposons, comme annoncé plus haut et au §4, que la restriction de $f: U \to X$ est une fibration (au moins au voisinage des points de $U \cap \operatorname{supp} M$).

On vérifie en fait aisément que la condition est locale sur X, et qu'elle ne dépend pas du choix d'une bonne filtration de M. Remarquons que les sections de $\mathcal{D}_{H/T^*Y}$ sont homogènes (polynomiales) dans les fibres de T^*Y, de sorte que $\mathrm{gr}^v M$ est relativement elliptique au dessus de T^*Y si et seulement s'il l'est le long de la section nulle de T^*Y. La condition d'ellipticité relative équivaut donc aussi bien à la suivante:

(ER) *(f submersive) M est relativement elliptique si pour tout $\xi \in \mathrm{SS}\, U$ non nul et tout germe de section $s \in M_x$, défini au voisinage du point de base x de ξ, il existe $P \in \mathcal{D}_X$ vertical, elliptique en ξ, tel que $sP = 0$.*

De façon équivalente.

(ER) bis *(f submersive) M est relativement elliptique si, au voisinage de tout point de X, il existe un $\mathcal{D}_{X/Y}$-module cohérent N relativement elliptique (pour f et U) tel que M soit isomorphe à un quotient de $N \otimes_{\mathcal{D}_{X/Y}} \mathcal{D}_X$.*

Un tel N existe semi-globalement, i.e. au voisinage de tout compact, si X est de Stein.

Comme dans le cas absolu, l'ellipticité relative n'est pas stable par immersion fermée, mais elle implique la presqu'ellipticité relative, qui elle est stable:

Définition 5.2. *M est relativement presqu'elliptique s'il existe une famille adaptée U_ε de voisinages tubulaires de U telle que M soit relativement elliptique sur U_ε pour $\varepsilon > 0$ assez petit.*[21]

Dans le cas général, si $f = p \circ i$ avec i immersion fermée et p submersive, on dira que M est relativement presqu'elliptique (pour f et U) si i_+M est presqu'elliptique au sens précédent pour p et $i(U)$ (ceci ne dépend pas du choix de i et p).

5.3 Finitude.

Nous supposons toujours f submersive. Nous complétons comme suit les notations ci-dessus:

(5.4)
$$\begin{cases} F^v: H \longrightarrow T^*(H/T^*Y) = T^*(X/Y) \times_Y T^*Y, & \text{la section nulle du fibré cotangent relatif de } H \text{ sur } T^*Y, \text{ et } Z^v \\ Z^v = \mathrm{car}(\mathrm{gr}^v M) \subset T^*(H/T^*Y), & \text{la variété caractéristique relative de } \mathrm{gr}^v M \\ U^v = U \times_Y H & \end{cases}$$

$\mathrm{SS}(U^v/T^*Y)$ est la projection de $\mathrm{SS}\, U$ dans $T^*(H/T^*Y) = T^*(X/Y) \times_Y T^*Y$.

Proposition 5.3. *Soit M un $\mathcal{D}_Y$-module bien filtré, relativement elliptique, et posons $Z'' = \overline{f}F^{v-1}(Z^v \cap \mathrm{SS}\, U^v)$. Supposons en outre f propre sur $\mathrm{supp}\, M \cap U$. Alors*

1. *$\mathrm{R}f_*(\mathrm{gr}^v M_{|U^v})$ est à cohomologie cohérente et bien filtrée sur T^*Y (et homogène dans les fibres de T^*Y).*
2. *On a $\mathrm{supp}\, \mathrm{R}f_*(\mathrm{gr}^v M_{|U^v}) \subset Z''$.*
3. *Si Z^v est de base compacte on a $[\mathrm{R}f_*(\mathrm{gr}^v M)]^{\mathrm{top}}_{Z''}$ est l'image K-théorique de la restriction $[\mathrm{gr}^v M]^{\mathrm{top}}_{Z^v}|_{\mathrm{SS}(U^v/T^*Y)}$ par la projection $\mathrm{SS}(U^v/T^*Y) \to T^*Y$.*

C'est aussi bien l'image K-théorique du produit $[\mathrm{gr}^v M]^{\mathrm{top}}_{Z^v} \cdot [\mathcal{O}_{U^v}]_{\mathrm{SS}(U^v/T^*Y)}$.

---

[21] Nous n'entrons pas ici dans le détail de ce "grossissement": si U est un produit $U' \times Y$ il suffit de grossir U' en U'_ε; en général et en particulier si Y est réel il est prudent de grossir aussi Y)

Comme au §4.4 on se ramène au cas où U est une fibré en ellipsoïdes: $U = B_\epsilon \times Y$. Alors $\mathrm{gr}^v M$ est un complexe de $\mathcal{D}_{H/T^*Y}$-modules verticaux sur $H \sim B_\epsilon \times T^*Y$, paramétré par T^*Y. Si M est relativement elliptique pour f et U, $\mathrm{gr}^v M$ est elliptique pour la projection $\overline{f} : B_\epsilon \times T^*Y \to T^*Y$, et la proposition 5.3 est une répétition du théorème d'indice du §4.4, en tenant compte en plus de l'homogénéité dans les fibres de T^*Y, qui définit la graduation de $\mathrm{R}f_*(\mathrm{gr}^v M_{|U^v})$.

5.4 Ellipticité relative géométrique — énoncé du théorème.

Notons que dans le cas absolu, resp. avec paramètres du §4.4, l'ellipticité équivaut à la condition géométrique $\mathrm{car}\, M \cap \mathcal{O}_U \subset \{0\}$, (resp. $\mathrm{car}\, M/Y \cap \mathrm{SS}(U/Y) \subset \{0\}$). Dans le cas relatif l'ellipticité relative s'écrit donc (en supposant toujours f submersive et que $f_{|U}$ est une fibration)

$$Z^v \cap \mathrm{SS}(U^v/T^*Y) \subset \{0\} \tag{5.5}$$

et comme on a remarqué, il faut et il suffit pour cela que ce soit vrai au dessus de la section nulle de T^*Y, parce que Z^v est bihomogène.

Lemme. *On a* $F^{v-1}(Z^v) \subset F^{-1}(Z)$.

En effet si ξ est un covecteur non nul horizontal dans T^*X et P un opérateur différentiel d'ordre m elliptique en ξ, il est immédiat que $\mathrm{gr}^v P$ homogène de degré m dans les fibres de T^*Y, et d'ordre 0 comme opérateur différentiel, et qu'il est aussi elliptique au point ξ. Si alors $\xi \notin Z$, i.e. pour tout $s \in M_x$ (x base de ξ) il existe P elliptique en ξ tel que $sP = 0$, le "symbole" $\mathrm{gr}^v P$ annule aussi $\mathrm{gr}^v s = 0$, autrement dit $\xi \notin Z^v$.

Compte tenu de ceci si M est relativement elliptique pour (f et U), la variété caractéristique $Z = \mathrm{car}\, M$ satisfait à la condition géométrique suivante d'ellipticité:

(EG) $F^1_\epsilon(Z)$ est contenu dans la section nulle de U_ϵ.

de façon équivalente: Z ne contient pas de covecteur de la forme $\eta + F(\xi)$, avec $\xi \in H$, $\eta \in \mathrm{SS}\, U$, $\eta \neq 0$.

Ceci suffit pour donner un sens à l'expression K-théorique qui figure dans la formule d'indice; en effet on a alors $Z' = \overline{f}_\epsilon F_\epsilon^{-1}(Z) \supset Z''$, et si de plus Z' est de base compacte l'image K-théorique est bien définie:

$$\overline{f}_{\epsilon *} F_\epsilon^{-1}([M]_Z^{\mathrm{top}}) \in K_{Z'}(T^*Y) \tag{5.6}$$

Notons néanmoins que la condition d'ellipticité géométrique ne suffit en général pas à assurer la finitude de $f_+ M$, parce qu'elle ne fait intervenir en gros que les termes de plus haut degré du système différentiel correspondant à M; seule la condition plus haut, qui fait intervenir le gradué $\mathrm{gr}^v M$ permet de prendre en compte l'effet de termes verticaux de trop bas degré donne une condition correcte pour la finitude. Par exemple le lecteur vérifiera simplement dans le cas où $Y = \mathbf{R}$, $X = \mathbf{R} \times W$ où W est une variété Riemannienne compacte, que les deux opérateurs $A = \Delta_W$, $B = \Delta_W + \partial/\partial y$ ont le même symbole, mais que le premier est relativement elliptique, et pas le second: l'image directe du $\mathcal{D}_X$-module correspondant à B n'est pas de type fini sur $\mathcal{D}_Y$.

On peut alors compléter comme suit la proposition 5.3:

Théorème 5.4. *Avec les hypothèses et les notations du théorème 5.3, et en supposant vérifiée la condition d'ellipticité géométrique, posons* $Z' = \overline{f}_\epsilon F_\epsilon^{-1}(Z)$, *et supposons* Z' *de base compacte. Alors*

1. $f_+(M)$ *est à cohomologie cohérente et bien filtrée.*

2. *On a* car $f_+(M) \subset Z'$.
3. *Si* Z' *est de base compacte on a* $[f_+(M)]^{\mathrm{top}}_{Z'} = \overline{f}_{e*}F_e^{-1}[M]^{\mathrm{top}}_Z$ *(image* K*-théorique).*

(On a aussi $Z' = \overline{f}(Z \cap \mathrm{SS}\, U)$). Voici une idée de la fin de la démonstration: ici encore l'assertion de cohérence et la formule d'indice relative sont stables par immersion fermée, et on se ramène, comme au §4.4 et ci dessus, au cas où U est une fibré en ellipsoïdes: $U = B_\varepsilon \times Y$. On a déjà montré que $\mathrm{R}f_*(\mathrm{gr}^v M_{|U^v})$ est cohérent (homogène) sur T^*Y. Or c'est le premier terme de la suite spectrale naturelle pour l'image directe $f_+(M_{|U})$. Comme ce premier terme est cohérent, la suite spectrale converge, par cohérence noethérienne, vers $\mathrm{gr}\, f_+(M_{|U})$ de sorte que $\mathrm{gr}\, f_+(M_{|U})$ et $f_+(MU)$ sont cohérents. Ceci démontre la première assertion. La seconde est évidente car on a $Z'' \subset Z'$. Les assertions 1) et 2) ont été démontrées par Houzel et Schapira.

Si de plus Z' est à base compacte, l'élément K-théorique $[f_+M]_{Z'}$ est bien défini. Il est associé à l'aboutissement de la suite spectrale est donc égal à l'élément K-théorique $[R\overline{f}_*\, \mathrm{gr}^v M]_{Z'}$ associé au premier terme de cette suite. D'après la proposition 5.3, c'est donc l'image K-théorique de la restriction $[\mathrm{gr}^v M]^{\mathrm{top}}_{Z'}{}_{|\mathrm{SS}(U^v/T^*Y)}$ (modulo l'agrandissement $Z'' \subset Z'$). Il reste encore à comparer à l'élément cette image K-théorique et celle du théorème.

Il est assez facile de voir que $[M]_Z$ a même image K-théorique dans $K_{Z'}(T^*X)$ que sa "déformation au cône tangent" $[\mathrm{gr}^v\, \mathrm{gr}\, M]_{CZ}$, où $CZ \subset H \times_Y T^*X/Y$ désigne le cône tangent de Z le long de H; il reste alors à montrer que les éléments $[\mathrm{gr}\, \mathrm{gr}^v M]$ et $[\mathrm{gr}^v\, \mathrm{gr}\, M]$ ont la même image K-théorique dans $K_{Z'}(T^*X)$. Ce dernier point n'est pas évident parce qu'il met en jeu deux filtrations dont la comparaison n'est pas immédiate. Nous ne reproduirons pas ici sa démonstration, qui utilise une astuce de géométrie algébrique exploitée par Y. Laurent pour l'étude de la seconde microlocalisation, et que le lecteur trouvera décrite en détail dans l'article de Malgrange pour le cas qui nous intéresse ici.

BIBLIOGRAPHIE

Atiyah M. F., *K-theory*, Benjamin, Amsterdam.

Atiyah, M. F., Bott, R., *On the periodicity theorem for complex vector bundles*, Acta Math. **112** (1964), 229–247.

Atiyah M. F., Bott R., Schapiro A., *Clifford modules*, Topology (1964), no. 3, supplément, 3–83.

Atiyah M. F., Hirzebruch F., *Vector bundles and homogeneous spaces*, Diff. Geometry, Proc. Symp. Pure Math., Amer. Math. Soc., Providence (1961), 7–38.

Atiyah M. F., Segal G. B., *The index of elliptic operators II*, Ann. Math. **87** (1968), 531–545.

Atiyah M. F., Singer I. M. [1], *The index of elliptic operators on compact manifolds*, Bull. Amer. Math. Soc. **69** (1963), 422–433.

— [2], *The index of elliptic operators I*, Ann. Math. **87** (1968), 484–530.

— [3], *The index of elliptic operators III*, Ann. Math. **91**, 546–604.

— [4], *The index of elliptic operators IV*, Ann. Math. **92** (1970), 119–138.

Baum P., Fulton W., Mac Pherson R., *Riemann-Roch and topological K-theory for singular varieties*, Acta Math. **143** (1979), no. 3-4, 155–192.

Bernstein I. M., Gelfand S. I., *Meromorphy of the function* P^λ, Funkc. Anal. i Prilozen **3** (1969), 84–85; Funct. Anal. appl. **3** (1969), 68–69.

Bernstein I. N., *Modules over rings of differential operators. An investigation of the fundamental solution of equations with constant coefficients*, Funkc. Anal i Prilozen **5** (1971), no. 2, 1–16; Funct. Anal. appl. **5** (1971), 89–101.

Björk J. E., *Rings of Differential Operators*, North Holland, 1979.

Borel A. et al. [1], *Intersection cohomology*, vol. 50, Progress in Math., Birkhäuser, 1984.

— [2], *Algebraic D-modules*, vol. 2, Perspect. in Math., Academic Press, 1987.

Boutet de Monvel L. [1], *On the index of Toeplitz operators of several complex variables*, Inventiones Math. **50** (1979), 249–272; Séminaire EDP 1979, Ecole Polytechnique.

— [2], *Systèmes presqu'elliptiques: une autre démonstration de la formule de l'indice*, Astérisque **131** (1985), 201–216.

— [3], The index of almost elliptic systems, E. de Giorgi Colloquium, vol. 125, Research notes in Math., Pitman, 1985, pp. 17–29.

Boutet de Monvel L., Lejeune M., Malgrange B., *Opérateurs différentiels et pseudodifférentiels*, Séminaire, Grenoble (1975-76).

Boutet de Monvel L., Malgrange B., *Le théorème de l'indice relatif*, Ann. Scientifiques de l' E.N.S. **23** (1990), 151–192.

Boutet de Monvel L., Sjöstrand J., Sur la singularité des noyaux de Bergman et de Szegö, Astérisque, vol. 34-35, 1976, pp. 123–164.

Brylinski J. L., Dubson A., Kashiwara M., *Formule de l'indice pour les modules holonômes et obstruction d'Euler locale*, C.R. Acad. Sci. **293** (1981), 573–576.

Cornalba H., Griffiths P., *Analytic cycles and vector bundles in non compact algebraic varieties*, Invent. Math. **28** (1975), 1–106.

Godement R., *Topologie algébrique et théorie des faisceaux.*, Activités scientifiques et industrielles, Hermann Paris, 1958.

Grauert H., *Ein theorem der analytischen Garben-theorie und die modulräume komplexe Structuren*, IHES Sci. Publ. Math. **5** (1960).

Grothendieck A., *SGA 5, Théorie des intersections et théorème de Riemann-Roch*, vol. 225, Lecture Notes in Math., Springer Verlag, 1971.

Hirzebruch F., *Neue topologische Methoden in der algebraiche geometrie*, Springer Verlag, Berlin.

Hörmander L., *The Analysis of Linear Partial Differential Operators, vol. III et IV*, vol. 124, Grundlehren der Math. Wiss..

Houzel Ch., Schapira P., *Images directes de modules différentiels*, C.R.A.S. **298** (1984), 461–464.

Hurewicz W., Wallman H., *Dimension theory*, vol. 4, Ann. of Math. series, Princeton University Press, 1941.

Jänich K., *Vektorraumbündel und das Raum der Fredholm operatoren*, Math. Ann. **161** (1965), 129–142.

Kashiwara M. [1], *Index theorem for a maximally overdetermined system of linear differential equations*, Proc. Jap. Acad. **49-10** (1973), 803–804.

— [2], *b-fonctions and holonomic systems*, Invent. Math. **38** (1976), 33–54.

— [3], *Analyse microlocale du noyau de Bergman*, Séminaire Goulaouic-Schwartz, Ecole Polytechnique, exp. n. 8 (1976-77).

— [4], Introduction to the theory of hyperfunctions, Seminar on microlocal analysis, Princeton University Press, 1979, pp. 3–38.

— [5], *Systems of microdifferential equations*, vol. 34, Progress in Math., Birkhäuser, 1983.

Kashiwara M., Kawai T., Kimura T., *Foundations of algebraic analysis*, vol. 37, Princeton Math. Series, Princeton University Press, Princeton N.J., 1986.

Kashiwara M., Kawai T., Sato M., *Microfunctions and pseudodifferential equations*, Lecture Notes, Springer-Verlag **287** (1973), 265–524.

Kashiwara M., Schapira P. [1], *Microlocal study of sheaves*, vol. 128, Astérisque, 1985.

— [2], *Sheaves on manifolds*, vol. 292, Grundlehren der mathematischen Wissenschaften, Springer, 1990.

Laumon G., *Sur la catégorie dérivée des $\mathcal{D}$-modules filtrés*, thèse, Orsay (1983).

Laurent Y., *Théorie de la deuxième microlocalisation dans le domaine complexe*, vol. 53, Progress in Math., Birkhäuser, 1985.

Levy R. N., *Riemann-Roch theorems for complex spaces*, Acta Math. **158** (1987), 149–188.

Malgrange B., *Sur les images directes de $\mathcal{D}$-modules*, Manuscripta Math. **50** (1985), 49–71.

Melin A., Sjöstrand J., *Fourier Integral operators with complex valued phase functions*, Lecture Notes **459** (1974), 120–223.

Ohana M., *Ellipticité et K-théorie*, Note aux C.R.A.S. (à paraître).

Pham F., *Progress in Math.*, vol. 2, Birkhäuser, 1980.

Schapira P., *Microdifferential systems in the complex domain*, vol. 269, Grundlehren der mathematischen Wissenschaften, Springer, 1985.

Schapira P., Schneiders J.P. [1], *Paires elliptiques I - Finitude et dualité*, C.R. Acad. Sci. **311** (1990), 83–86.

— [2], *Paires elliptiques II - Classes d'Euler et indice*, C.R. Acad. Sci. **312** (1991), 81–84.

Segal G., *Fredholm complexes*, Quat. J. Math., Oxford Series **21** (1970), 385–402.
Steenrod N., *The topology of fibre bundles*, Ann. of Math. series, Princeton University Press **144** (1951).
Verdier J. L., Catégories dérivées, état 0, SGA $4\frac{1}{2}$, vol. 569, Springer Lecture Notes in Math., 1977, pp. 262–311.

QUANTUM GROUPS

C. De Concini S.N.S Pisa, C. Procesi Univ. di Roma

INDEX

Introduction.

The theory of Quantum groups, although rather young, since the expression Quantum group seems to appear only with the work of Drinfeld if the late 70's, has already produced a very large number of papers treating a variety of ideas and applications. From the purely algebraic theory to the theory of Yang-Baxter equations (motivated in turn by the exactly solvable planar lattice models of statistical mechanics), from knot theory and invariants of 3-manifolds to the fusion rules of conformal field theory or the applications to modular representations. This is thc whole more remarkable since in a way there is no unified theory nor a unified definition but rather several very interesting examples.

As may already be clear from the previous remarks a comprehensive treatment of this theory is at this moment untimely and perhaps not possible. In these notes we restrict ourselves to present a very particular part of the theory, that which refers mostly to certain *quantizations* of remarkable Poisson groups, by this one means a Hopf algebra depending from a parameter q which for a special value of q is commutative and hence the coordinate ring of an algebraic group (which acquires an extra Poisson structure from the deformation) (cf. §10,11), in particular we will concentrate on the case when the value of the natural parameter q is a root of 1. The choice is exclusively due to the fact that we thought useful to try to present in a unified way 4 articles, the first by De Concini Kac and the remaining by De Concini Kac Procesi. In fact these notes are a reelaboration of these papers with a minimum amount of introductory material. Not everything here is proved and we refer to several papers specially for some foundational material and for the case q generic.

With respect to the papers there are several improvements (which in part can be considered as new results) and corrections of mistakes. In particular we should point out the introduction of a suitable form of the quantum group, the algebra A introduced in §12, and the corresponding discussion of the Frobenius map which we hope is more complete here.

Unfortunately the theory as it stands now is based on a heavy computational machinery which only at the end reveals its structural features. This is certainly due to the present very algebraic definitions, a more geometric definition of the objects we study should be possible and it may eliminate some of the most annoying computations.

CHAPTER 1

Hopf algebras

§1 Hopf Algebras.

1.1 The theory of *Quantum groups* is still in a state of development so that in reality there is not yet a completely satisfactory definition of what a quantum group should be; there are nevertheless several rather interesting examples and some of these will be the objects of these notes.

There are several inputs to the theory and we will be able only to discuss some, we may recall the main ones:

Poisson groups and classical Yang-Baxter equations.
Quantization of Poisson structures.
Universal R–matrices and Yang-Baxter equations.
Applications of q–analogues to Lie algebras and groups.
Applications to knots and 3-dimensional manifolds.
Tensor categories and fusion rules.

We will discuss several kinds of algebras, we will assume to have fixed a base field F, algebras will be assumed to be over F as well as tensor products, in some sections F will be the field of rational functions $\mathbb{C}(q)$.

1.2 Representation theory in a rather abstract form may be viewed as the theory of modules over a given algebra, Representations of groups or of Lie algebras can be presented in this setting. For a group G we have the group algebra $F[G]$ with the same module theory, for a Lie algebra $\mathfrak{g}$ instead one constructs its: *Universal enveloping algebra.* This is an associative algebra $U(\mathfrak{g})$ containing $\mathfrak{g}$ as a Lie subalgebra and universal with respect to maps of $\mathfrak{g}$ into associative algebras. By its very definition the representation theory of $\mathfrak{g}$ is completely equivalent to the theory of modules for $U(\mathfrak{g})$. There is though an important extra feature in this theory; given two representations M, N of a group, or of a Lie algebra, also $M \otimes N$ is a representation, setting $g(m \otimes n) := gm \otimes gn$ in the group case and $a(m \otimes n) := am \otimes n + m \otimes an$ for Lie algebras. Moreover the dual M^* is also a representation by $< g\phi|m >:=< \phi|g^{-1}m >$ for groups and $< a\phi|m >:=< \phi| -am >$ for Lie algebras, and finally among all representation we have the trivial (1 dimensional) representation, for a group the identity and for a Lie algebra the 0 action.

Of course these definitions are tied together. For a Lie group G with Lie algebra $\mathfrak{g}$ let $g = exp(ta)$ be a 1-parameter subgroup, that the action of a is the infinitesimal action induced from the one of g, one easily sees then that the formulas given for the Lie algebra actions are just the infinitesimal formulations of the formulas valid for the group.

This extra structure in the representation theory is reflected in extra structure for $F[G]$ or $U(\mathfrak{g}$. An axiomatization of this leads to the notion of Hopf algebras.

Let us go through the main steps. Given two representations M, N of an algebra R, $M \otimes N$ is a representation of $R \otimes R$, setting $a \otimes b(m \otimes n) := am \otimes bn$. If we want to have an R module structure on $M \otimes N$ it is natural then to require the existence of a homomorphism:

$$\Delta : R \to R \otimes R. \text{ a comultiplication.}$$

For a group algebra it is given by:

$$\Delta(g) := g \otimes g, \; g \in G \otimes G.$$

For a Lie algebra $\mathfrak{g}$ we set:

$$\Delta(a) := a \otimes 1 + 1 \otimes a, \; a \in \mathfrak{g}.$$

These definitions give rise to the previously described module structures. It is convenient to express this by a:

Definition. *For a given vector space A a linear map $\Delta : A \to A \otimes A$ is called* **a comultiplication** *and A* **a coalgebra**.

It should be clear how to define homomorphisms of coalgebras, tensor product of two coalgebras and finally the property, for the comultiplication, to be co-associative or co-commutative.

1.3 If A is an algebra and a coalgebra it is natural to consider compatibility properties between the two operations.

Definition. *We say that A is a* **bialgebra** *if the comultiplication is a algebra homomorphism while the multiplication is a coalgebra homomorphism.*

We stress again that, giving on A a comultiplication allows us to perform the tensor product of two A modules M, N. Since obviously $M \otimes N$ is an $A \otimes A$ module we consider it as an A module setting $av := \Delta(a)v$ (we say that we *restrict* the $A \otimes A$ module to A, via the map Δ).

1.4 It is usually necessary, to have a reasonable theory, to assume that the tensor product construction is canonically associative, i.e. that for 3 modules $(M \otimes N) \otimes P \simeq M \otimes (N \otimes P)$. This is insured if the map Δ is *coassociative* i.e. if the 2 maps from A to $A \otimes A \otimes A$ given by $(\Delta \otimes 1)\Delta$ and $(1 \otimes \Delta)\Delta$ coincide. In commutative diagram form:

$$\text{(1.4.1)} \qquad \begin{array}{ccc} A & \xrightarrow{\Delta} & A \otimes A \\ {\scriptstyle \Delta}\downarrow & & \downarrow{\scriptstyle \Delta \otimes 1} \\ A \otimes A & \xrightarrow[1 \otimes \Delta]{} & A \otimes A \otimes A \end{array}$$

It is quite obvious that the comultiplication defined for group algebras or enveloping algebras is coassociative (it is enough to check such a property on a set of generators, in this case the group or the Lie algebra itself).

Let us make a side remark useful for the theory of enveloping algebras:

Definition. *An element $a \in A$ of a coalgebra is called* **primitive** *if*

$$\Delta(a) = a \otimes 1 + 1 \otimes a.$$

It is an immediate exercise that:

Proposition. *The primitive elements form a Lie subalgebra.*

Less obvious but not very difficult is (cf. Inserire):

Theorem. *The primitive elements of $U(\mathfrak{g})$ are exactly $\mathfrak{g}$.*

Thus in a sense, which is easy to make precise, to give the Lie algebra $\mathfrak{g}$ is equivalent to give its enveloping algebra with its comultiplication.

1.5 In the module theory of groups or Lie algebras the dual of a module can be also given a module structure. In general, given an A module M we may define, by transposition, an action $< a * \phi | m >:=< \phi | am >$. This is *not* a module structure for A but rather for its opposite algebra A^0, due to the fact that transposition is an antihomomorphism. Thus, in order to have an A module structure on M^*, it is necessary to give in A a antihomomorphism S. Then we may define:

$$< a\phi | m >:=< \phi | S(a)m > .$$

In the theory of Hopf algebras such a antihomomorphism is called an ***antipode***.

For the group algebra the antipode is the unique antihomomorphism which extends the map $g \to g^{-1}$ in G. For the enveloping algebra the antipode is the unique antihomomorphism which extends the map $a \to -a$ in $\mathfrak{g}$.

There are several extra conditions that one may want to impose to an antipode. The most natural is the one that insures that, given two modules M, N, the natural map

$$M^* \otimes N^* \to (M \otimes N)^*$$

is a module homomorphism. It is easily seen that the natural requirement is that S should be a ***coalgebra homomorphism*** or that the diagram:

(1.5.1)
$$\begin{array}{ccc} A & \xrightarrow{\Delta} & A \otimes A \\ {\scriptstyle S}\downarrow & & \downarrow{\scriptstyle S\otimes S} \\ A & \xrightarrow[\Delta]{} & A \otimes A \end{array}$$

is commutative.

1.6 We are approaching the definition of Hopf algebra, we still need one small data, the trivial representation. In the language of Hopf algebras this is a homomorphism $\varepsilon : A \to F$ to the base field, called the *counit*.

For the group algebra it is the one which maps the group G to 1. For the enveloping algebra it is the one which maps the Lie algebra $\mathfrak{g}$ to 0.

If we wish to have that the trivial representation acts as a unit element under tensor product we should assume that: the counit is compatible with comultiplication, and it behaves as a counit element. That is we have a map which we may call $(\varepsilon \otimes \varepsilon) : A \otimes A \to F$ given by $(\varepsilon \otimes \varepsilon)(a \otimes b) := \varepsilon(a)\varepsilon(b)$. We need

$$(\varepsilon \otimes \varepsilon)\Delta = \varepsilon.$$

Furthermore map $A \otimes A$ to $A = F \otimes A$ by $\varepsilon \otimes 1$. We need $(\varepsilon \otimes 1)\Delta = 1_A$ (left counit), similarly on the right. In diagram form:

(1.6.1)
$$\begin{array}{ccc} A & \xrightarrow{\Delta} & A \otimes A \\ & {\scriptstyle 1_A}\searrow & \downarrow{\scriptstyle \varepsilon\otimes 1} \\ & & A \end{array}$$

The final useful requirement is made in order to insure that, for a finite dimensional module M the identity map $1 : F \to End(M) = M^* \otimes M$ and the trace $tr : M \otimes M^* \to F$ are module homomorphisms.

The antipode is like an *inverse* in a group. Formally the requirement is that, composing Δ with $1 \otimes S$ and with the multiplication m, we have the identity map:

(1.6.2)
$$\begin{array}{ccc} A & \xrightarrow{\Delta} & A \otimes A \\ {\scriptstyle 1_A}\downarrow & & \downarrow{\scriptstyle 1\otimes S} \\ A & \xleftarrow[m]{} & A \otimes A \end{array}$$

similarly with $S \otimes 1$.

Finally we can summarize:

Definition. *A* **Hopf Algebra** *is a bialgebra with counit and antipode satisfying all the extra conditions imposed by the commutative diagrams considered.*

1.7 A different way to understand these definitions is to think of duality. If A is an algebra its multiplication map induces a transpose map $A^* \to (A \otimes A)^*$. Of course in general $(A \otimes A)^*$ is not isomorphic to $A^* \otimes A^*$. Nevertheless this is true certainly in the finite dimensional case and it can sometimes be restored by restricting to a suitable subspace of the dual. For instance if A is graded with finite dimensional homogeneous parts we may replace A^* with $\sum A_k^*$. Then it is clear that the dual of a Hopf algebra can often be defined as a new Hopf algebra. Multiplication and comultiplication are exchanged by duality. For instance the comultiplication in the algebra of functions on a group gives by duality the convolution.

We want to point out a useful construction. Given an algebra A, it can be considered as an $A \otimes A^0$ module setting:

$$(a \otimes b)c := acb.$$

The map

(1.7.1)
$$A \xrightarrow{\Delta} A \otimes A \xrightarrow{1\otimes S} A \otimes A^0$$

is clearly a homomorphism and so it induces a module structure of A on itself. This action is called the *Adjoint action* and will be indicated by

$$ad(a)(b) := \sum_i u_i b S(v_i), \quad \text{if } : \Delta(a) = \sum_i u_i \otimes v_i.$$

§2 Categories, functors etc...

2.1 We have promised to explain in a different way the theory. This is a digression and not essential for the sequel but it is useful to establish a point of view. Let us first describe in a more concrete way commutative Hopf algebras (since there are so many operations let us stress that, when we speak of a commutative Hopf algebra we assume that the multiplication is commutative).

The standard way to construct such a Hopf algebra is the following.

Let G be a group and $A := F[G]$ be the algebra of functions on G, this is not the group algebra where multiplication is convolution, but the ordinary algebra of functions. In general for an infinite group we may want to restrict to some special functions connected with some extra structure on G (continuous, L^1, algebraic etc.). Let us stick to the algebraic case which is what we are exactly treating. Thus G indicates an affine algebraic group and A its algebraic coordinate ring (the ring of regular algebraic functions on G).

The coordinate ring of $G \times G$ is then $A \otimes A$. The multiplication map $G \times G \to G$ induces Δ as $\Delta(f)(x,y) := f(xy)$, the antipode is $S(f)(x) := f(x^{-1})$ the counit $\varepsilon(f) = f(1)$. The various axioms required are just the translations, for the functions on G, of the properties of groups. This special example extends easily to the general case, provided that we use the categorical language.

2.2 Let us recall that a category C consists in giving:

i) A class of elements, called the *objects*.

ii) For each pairs of objects A, B a set $hom_C(A, B)$ of elements called *morphisms*.

iii) For any 3 objects A, B, C a *composition of morphisms*:

$$hom_C(A, B) \times hom_C(B, C) \to hom_C(A, C)$$

These data are subject to the following simple axioms:

a) Composition is associative whenever defined.

b) For every object A there is a morphism $1_A \in hom_C(A, A)$ which behaves as a unit element under compositions:

$$1_A f = f \text{ and } g 1_A = g, \text{ when defined.}$$

It is often convenient to indicate by $f : A \to B$ instead of $f \in hom_C(A, B)$, also if there is no ambiguity we will write simply $hom(A, B)$ instead of $hom_C(A, B)$.

2.3

Definition. *Given two categories $\mathcal{A}$, $\mathcal{B}$ a* **covariant functor** $F : \mathcal{A} \to \mathcal{B}$ *consists in a map that, to each object $A \in \mathcal{A}$ assigns an object $F(A) \in \mathcal{B}$ and to a map $f : A \to B$ assigns a map $F(f) : F(A) \to F(B)$.*

Again we assume the simple axioms:
a) F preserves composition, i.e. $F(fg) = F(f)F(g)$.
b) F preserves the identities, i.e. $F(1_A) = 1_{F(A)}$.

2.4 Functors can be made into a category by defining *natural transformations.*

Definition. *Given two functors $F, G : \mathcal{A} \to \mathcal{B}$ a* natural transformation *consists in giving, for each object $A \in \mathcal{A}$ a map $\eta_A : F(A) \to G(A)$ such that, for any $A, C \in \mathcal{A}$ and any mapping $f : A \to C$ the diagram:*

(2.4.1)
$$\begin{array}{ccc} F(A) & \xrightarrow{\eta_A} & G(A) \\ {\scriptstyle F(f)}\downarrow & & \downarrow{\scriptstyle G(f)} \\ F(C) & \xrightarrow[\eta_C]{} & G(C) \end{array}$$

is commutative.

The set of all natural transformations between two functors is usually denoted by $Nat(F, G)$. It is easy to see that in this way the functors become a new category.

REMARK We are being somewhat sloppy at the level of foundations, one should distinguish between sets and classes and perhaps talk about small categories. In fact there are several ways of avoiding to fall in the dangerous pitfalls of set theory but we will ignore all these things which for our purposes are of little use.

Sets, groups, algebras, topological spaces, differentiable manifolds, analytic spaces, algebraic manifolds, modules, sheaves etc., are all examples of categories if one uses as morphisms the appropriate maps. Functors, on the other hand are obtained once we perform canonical constructions on these objects, (the free group on a set, the singular cohomology, the *Ext* and *Tor* functors etc.), natural transformations appear when we compare different canonical constructions.

Together with covariant functors there appear naturally *contravariant functors* which reverse composition. From a very abstract point of view one can treat them in the same way, introducing the opposite $\mathcal{C}^0$ of a category $\mathcal{C}$. $\mathcal{C}^0$ has the same objects as $\mathcal{C}$ but we set

$$hom_{\mathcal{C}^0}(A, B) := hom_{\mathcal{C}}(B, A).$$

2.5 Given a category $\mathcal{C}$ a set valued covariant functor F on $\mathcal{C}$ is called *representable* if there exists an object $A \in \mathcal{C}$ such that F is naturally isomorphic to $hom(A, -)$. To keep our notations not too complicated let us indicate by F_A the functor $hom(A, -)$. The main idea in the language of representable functors is given by Yoneda's Lemma.

Lemma. *Given two objects A, $B \in \mathcal{C}$ we have a canonical identification of $Nat(F_A, F_B)$ with $hom(B, A)$.*

Proof. If $g \in hom(B, A)$, it induces a natural transformation of functors given by $\eta^g(f) := fg$. Of course $g = \eta^g(1_A)$. Conversely if η is a natural transformation and $g := \eta_A(1_A)$ we have, for any given mapping $f : A \to C$, the commutative diagram

(2.5.1)
$$\begin{array}{ccc} F_A(A) & \xrightarrow{\eta_A} & F_B(A) \\ {\scriptstyle F_A(f)}\downarrow & & \downarrow{\scriptstyle F_B(f)} \\ F_A(C) & \xrightarrow[\eta_C]{} & F_B(C) \end{array}$$

since $f = f1_A = F_A(f)(1_A)$ we have that $\eta_C(f) = F_B(f)(\eta_A(1_A)) = fg$, as desired.

A rather fancy way of expressing this lemma is:

Proposition. *The Category C^0 opposite to C is isomorphic to the category of representable functors, a full subcategory of the functors.*

This rather formal lemma and proposition have deep implications if one is willing to adjust to the functorial language. The first idea is the following, all the possible constructions we are used to make with sets we can do with set valued functors, by just applying them to the values of the functors. Whenever the resulting functor is representable we have thus a construction, analogous to the set theoretical one considered, but in the given category C.

In fact for coherence, since we identify the opposite of the category with the functors we should really think of the dual construction.

For instance one starts by saying that two objects A, B have a *coproduct* if the functor $F_A \times F_B$ is representable, then we write $A \sqcup B$ for an object representing it, (unique up to unique isomorphism by Yoneda's Lemma). Then it is convenient to recall that the product of two sets is equipped with the two canonical projection maps. In our setting these are two natural transformations and so correspond (again by Yoneda's Lemma) to two maps of A, B to $A \sqcup B$ wich satisfy the *universal property*, given any object C and two maps of A, B to C there is a unique map $A \sqcup B \to C$ that factors the two given maps.

For our purposes the main remark is that, in the category of commutative algebras, the copruduct is the usual tensor product.

2.6 Now comes the second idea, we know how to define algebraic structures giving operations and axioms on these operations, for instance groups. We can try to perform similar constructions in a category by considering functors.

So assume that in C we are given a *group valued* functor F. Assume now that F is representable by an object A and that we also have the copruduct $A \sqcup A$. Then the group axioms translate immediately by Yoneda's lemma in a series of facts for A. The group multiplication gives a map $\Delta : A \to A \sqcup A$ and the associative law gives the coassociativity of this map. The inverse map of groups induces the antipode and the unit element induces the counit map if we assume that the *point* is a representable functor (then a representing object is called an *initial object*). The point is the constant functor that to each object associates a fixed set with 1 element.

An expressive way of thinking is to consider the object F as a group in the category of functors and A as a group in C^0, identified to the category of representable functors.

In particular we think of the opposite of the category of commutative algebras as the *Affine schemes.* If A is an algebra we denote $Spec(A)$ the same object in the opposite category. A morphism $A \to B$ is thought as a point of $Spec(A)$ with coordinates in B, or a point *rational* over B. Finally an *affine group* is a group in the category of affine schemes. Its rational points over B form a group, functorial in B. Its *coordinate ring* A is then a commutative Hopf algebra. We rather leave the details to the reader, but we want to point out some interesting representable (to be checked) functors on the category of commutative algebras, which then correspond to commutative Hopf algebras.

The General Linear Group $GL(n, A)$ is the group of invertible matrices of A. The Orthogonal, Special orthogonal and Symplectic groups can be defined in the usual way for any algebra and are functorial. Finally one which may be a little less usual. Suppose we have a finite dimensional algebra R over F, for each commutative algebra A set $R_A := R \otimes A$. The group of A algebra automorphisms of R_A is representable. For instance when $R = M_n(F)$, the $n \times n$ matrices, we get the Projective linear group. But also if R is a Lie algebra this defines a group closely connected to the usual Lie group construction.

2.7 The theory, as we have presented it, does not apply directly to general (not necessarily commutative) Hopf algebras. The reason being that in the category of associative algebras the coproduct is the *free product* and not the tensor product. If we want to retain at least in part the suggestive language of functors we can still do so paying a small price.

Definition *We say that two maps* $f_1 : A_1 \to B$, $f_2 : A_2 \to B$ *commute if* $f_1(a_1)$ *commutes with* $f_2(a_2)$ *for all* $a_1 \in A_1$ *and* $a_2 \in A_2$

Let us denote by $F_A(B)$ the set of all maps from A to B and by $(F_{A_1}(B) \times F_{A_2}(B))_c$ the pairs of commuting maps. It is then easy to see that this functor is represented by the tensor product $A_1 \otimes A_2$.

Then a bialgebra will consist in giving functorially a *partial group law* $(F_A(B) \times F_A(B))_c \to F_A(B)$. Similar considerations hold for the antipode and the counit and we leave them to the reader.

§3 Complete reducibility.

3.1 Hopf algebras with the property that their finite dimensional representations are always completely reducible seem of particular interest and we want to start their study.

Definition. *A Hopf algebra for which finite dimensional representations are always completely reducible will be called* **reductive**.

First of all some generalities. If A is a Hopf algebra M, N two modules, we have a natural module structure on $M \otimes N^*$ given by $a(m \otimes \phi) := \sum a_i m \otimes b_i \phi$ where $\Delta a = \sum a_i \otimes b_i$ and $< c\phi | n > = < \phi | S(c) n >$. If N is finite dimensional we identify $M \otimes N^*$ to $hom(N, M)$ and we see that the action of a on a homomorphism f is:

$$(af)(n) = \sum a_i f(S(b_i)n).$$

For any module M we define $M^A := \{m \in M | am = \varepsilon(a)m, \forall a \in A\}$. Borrowing the notion from group representations we may consider M^A as the subspace of *invariants.* Let us assume, for the rest of this paragraph, that all modules under consideration are finite dimensional, we also assume for the Hopf algebra that the antipode S is invertible. Then we see that:

Lemma. $hom_A(N, M) = hom(N, M)^A$.

Proof. In fact if $f(an) = af(n)$ we have $(af)(n) = \sum a_i f(S(b_i)n) = \sum a_i S(b_i) f(n) = \varepsilon(a) f(n)$.

As for the converse notice that $A \otimes A$ acts on $hom(N, M)$ by ${}^{a \otimes b} f = af S(b)$. From the coassociativity of the comultiplication we get $\forall a \in A$ that, setting $\Delta(a) = \sum_i a_i \otimes b_i$, $\Delta(a_i) = \sum_j a_{ij} \otimes b_{ij}$ and $\Delta(b_i) = \sum_h c_{ih} \otimes d_{ih}$ we have:

$$\sum_{ij} a_{ij} \otimes b_{ij} \otimes b_i = \sum_{ih} a_i \otimes c_{ih} \otimes d_{ih}$$

in $A \otimes A \otimes A$. Applying the operator $(1 \otimes S^{-1})(1 \otimes m)(1 \otimes S \otimes 1)$ to the above identity and using the fact that $\sum_h S(c_{ih}) d_{ih} = \varepsilon(b_i)$ we get

$$\sum_{ij} a_{ij} \otimes S^{-1}(b_i) b_{ij} = \sum_i a_i \otimes \varepsilon(b_i) 1 = a \otimes 1$$

From this is it clear that $\forall n \in N$, $f \in hom(N, M)$ we get

$$af(n) = \sum_{ij} a_{ij} f(S(b_{ij}) b_i n).$$

If $f \in hom(N,M)^A$ then $\sum_j a_{ij} f S(b_{ij}) = \varepsilon(a_i) f$ so we get

$$af(n) = \sum_{ij} a_{ij} f(S(b_{ij}) b_i n) = f(\sum_i \varepsilon(a_i) b_i n) = f(an).$$

Lemma. *If every exact sequence $0 \to M \to N \to P \to 0$ splits, when P is the trivial 1 dimensional representation then A is reductive.*

Proof. The usual proof as for Lie algebras works. Let $M \subset N$ be modules, let us consider the restriction map $\pi : hom(N,M) \to hom(M,M)$. It is an A module homomorphism. In $hom(M,M)$ the multiples of the identity form a (1 dimensional submodule) P isomorphic to the trivial representation (given by ε). Let R be the submodule of $hom(N,M)$ mapping to P. From our hypothesis the projection of R to P splits and we thus can find an element $f \in hom(M,N)^A$ restricting to the identity of M. Thus we have that f is a module homomorphism and its kernel is a complement for M.

We finally need a sufficient criterion to satisfy the hypothesis of the lemma. As usual one can by induction restrict to the sequences $0 \to M \to N \to P \to 0$ in which M is irreducible. If M is also trivial, a simple sufficient condition is to assume taht if $J := Ker(\varepsilon)$ is the augmentation ideal, we have that the ideal generated by the commutators $[J,J]$ equals J (which is verified for the enveloping algebra of a semisimple Lie Algebra).

For non trivial M the following is sufficient. Let Z be the center of A, for every irreducible representation M the center acts as scalars on M, we call the corresponding homomorphism the *character* of M. It is then enough to assume that, if the representation is not trivial the character should be different from the trivial character. In fact if $c \in Z$ acts on M by a scalar b different from $\varepsilon(c)$ it is clear that the operator $c - b$ has a 1 dimensional kernel which provides the required complement.

For the enveloping algebra of a semisimple Lie Algebra the Casimir element gives this required element.

Summarizing:

Theorem. *Given a Hopf algebra A, if:*

1) Setting $J := Ker(\varepsilon)$ the augmentation ideal, we have that the ideal generated by the commutators $[J,J]$ equals J.

2) For a non trivial irreducible representation the central character is different from the trivial character.

Then A is reductive.

CHAPTER 2

FINITE DIMENSIONAL REPRESENTATIONS

§4. Finite dimensional representations of algebras and filtrations.

4.1. In this section we will discuss the theory of finite dimensional representations of algebras. Let A be an associative algebra with a unit element 1 over a field $\mathbb{F}$ and let us denote by $\overline{\mathbb{F}}$ the algebraic closure of $\mathbb{F}$. For an algebra A we denote by Spec A the set of all equivalence classes of finite dimensional irreducible representations over $\overline{\mathbb{F}}$ so that, if A is a finitely generated commutative algebra over $\overline{\mathbb{F}}$ we are in fact thinking of its maximal spectrum.

If Z is the center of A , then (by Schur's lemma) we have a canonical map (the central character map)

$$\text{Spec } A \xrightarrow{\chi} \text{Spec } Z. \tag{4.1.1}$$

A good theory of finite dimensional representations can be developed when the algebra A is:

1) finitely generated over $\mathbb{F}$,

2) a finite module over its center Z (this already implies that every irreducible module is finite dimensional)

3) It has a suitable reduced trace map.

(cf. [A] and [P1-2-3-4]).

For 3) we mean the following:

Definition. *A trace map in an algebra R is a linear map $tr : R \to R$ satisfying the following axioms:*

For all pairs of elements $a, b \in R$:

i) $tr(ab) = tr(ba)$

ii) $tr(a)b = b\, tr(a)$

iii) $tr(tr(a)b) = tr(a)tr(b)$

Notice that the values of the map tr form a subalgebra of the center.

The definition has been set is such a way that it is suitable for universal algebra. Algebras with trace form a category, where morphisms are algebra morphisms which are compatible with the trace.

An ideal I in an algebra A with trace is an ordinary ideal closed under trace, so that A/I inherits a trace.

In this category we can construct a free algebra (with trace) on a given set of generators x_i. By definition a free algebra R on elements x_i in a category of algebras is an algebra with elements x_i such that, for any algebra A in the category and elements $a_i \in A$, there is a unique morphism of $\pi : R \to A$ with $\pi(x_i) = a_i$.

To construct a free algebra with trace consider the ordinary free algebra in the x_i with coefficients in a polynomial ring in formal variables $tr(M)$ as M runs over all non commutative monomials in the x_i considered up to cyclic order equivalence, plus a further indeterminate t which stands for the trace of 1. The trace is then defined in the obvious way and it is fairly evident that this object is a free algebra in our category.

Let us indicate by:

(4.1.2) $$F < x_i >_{i \in I}$$

the free algebra with trace in a set of variables x_i, $\{i \in I\}$.

In practice it is often important to restrict the trace of 1 to be a given number and in the theory to be developed a positive integer d, we will assume from now on to be in characteristic 0.

4.2 To proceed further towards a theory of finite dimensional representations we introduce a particular category of algebras with trace as in ([P4]).

Once we have a trace map in an algebra in characteristic 0 we want to define, for any given element a, elements $\sigma_k(a)$ (to be thought formally as the elementary symmetric functions in the *eigenvalues* of a) by declaring that $tr(a^k)$ should be the sum of the kth powers of the eigenvalues.

To do this recall that in the ring $\mathbb{Q}[x_1, x_2, \ldots, x_m]$ one defines the elementary symmetric functions by the identity

$$\prod(t - x_i) := \sum_{i=0}^{d}(-1)^i \sigma_i t^{d-i}$$

and the power sums functions $\psi_k := \sum x_i^k$.

It is easy to prove the existence, for every $k \leq m$, of a polynomial $p_k(y_1, \ldots, y_k)$ with rational coefficients, independent of m and such that:

$$\sigma_k = p_k(\psi_1, \psi_2, \ldots, \psi_k)$$

We then set

$$\sigma_k(a) := p_k(tr(a), tr(a^2), \ldots, tr(a^k)).$$

Next we can formally define for every element a in A and for every integer d a dth*–characteristic polynomial*:

$$\chi_{d,a}[t] := \sum_{i=0}^{d}(-1)^i \sigma_i(a) t^{d-i}.$$

This formal polynomial is useful in representation theory if we have the formal Cayley–Hamilton theorem:

Definition. *i) We say that an algebra R with trace satisfies the $d-th$ formal Cayley-Hamilton theorem if $\chi_{d,a}[a] = 0$ for every $a \in R$.*

ii) We say that R has **degree** *d if it satisfies the $d-th$ formal Cayley-Hamilton theorem and $tr(1) = d$.*

It can be shown that condition i) alone implies that $tr(1)$ must be a positive number $\leq d$.

4.3 Algebras with trace of degree d form an interesting category of algebras which will be denoted by C_d. We shall give now some examples of algebras in this category:

First we may consider the algebras $M_d(A)$ with A a commutative algebras with the ordinary trace. In particular we single out $M_d\overline{\mathbb{F}}$, the matrices over the algebraic closure of the base field. This algebra plays the same role as $\overline{\mathbb{F}}$ does for commutative algebras and we stress it with a definition (R is an algebra with trace):

Definition. *A $d-$dimensional representation of an algebra with trace R with values in a commutative algebra A is a homomorphism $\varrho : R \to M_d(A)$ compatible with the trace map. If $A = \overline{\mathbb{F}}$ we think of this representation as a* **geometric point**.

One can start building some of the theory from this example using the language of category theory.

Let us think to $A \to M_d(A)$ as a functor from the category of commutative algebras to $\mathcal{C}_d$. If $j : A \to B$ is a mapping of commutative algebras we indicate by j_d the induced morphism on matrices.

We ask whether this functor admits an adjoint, i.e. a functor $\mathcal{F}(R)$ going in the opposite direction with a natural isomorphism:

$$Hom_{\mathcal{C}_d}(R, M_d(A)) = Hom(\mathcal{F}(R), A). \tag{4.3.1}$$

The right hand side of 4.3.1 is the set of d dimensional representations of R with values in A while the left hand side is the set of commutative algebra homomorphisms.

This is a typical *universal problem* as explained in 2.3. The identity map $1_{\mathcal{F}(R)}$ must correspond to a canonical representation:

$$i : R \to M_d(\mathcal{F}(R))$$

such that, any other representation $j : R \to M_d(A))$ there is a unique map $\overline{j} : \mathcal{F}(R) \to A$ such that $j = \overline{j}_d i$ or in diagram form:

$$\begin{array}{ccc} R & \xrightarrow{\ i\ } & M_d(\mathcal{F}(R)) \\ & {\scriptstyle j} \searrow & \big\downarrow {\scriptstyle \overline{j}_d} \\ & & M_d(A)) \end{array} \tag{4.3.1}$$

We think of i as a generic point and as Spec$(\mathcal{F}(R))$ as the scheme of $d-$dimensional representations.

The solution of this problem is quite simple, first we solve it for the free algebra (with trace) $F < x_i >_{i \in I}$. It is fairly clear that the solution is the following:

Consider a set of variables $\xi_{hk}^{(i)}$ as $i \in I,\ h, k = 1, \dots, d$. Let $F[\xi_{hk}^{(i)}]$ be the usual (commutative) polynomial ring in this set of variables. Consider the matrices $\xi^{(i)} := (\xi_{hk}^{(i)})$ with entries the variables $\xi_{hk}^{(i)}$, for given i, and the map:

$$i : F < x_i >_{i \in I} \to M_d(F[\xi_{hk}^{(i)}])$$

given by $i(x_i) := \xi^{(i)}$. It is well defined since $F < x_i >_{i \in I}$ is free over the x_i and it is easily seen that it is universal. In particular $\mathcal{F}(F < x_i >_{i \in I}) = F[\xi_{hk}^{(i)}]$.

Now take any algebra R and present it as $R = F < x_i >_{i \in I} / J$. A $d-$ dimensional representation of R is just a $d-$ dimensional representation of $F < x_i >_{i \in I}$ vanishing on the ideal J. Let us consider the two sided ideal generated by $i(J)$ in $M_d(F[\xi_{hk}^{(i)}])$. As every two sided ideal in a matrix algebra it has the form $M_d(\overline{J})$ for an ideal $\overline{J}$ of $F[\xi_{hk}^{(i)}]$. It is then clear that the composition:

$$i : F < x_i >_{i \in I} \to M_d(F[\xi_{hk}^{(i)}]) \to M_d(F[\xi_{hk}^{(i)}]/\overline{J})$$

factors through R giving rise to the universal map for R.

There is an important feature of this construction which connects it with invariant theory. Let $G := Aut(M_d(F))$ be the group of algebra automorphisms. Since every automorphism is inner $G = GL(d,F)/F^* = PGL(d,F)$ is the projective linear group. Since $M_d(A) = A \otimes_F M_d(F)$ the group G acts as automorphisms of the A algebra $M_d(A)$ and in fact it acts on the functor $M_d(A)$, an element g acting by $1 \otimes g$. Thus by the universal property, for any algebra R for every $g \in G$ we have an induced automorphism $\varphi(g) : \mathcal{F}(R) \to \mathcal{F}(R)$ so that we have the commutative diagram:

(4.3.2)
$$\begin{array}{ccc} R & \xrightarrow{\ i\ } & M_d(\mathcal{F}(R)) \\ {\scriptstyle i}\downarrow & & \downarrow{\scriptstyle \varphi(g)_d} \\ M_d(\mathcal{F}(R)) & \xrightarrow{1\otimes g} & M_d(\mathcal{F}(R)) \end{array}$$

Or $(1 \otimes g)i = \varphi(g)_d i$, hence for two group elements we have:

$$\varphi(gh)_d i = (1\otimes gh)i = (1\otimes g)(1\otimes h)i = (1\otimes g)\varphi(h)_d i = \varphi(h)_d(1\otimes g)i = \varphi(h)_d\varphi(g)_d i = (\varphi(h)\varphi(g))_d$$

By the universal property we have:

(4.3.3)
$$\varphi(gh) = \varphi(h)\varphi(g)$$

hence setting

$$ga := \varphi(g^{-1})a$$

we have a G action on $\mathcal{F}(R)$.

Now with this G–action the formula 4.3.2 takes a new light. In fact it can be formulated as follows:

Define an action of G on $M_d(\mathcal{F}(R)) = \mathcal{F}(R) \otimes M_d(F)$ as the diagonal action associated to the 2 actions described, then the formula 4.3.2 is equivalent to the following:

Lemma. *The map i has its image contained in the ring of invariants $M_d(\mathcal{F}(R))^G$.*

In fact one has a much more precise statement:

Theorem. a) *The kernel of i is the minimal ideal J such that $R/J \in C_d$.*

b) *The image of i* **equals** *the ring of invariants $M_d(\mathcal{F}(R))^G$.*

c) $i(tr(R)) = \mathcal{F}(R)^G$.

Before giving a sketch of the proof of this theorem let us remark its implications for algebras in C_d. The functor $\mathcal{F}(R)$ is now to be thought as a functor from the category C_d to the category of commutative algebras with a G action. We have a *left inverse* to this functor, if A is a commutative algebra with a G action it is given by $M_d(A)^G$ (relative to the diagonal action). It is easily seen that under these functors the property of being finitely generated is preserved. It is therefore possible to translate problems of representation theory into problems of algebraic varieties with G actions. The functor is not an equivalence, to find an example the reader may look as an exercise at the closures of conjugacy classes of nilpotent matrices and their corresponding non commutative algebras.

Sketch of proof. One takes 2 steps, first we prove the theorem for the free algebra. In this case we the fact that, in characteristic 0, we can interpret the classical first and second fundamental theorem of invariant theory for $GL(d)$ as saying that all invariants of k–tuples of matrices are generated by traces of monomials and their relations can be deduced from the Cayley-Hamilton theorem.

The second step is done by taking an algebra $R = F_d[x_i]_{i\in I}/J$ where $F_d[x_i]_{i\in I}$ now denotes a free algebra in the category $\mathcal{C}_d$. Using the universal embedding one has to prove that the ideal $\overline{J}$ generated by J in $M_d(F[\xi_{hk}^{(i)}])$ intersects $F_d[x_i]_{i\in I}$ back into J. This is a statement that for commutative rings and rings of invariants of linearly reductive groups is proved using the Reynolds operator. In our non commutative situation one proceeds similarly but there is a little twist in which the use of the trace is essential and allows to make a reduction to the commutative method.

We go back to further examples.

Next let us consider A to be an order in a finite dimensional central simple algebra D. This means that the center Z of A is a domain, A is torsion free over Z and, we have $D = A \otimes_Z Q(Z)$ where $Q(Z)$ is the field of fractions of Z.

In other words A embeds naturally in D which is its *ring of fractions* (cf. [P1] for a more abstract approach [Posner's Theorem]).

If $\overline{Q(Z)}$ denotes the algebraic closure of $Q(Z)$ we have that $A \otimes_Z \overline{Q(Z)}$ is the full ring $M_d(\overline{Q(Z)})$ of $d \times d$ matrices over $\overline{Q(Z)}$. Hence we have on D (and on A) the usual reduced trace map $tr: D \to Q(Z)$. It is well-known (and also easy) that $tr(A) = Z$ if, A is a finite Z module, Z is integrally closed and the characteristic is 0. So under these hypotheses A is an algebra of degree d.

The third example is also rather interesting and it is the free algebra $F_d[x_1, \dots, x_k]$ in k generators $x_1, \dots, x_k$ in the category $\mathcal{C}_d$. This is constructed in the following way:

Consider the free algebra with trace R and in R the ideal J (closed under trace) and generated by the elements:

$$tr(1) - d, \ \chi_{d,a}[a] \forall a \in R.$$

J is the minimal ideal such that $R/J \in \mathcal{C}_d$ and thus R/J is the free algebra in $\mathcal{C}_d$.

According to the description of the universal map for a free algebra and Theorem 4.3 we see that the free algebra in the category $\mathcal{C}_d$ can be described by invariant theory as the algebra of polynomial maps from the space $X_{k,d}$ of $k-$tuples of $d \times d$ matrices to the algebra of matrices which are equivariant under the conjugation action of the group $Gl(d)$ of invertible matrices. The generators x_i corresponding then to the *coordinate* functions.

The algebra of polynomial maps from the space of $k-$tuples of $d \times d$ matrices to the algebra of matrices, without any equivariance restriction, is just the algebra of $d \times d$ matrices with entries the ring A of polynomial functions on the space $X_{k,d}$ which is an affine space of dimension kd^2. That is is the coordinate ring of the universal embedding:

$$i : F_d[x_i]_{i\in I} \to M_d(F[\xi_{hk}^{(i)}])$$

so any formal identity which can be deduced for matrices is valid in $F_d[x_1, \dots, x_k]$ and then by specialization in any algebra R in the category $\mathcal{C}_d$.

As application let us define the *norm* in such an algebra setting:

$$N(a) := \sigma_d(a). \tag{4.3.1}$$

We have, as a corollary of the previous discussion:

$$N(1) = 1, N(ab) = N(a)N(b), \ \forall a, b \in R. \tag{4.3.2}$$

Furthermore the norm extends to a polynomial map on $R[t]$ and $\chi_{d,a}[t] = N(t-a)$, in particular the trace of a is obtained by polarization of the norm.

The fourth example is important for representation theory.

Let $A = \oplus_{i=1}^n M_{d_i}(F)$ be an algebra with trace in the category $\mathcal{C}_d$, such that the values of the trace lie in the field F then:

Proposition. *There exist positive integers* $k_1, \dots, k_n$ *such that* $d = \sum k_i d_i$ *and if* $a = (A_1, \dots, A_n)$ *with* $A_i \in M_{d_i}(F)$ *matrices, we have* $tr(a) = \sum k_i Tr(A_i)$*, where* $Tr(A_i)$ *is the ordinary trace of matrices.*

Proof. Consider the norm map N restricted to the group of invertible elements:

$$N : A' = \prod_{i=1}^{n} Gl(d_i, F) \to F'$$

it is a character and so there are integers $k_1, \dots, k_n$ such that, if $a = (A_1, \dots, A_n)$ with $A_i \in M_{d_i}(F)$ are invertible matrices, we have $N(a) = \prod det(A_i)^{k_i}$, where $det(A_i)$ is the ordinary determinant of matrices. Since N extends to a polynomial map on A we must have that the $k_i \geq 0$ and the statement $tr(a) = \sum k_i Tr(A_i)$ follows by polarization. We need only show that no $k_i = 0$. In fact otherwise in one of the blocks $M_{d_i}(F)$ we would have that the trace is identically 0, then the Cayley-Hamilton polynomial for any element in this block would be t^d which is manifestly absurd.

The representations of A are all of the form $\oplus m_i N_i$ where $N_i = F^{d_i}$ is the standard representation of $M_{d_i}(F)$. It is then easy to verify that A has a unique d–dimensional representation (according to our definitions) up to equivalence, the one in which the multiplicity $m_i = k_i$.

According to Theorem 4.3 this algebra corresponds to an algebraic scheme with a G action at it is possible to see that this is a homogeneous space the set of equivalent d–dimensional representations which is isomorphic to G/H where H is the automorphism group of the representation which is easily seen to be $\prod_i GL(k_i)$.

More generally let A be a finite dimensional algebra over F in the category $\mathcal{C}_d$, such that the values of the trace lie in the field F. Let J be the radical of A. Since the norm $N : A' \to F^*$ is a character it is 1 on the unipotent group $1 + J$ and so $tr(J) = 0$. We can thus factor the trace modulo J and A/J is again in the category $\mathcal{C}_d$.

If F is algebraically closed we are then in the previous case, otherwise by extension of scalars we can reduce to that case always. Now A may have several d–dimensional representations but only one semisimple, that one which factors through A/J.

4.4 The fifth example is given by Azumaya algebras. Recall that:

Definition. *An algebra* R *over a commutative ring* A *is called an Azumaya algebra of degree* d *over* A*, if there exists a faithfully flat extension* B *of* A *such that* $R \otimes_A B$ *is isomorphic to the algebra* $M_d(B)$.

In this case it is easy to show that the ordinary trace maps R into A.

Remark. *In the correspondence between algebras and* G*–varieties Azumaya algebras correspond to* **principal bundles** *(called also Brauer Severi varieties); in particular Azumaya algebras can be split also by etale maps.*

A very general connection of this notion with representation theory may be guessed from the following theorem of M. Artin [A]:

Theorem. *A ring* R *is an Azumaya algebra of degree* d *over its center if and only if it satisfies the polynomial identities of* $d \times d$ *matrices and has no representation of dimension* $< d$,

4.5 Let us go back to theorem 4.3 for a finitely generated algebra $R = \mathbb{F}[a_1, a_2, \dots, a_n]$ and, in order to simplify the treatment and stick to a geometric language assume F is algebraically closed.

We want to study d–dimensional representations of R and, from the universal map $i : R \to M_d(\mathcal{F}(R))$ we reduce to the case $R \in \mathcal{C}_d$ so that we identify

$$R = M_d(\mathcal{F}(R))^G, tr(R) = \mathcal{F}(R)^G.$$

Let us indicate by V_R the variety associated to $\mathcal{F}(R)$. By the very definition of the universal map V_R parametrizes d–dimensional representations of R.

Theorem 4,3 c) implies that $tr(R) = \mathcal{F}(R)^G$ so that, if we denote by W_R the variety associated to it we can apply geometric invariant theory.

The projection map $\pi : V_R \to W_R$ is a *geometric quotient*.

The points of W_R parametrize closed orbits of representations.

It is known (cf. [A]) that the closed orbits of this action are the semisimple representations, and that, given any representation V a Jordan Hölder series $V \supset V_1 \ldots \supset V_n \supset 0$, the associated graded semisimple representation is in the closure of its orbit by applying a 1-parameter subgroup.

We can thus complete the analysis:

Theorem. *Assume that $R \in \mathcal{C}_d$ is a finitely generated algebra. Set $T = tr(R)$.*

(a) T is a finitely generated algebra, and R is a finite module over T.

(b) The points of Spec T parametrize equivalence classes of d–dimensional semisimple representations of R.

(c) The canonical map Spec $R \xrightarrow{X}$ Spec T is surjective and each fiber consists of all those irreducible representations of R which are irreducible components of the corresponding semisimple representation. In particular each irreducible representation of R has dimension at most d.

(d) Spec(R) (the set of equivalence classes of irreducible representations) is in bijective correspondence (by associating to a representation its kernel) with the set of maximal two sided ideals of R.

(e) The set

$$\Omega_R = \{a \in \text{Spec } T \mid \text{ the corresponding semisimple representation is irreducible } \}$$

is a Zariski open set. This is exactly the part of Spec T over which A is an Azumaya algebra of degree d. □

Proof. We sketch the main steps still missing in the proof. Let us assume that $F = \mathbb{C}$, the complex numbers, to simplify one of the steps (cf. [P1]).

(a) T is a finitely generated algebra by Hilbert theorem on invariants and every element of R satisfies its characteristic polynomial hence it is integral over T. To prove that R is a finite module one has several options, there is a general theorem of Shirshov on integral PI-rings that one can invoke or one can reduce to the free algebra where the theorem is easily proved by invariant theory.

(b) Has been explained by Invariant theory but it can be further explicited. Let m be a maximal ideal in T, then $T/m = \mathbb{C}$ and (from part (a)) R/mR is a finite dimensional algebra in the category $\mathcal{C}_d$ with trace values in $\mathbb{C}$. We are thus in the example 4 of the previous paragraph where we have described the unique d–dimensional semisimple representation of such an algebra. This is the explicit correspondence between points in the spectrum of T and semisimple representations of R.

(c) Let us consider an arbitrary irreducible representation M, it can be presented as R/I, I a left ideal and hence its dimension over $\mathbb{C}$ is countable. By Schur's lemma its centralizer is a division algebra of countable dimension over $\mathbb{C}$ thus it must equal $\mathbb{C}$. Thus T acts on M by scalars so there is a maximal ideal m of T annihilating M and M is an irreducible R/mR module. We are thus back in the analysis performed in (b) and (c) follows.

(d) Is a simple consequence of the previous steps.

(e) These are the points where the quotient map is a principal fibration.

Remark. *i) If R is an order in a central simple algebra of degree d then T equals the center of R (and will be usually denoted by Z), furthermore since the central simple algebra splits in a d-dimensional matrix algebra which one may consider as a generic irreducible representation, it is easily seen that the open set Ω_R is non empty.*

ii) If T is a finitely generated module over a subalgebra Z_0, we can consider the finite surjective morphism

$$Spec\ T \xrightarrow{\tau} Spec\ Z_0.$$

Then by the properness of τ we get that the set $\Omega_A^0 := \{a \in SpecZ_0 | \tau^{-1}(a) \subset \Omega_R\}$ is a Zariski dense open subset of $SpecZ_0$.

We will use this remark in the theory of quantum groups where there is a natural subalgebra Z_0 which appears in the picture.

Further consequences of this picture are the following. Suppose A is as before an algebra of degree d.

Fix a positive integer k and define a new trace $tr_k(a) := ktr(a)$.

It is easy to see that, if A satisfies the dth characteristic polynomial under the original trace, it does satisfy the kd^{th} characteristic polynomial under the new trace. In fact we have seen that such an algebra embeds into $d \times d$ matrices over a commutative ring in such a way that the trace extends to the ordinary trace and then it is enough to embed $M_d(A)$ into $M_d(A) \otimes M_k(F) = M_{dk}(A)$. Then the same algebra C parametrizes equivalence classes of kd-dimensional semisimple representations and (it is easy to see using the example 4 of 4.2) these are just obtained from the previous d-dimensional representations by considering each such representation with multiplicity k.

From a geometric point of view the embedding of $M_d(F)$ into $M_d(F) \otimes M_k(F) = M_{dk}(F)$ embeds $PGL(d)$ in $PGL(kd)$ hence, given a variety V with $PGL(d)$ action we get one V_k with $PGL(kd)$ action by change of groups:

$$V_k := V \times_{PGL(d)} PGL(kd)$$

If V is associated to A as algebra in C_d so is V_k associated to A as algebra in C_{kd}

A second consequence is the following. Suppose that $Z \subset C$ is a subring and that C is a finite free extension of Z of degree h. Consider the reduced trace $tr_{C/Z} : C \to Z$ and the composition $tr_{A/Z}(a) := tr_{C/Z}(tr(a))$.

Under the same hypotheses as before the algebra A equipped with this new trace satisfies the hd^{th} characteristic polynomial and $Spec\ Z$ parametrizes equivalence classes of hd-dimensional representations.

The picture is the following. Call $\pi :\ Spec\ C \to\ Spec\ Z$ the finite map of spectra. Given a pont $P \in Spec\ Z$ one defines $\pi^{-1}(P) = \sum h_i P_i$ as a positive 0 cycle.

Proposition. *Each P_i corresponds to a semisimple representation ρ_i of A of dimension d and P corresponds to $\sum h_i \rho_i$.*

Proof. This is easy to justify again using example 4 of 4.2. Let m be a maximal ideal of Z and m_i the maximal ideal of P_i in C. Decompose $C/mC = \oplus \overline{C}_i$ where $\overline{C}_i$ is the coordinate ring of the fiber in the point P_i. In particular $\dim_F \overline{C}_i = h_i$ and for an element $a \in C/mC$ its trace in $F = Z/m$ is $\sum h_i a_i$ where a_i is the value of a in the point P_i. If e_i denotes the central idempotent of the factor $\overline{C}_i$ we have $A/mA = \oplus A/mAe_i$. If $\overline{A}$, resp $\overline{A}_i$ denote the semisimple algebras associated to A/mA and A/m_iA then $\overline{A} = \oplus \overline{A}_i$ and the trace in $\overline{A}$ is the sum of the traces in the factors $\overline{A}_i$ with coefficients h_i. From example 4 we get the claim.

Remark. *One can give also a geometric interpretation of this statement in terms of G varieties.*

There is a more general notion of degree in non commutative algebras which comes from the theory of polynomial identities. Onc may say that:

Definition. *An algebra has degree $\leq d$ if it satisfies all polynomial identities of $d \times d$ matrices.*

The relation with the previous definition is the following, an algebra in the category C_d satisfies the polynomial identities of $d \times d$ matrices since it can be embedded in matrices over a commutative ring, the converse is not true. In fact we will essentially use this notions for orders in a simple algebra D where the degree is the square root of the dimension of D over its center.

4.6 For a given algebra A in C_d the problem of the study of its spectrum can be thus naturally divided in two steps. Let $Z = tr(A)$, which for an order in a simple algebra of degree d will coincide with the center. First one has to develop a geometric description of Spec Z, then for each point of Spec Z we need a description of the corresponding semisimple representations, i.e. of its irreducible components and multiplicities.

We have already stressed the importance of orders in simple algebras, an important special case is the notion of maximal order which in non commutative algebra replaces the one of integrally closed domain.

Before giving this definition let us make some general remarks (cf. [MC,R] or [Ch], [vB-vO],[Sm]).

Given an order R in a central simple algebra D an element $a \in R$ is a non zero divisor in R if and only if it is invertible in D, such an element is called a *regular* element.

Given two orders R_1 and R_2 let us consider the following condition:

There exist regular elements $a, b \in R_1$ such that $R_1 \subset aR_2b$. This relation generates an equivalence of orders and a *maximal order* is one which is maximal with respect to this equivalence.

Definition. *An order R in a central simple algebra D is called a maximal order if given any central element $c \in R$ and an algebra S with $R \subset S \subset 1/cR$ we have necessarily that $R = S$.*

We shall encounter maximal orders in our study of quantum groups, for the moment we remark an important property of maximal orders:

Remark. *The center Z of a maximal order R is integrally closed. If R is a finitely generated algebra over a field F then R is a finite module over Z.*

Corollary. *A maximal order R in a central simple algebra D of degree d is closed under the reduced trace and hence it is an algebra in the category C_d.*

4.7 In our study of representation theory we will need methods of simplifications of our given algebras which will allow us to compute their degrees and give us some information on their spectrum, a standard tool is that of filtered algebras and their associated graded algebras, let us recall some main ideas.

Definition. *An algebra A is called $(\mathbb{Z}_+)$ filtered if $A = \bigcup_{j \in \mathbb{Z}_+} A_j$ is a union of $\mathbb{F}$-submodules A_j such that the following two properties hold:*

(4.7.1) $$1 \in A_0 \subset A_1 \subset A_2 \subset \ldots,$$

(4.7.2) $$A_i A_j \subset A_{i+j}.$$

There is a more general notion of filtered algebra relative to an *ordered monoid.* With this we mean a commutative semigroup S with an order compatible with its sum. Then an S filtration in A means to give subspaces A_s, $s \in S$, such that $A_s A_t \subset A_{s+t}$, $A = \bigcup A_s$ and $A_s \subset A_t$ whenever $s \leq t$. This notion will be used for $S = \mathbb{N}^k$ with the degree-lexicographic order (cf. 3.5).

For a filtered algebra A one can construct $\overline{A} = \oplus_{j\in\mathbb{Z}_+}(A_j/A_{j-1})$ the *associated graded algebra.*

Given $a \in A$, we let $\deg a$ be the minimal j for which $a \in A_j$, and let $\overline{a}$ be the image of a in A_j/A_{j-1}. For a subset S of A we let $\overline{S} = \{\overline{a} \in \overline{A}$ where $a \in S\}$. For an ideal I we will, by abuse of notations, indicate by $\overline{I}$ not just the previously defined set of homogeneous elements but also their (direct) sum, which is an ideal of $\overline{A}$.

Lemma. *Let A be a filtered algebra.*

(a) If $a, b \in A$ and $\overline{a}\overline{b} \neq 0$, then $\overline{ab} = \overline{a}\overline{b}$; in particular if $\overline{A}$ has no zero divisors then A has no zero divisors.

(b) Let B be a subalgebra of A with induced filtration. Let $a_1, a_2, \ldots \in A$ be such that $\overline{a}_1, \overline{a}_2, \ldots$ are homogeneous generators of the left $\overline{B}$-module $\overline{A}$. Then any element a of A can be written in the form

$$a = \sum_i b_i a_i, \text{ where } \deg a_i + \deg b_i \leq \deg a.$$

Proof. (a) is standard. In order to prove (b) note that we may write $\overline{a} = \sum_i \overline{b}_i \overline{a}_i$, where $\overline{b}_i$ are some homogeneous elements of $\overline{B}$ such that $\deg \overline{a}_i + \deg \overline{b}_i = \deg \overline{a}$. We have $a = \sum_i b_i a_i + a'$, where $\deg a' < \deg a$, and we apply the inductive assumption to a' (in degree 0 it is clear). □

4.8 Let A be a filtered algebra and let $A[t]$ (resp. $A[t, t^{-1}]$) denote the ring of polynomials (resp. Laurent polynomials) over A. The *Rees algebra* $\mathcal{R}(A)$ of A is the following subalgebra of $A[t]$ (cf. [A-O-B]:

$$\mathcal{R}(A) = \sum_{j\in\mathbb{Z}_+} A_j t^j.$$

The following properties of the algebra $\mathcal{R}(A)$ are obvious:

Lemma. *(a) If A has no zero divisors, then the same is true for $\mathcal{R}(A)$.*

(b) If $\overline{A}$ is generated by homogeneous elements $\underline{a}_1, \underline{a}_2, \ldots$ of degree $r_1, r_2, \ldots$, then $\mathcal{R}(A)$ is generated by the elements $t, t^{r_1}a_1, t^{r_2}a_2, \ldots$ where the a_i lift the $\underline{a}_i$.

(c) $A[t, t^{-1}] = \mathcal{R}(A)[t^{-1}]$.

(d) $\mathcal{R}(A)/(t) \simeq \overline{A}$.

(e) If $B \subset A$ is a subalgebra with induced filtration, then $\mathcal{R}(B) \subset \mathcal{R}(A)$. □

REMARK . It follows from Lemma 4.8c that

(4.8.1) $$\text{degree } \overline{A} \leq \text{ degree } \mathcal{R}(A).$$

From part (d) of the same lemma we deduce

(4.8.2) $$\text{degree } \overline{A} \leq \text{ degree } A.$$

The following proposition follows from Lemma 4.7b.

Proposition. *Let A be a filtered algebra, and let B be a subalgebra of A. Let $a_1, a_2, \ldots$ be elements of A of degrees $r_1, r_2, \ldots$ such that $\overline{A}$ is a left $\overline{B}$-module on generators $\overline{a}_1, \overline{a}_2, \ldots$. Then $\mathcal{R}(A)$ is a left $\mathcal{R}(B)$-module on generators $t^{r_1}a_1, t^{r_2}a_2, \ldots$.* □

4.9 We collect some general facts on ideals and their graded counterparts.

Lemma. *Let A be a filtered algebra and I an ideal, then $\overline{I}$ is an ideal in $\overline{A}$, and if H is the ideal of $A[t,t^{-1}]$ generated by I, in $\overline{A}$ we have:*

$$(H \cap \mathcal{R}(A) + t\mathcal{R}(A))/t\mathcal{R}(A) = \overline{I}.$$

Proof. Clear. □

In general it is difficult to determine generators for $\overline{I}$, therefore the next proposition is particularly useful when it can be applied.

Proposition. *Let A be a commutative filtered algebra and let $a_1, \dots, a_n \in A$ be such that $\overline{a}_1, \dots, \overline{a}_n$ is a regular sequence of $\overline{A}$. Let $I = (a_1, \dots, a_n)$ be the ideal of A generated by $a_1, \dots, a_n$. Then*

(a) $a_1, \dots, a_n$ is a regular sequence in A.

(b) The ideal $\overline{I}$ of $\overline{A}$ is generated by the elements $\overline{a}_1, \dots, \overline{a}_n$.

Proof. (a) Suppose that $a_1, \dots, a_n$ is not a regular sequence of A. Then there exists a $k \le n$ and $b_1, \dots, b_k \in A$ such that

(4.9.1) $$\sum_{j=1}^{k} a_j b_j = 0 \quad \text{and} \quad b_k \notin (a_1, \dots, a_{k-1}).$$

We may assume that $d := \max_i \deg a_i b_i$ is minimal possible for all such relations and (reordering if necessary) that for some $m \ge 1$:

(4.9.2) $$d = \deg a_1 b_1 = \dots = \deg a_m b_m \text{ and } d > \deg a_j b_j \quad \text{for} \quad j > m.$$

Then

(4.9.3) $$\sum_{i=1}^{m} \overline{a}_i \overline{b}_i = 0.$$

This reorders the a_i, but since $\overline{A}$ is graded and the elements $\overline{a}_1, \dots, \overline{a}_n$ are homogeneous we have that $\overline{a}_1, \dots, \overline{a}_m$ is a regular sequence in $\overline{A}$ and the corresponding first Koszul homology group $H_1(\overline{A}; \overline{a}_1, \dots, \overline{a}_m)$ vanishes [B]. This implies that there exists a skew-symmetric $m \times m$ matrix $\overline{B}$ with homogeneous entries over $\overline{A}$ such that

$$(\overline{b}_1, \dots, \overline{b}_m) = (\overline{a}_1, \dots, \overline{a}_m)\overline{B}.$$

Let B be a skew symmetric matrix over A whose image in $\overline{A}$ is $\overline{B}$. Let

(4.9.4) $$(b'_1, \dots, b'_m) = (a_1, \dots, a_m)B.$$

Then $\overline{b}'_i = \overline{b}_i$ and

(4.9.5) $$\sum_{i=1}^{m} a_i b'_i = 0 \text{ (since } B \text{ is antisymmetric).}$$

Let $b_i'' = b_i - b_i'$ with $b_i' = 0$ for $i > m$. We have: $\sum_{i=1}^k a_i b_i'' = 0$ by (4.9.1) and (4.9.5), and $\max_i \deg a_i b_i'' < d$. Since $b_i' \in (a_1, \ldots, a_{i-1}, a_{i+1}, \ldots)$ by (4.9.4) (recall that the diagonal entries of B are zero) we obtain a contradiction with (4.9.1).

The proof of (b) is similar. Let $x \in I$, if we can find an expression $x = \sum_{i=1}^n a_i b_i$ such that, setting $d = \max_i \deg a_i b_i$, we have $d = \deg x$ we are clearly done. Suppose this is not the case and choose an expression for which $d > \deg x$ is minimal. As in (4.9.2) assume $\deg a_i b_i = d$ for $i = 1, \ldots, m$ while $\deg a_i b_i < d$ for $i > m$. Thus we have $\sum_{i=1}^m \overline{a}_i \overline{b}_i = 0$ and as before we can find $b_1', \ldots, b_m'$ such that $\sum_{i=1}^m a_i b_i' = 0$ and $\deg a_i(b_i - b_i') < d$ for $i = 1, \ldots, m$, reaching a contradiction. □

4.10 Let A be an order closed under trace in a central simple algebra D, let Z be the center of A and $Q(Z)$ that of D as in §4.1.

Definition. *A trace filtration for A is a filtration A_i such that:*
(a) $tr(A_i) \subset A_i$.
(b) $\overline{A}$ is finitely generated.

Proposition. *Let A be an order in D closed under trace and with a trace filtration, then $\mathcal{R}(A)$ is a finitely generated order in $D(t)$ closed under trace.*

Proof. By Lemma 4.10c it follows immediately that $\mathcal{R}(A)$ is an order in $D(t)$, but by Lemma 4.10b both A and $\mathcal{R}(A)$ are finitely generated. The assumptions on the filtration imply that $\mathcal{R}(A)$ is closed under trace. □

The following simple lemma will be useful to analyze the spectrum of quantum groups.

Lemma. *Let A be a filtered algebra such that $\overline{A}$ is a finitely generated order of the same degree as A, and let Z_0 be a central subalgebra of A such that $\overline{Z}_0$ is finitely generated and $\overline{A}$ is a finitely generated module over $\overline{Z}_0$. Let I be an ideal of Z_0 and $\overline{I}$ the associated graded ideal of $\overline{Z}_0$. Let $\mathcal{O}$ (resp. $\mathcal{O}_1$) be the set of zeros of I (resp. $\overline{I}$) in Spec Z_0 (resp. Spec $\overline{Z}_0$). Suppose that $\mathcal{O}_1 \cap \Omega^0_{\overline{A}} \neq \emptyset$. Then $\mathcal{O} \cap \Omega^0_A \neq \emptyset$.*

Proof. Consider the Rees algebra $\mathcal{R}(A)$ of A. Its subalgebra $\mathcal{R}(Z_0)$ is central and by Proposition 4.10, $\mathcal{R}(A)$ is a finitely generated $\mathcal{R}(Z_0)$–module. We have seen that

(4.10.1) $$\text{degree } \mathcal{R}(A) = \text{degree } A$$

and by the hypothesis:

(4.10.2) $$\text{degree } (\overline{A}) = \text{degree } A,$$

hence

(4.10.3) $$\Omega^0_{\mathcal{R}(A)} \cap \text{Spec}\overline{Z}_0 = \Omega^0_{\overline{A}}.$$

Clearly

(4.10.4) $$\Omega^0_{\mathcal{R}(A)} \supset \Omega^0_{A[t,t^{-1}]} = \Omega^0_A \times \mathbb{F}^\times.$$

and by Proposition 4.9 we have:

$$\overline{(\mathcal{O} \times \mathbb{F}^\times)} \cap \text{Spec } \overline{Z}_0 = \mathcal{O}_1.$$

Hence $\Omega^0_{\mathcal{R}(A)} \cap \overline{(\mathcal{O} \times \mathbb{F}^\times)} \supset \Omega^0_{\overline{A}} \cap \overline{(\mathcal{O} \times \mathbb{F}^\times)} \neq \emptyset$ (where $\overline{\mathcal{O} \times \mathbb{F}^\times}$ stands for Zariski closure of $\mathcal{O} \times \mathbb{F}^\times$) since $\mathcal{O}_1 \cap \Omega^0_{\overline{A}} \neq \emptyset$ by the hypothesis. It follows that $\Omega^0_{\mathcal{R}(A)}$ intersects with $\mathcal{O} \times \mathbb{F}^\times$ in a non–empty open subset. But, obviously, this intersection is $(\mathcal{O} \cap \Omega^0_A) \times \mathbb{F}^\times$. It follows that $\mathcal{O} \cap \Omega^0_A \neq \emptyset$. □

§5. Twisted derivations and polynomial algebras.

5.1. Let A be an algebra and let σ be an automorphism of A.

Definition. *A twisted derivation of A relative to σ is a linear map $D: A \longrightarrow A$ such that:*

$$D(ab) = D(a)b + \sigma(a)D(b).$$

Example. An element $a \in A$ induces an inner twisted derivation $ad_\sigma a$ relative to σ defined by the formula:

$$(ad_\sigma a)b = ab - \sigma(b)a.$$

The following well-known fact is very useful in calculations with twisted derivations. (Here and further we use the usual "box" notation:

$$[n] = \frac{q^n - q^{-n}}{q - q^{-1}}, \; [n]! = [1][2]\dots[n], \; \begin{bmatrix} m \\ n \end{bmatrix} = \frac{[m][m-1]\dots[m-n+1]}{[n]!}.$$

One also writes $[n]_d$, etc. if q is replaced by q^d.)

Proposition. *Let $a \in A$ and let σ be an automorphism of A such that $\sigma(a) = q^2 a$, where q is a scalar. Then*

$$(ad_\sigma a)^m(x) = \sum_{j=0}^{m} (-1)^j q^{j(m-1)} \begin{bmatrix} m \\ j \end{bmatrix} a^{m-j}\sigma^j(x)a^j.$$

Proof. Let L_a and R_a denote the operators of left and right multiplications by a in A. Then

$$ad_\sigma a = L_a - R_a\sigma.$$

Since L_a and R_a commute, due to the assumption $\sigma(a) = q^2 a$ we have

$$L_a(R_a\sigma) = q^{-2}(R_a\sigma)L_a.$$

Now the proposition is immediate from the following well-known binomial formula applied to the algebra End A.

Lemma. *Suppose that x and y are elements of an algebra such that $yx = q^2xy$ for some scalar q. Then*

$$(x+y)^m = \sum_{j=0}^{m} \begin{bmatrix} m \\ j \end{bmatrix} q^{j(m-j)} x^j y^{m-j}.$$

Proof. is by induction on m using

$$\begin{bmatrix} m \\ j-1 \end{bmatrix} q^{m+1} + \begin{bmatrix} m \\ j \end{bmatrix} = \begin{bmatrix} m+1 \\ j \end{bmatrix} q^j,$$

which follows from

$$q^b[a] + q^{-a}[b] = [a+b]. \; \square$$

We note some consequence which will be used for ℓ a positive integer and q a primitive ℓ-th root of 1. Let $\ell' = \ell$ if ℓ is odd and $= \frac{1}{2}\ell$ if ℓ is even. Then, by definition, we have

$$\begin{bmatrix} \ell' \\ j \end{bmatrix} = 0 \text{ for all } j \text{ such that } 0 < j < \ell'.$$

This together with Proposition 5.1 implies

Corollary. *Under the hypothesis of Proposition 5.1 we have:*
i) If $yx = q^2xy$ *then* $(x+y)^\ell = x^\ell + y^\ell$.

(ii)) $$(ad_\sigma a)^{\ell'}(x) = a^{\ell'}x - \sigma^{\ell'}(x)a^{\ell'} \text{ if } q \text{ is a primitive } \ell\text{–th root of } 1.$$

Remark. Let D be a twisted derivation associated to an automorphism σ such that $\sigma D = q^2 D\sigma$. Then by induction on m one obtains the following well-known q-analogue of the Leibnitz formula:

$$D^m(xy) = \sum_{j=0}^{m} \begin{bmatrix} m \\ j \end{bmatrix} q^{j(m-j)} D^{m-j}(\sigma^j x) D^j(y).$$

It follows that if q is a primitive ℓ–th root of 1, then $D^{\ell'}$ is a twisted derivation associated to $\sigma^{\ell'}$.

5.2. Given an automorphism σ of A and a twisted derivation D of A relative to σ we define the *twisted polynomial algebra* $A_{\sigma,D}[x]$ in the indeterminate x to be the $\mathbb{F}$-module $A \otimes_{\mathbb{F}} \mathbb{F}[x]$ thought as formal polynomials with multiplication defined by the rule:

$$xa = \sigma(a)x + D(a).$$

When $D = 0$ we will also denote this ring by $A_\sigma[x]$. Notice that the definition has been chosen in such a way that in the new ring the given twisted derivation becomes the inner derivation $ad_\sigma x$.

Let us notice that if $a, b \in A$ and a is invertible we can perform the change of variables $y := ax + b$ and we see that $A_{\sigma,D}[x] = A_{\sigma',D'}[y]$. It is better to make the formulas explicit separately when $b = 0$ and when $a = 1$. In the first case $yc = axc = a(\sigma(c)x + D(c)) = a(\sigma(c))a^{-1}y + aD(c)$ and we see that the new automorphism σ' is the composition $(Ad\,a)\sigma$, so that $D' := aD$ is a twisted derivation relative to σ'. Here and further $Ad\,a$ stands for the inner automorphism:

$$(Ad\,a)x = axa^{-1}.$$

In the case $a = 1$ we have $yc = (x+b)c = \sigma(c)x + D(b) + bc = \sigma(c)y + D(b) + bc - \sigma(c)b$, so that $D' = D + ad_\sigma b$. Summarizing we have

Proposition. *Changing* σ, D *to* $(Ad\,a)\sigma, aD$ *(resp. to* $\sigma, D + D_b$*) does not change the twisted polynomial ring up to isomorphism.* □

We may express the previous fact with a definition: For a ring A two pairs (σ, D) and (σ', D') are *equivalent* if they are obtained one from the other by the above moves.

If $D = 0$ we can also consider the twisted Laurent polynomial algebra $A_\sigma[x, x^{-1}]$. It is clear that if A has no zero divisors, then the algebras $A_{\sigma,D}[x]$ and $A_\sigma[x, x^{-1}]$ also have no zero divisors.

5.3 The importance for us of twisted polynomial algebras will be clear in the section on quantum groups.

Remark. *Given a twisted polynomial algebra* $A_{\sigma,D}[x]$ *we can construct a natural filtration given by the degree of the polynomials, the associated graded algebra is clearly* $A_\sigma[x]$.

Definition. *We shall say that the algebra* $A_\sigma[x]$ *is a simple degeneration of* $A_{\sigma,D}[x]$.

We want to analyze a special case which will play an important role in the future. Let A be an algebra over a field $\mathbb{F}$ of characteristic 0, let $x_1, \ldots, x_n$ be a set of generators of A and let Z_0 be a central subalgebra of A. For each $i = 1, \ldots, k$, denote by A^i the subalgebra of A generated by $x_1, \dots, x_i$, and let $Z_0^i = Z_0 \cap A^i$. We assume that the following three conditions hold for each $i = 1, \dots, k$:

(a) $x_i x_j = b_{ij} x_j x_i + P_{ij}$ if $i > j$, where $b_{ij} \in \mathbb{F}$, $P_{ij} \in A^{i-1}$.

(b) Formulas $\sigma_i(x_j) = b_{ij} x_j$ for $j < i$ define an automorphism of A^{i-1}.

(c) Letting $D_i(x_j) = P_{ij}$ for $j < i$, we obtain $A^i = A^{i-1}_{\sigma_i, D_i}[x_i]$, as twisted polynomial algebra.

We may consider the twisted polynomial algebras $\overline{A}^i$ with zero derivations, so that the relations are $x_i x_j = b_{ij} x_j x_i$ for $j < i$. We call this the *associated quasipolynomial algebra* .

We can prove now the main theorem of this section.

Theorem. *Under the above assumptions, the quasipolynomial algebra $\overline{A}$ is obtained from A by a sequence of simple degenerations.*

Proof. We use this following remark. If there is an index h such that the elements $P_{ij} = 0$ for all $i > h$ and all j, then the monomials in the variables different from x_h form a subalgebra B and the algebra A is a twisted polynomial ring $B_{\sigma,D}[x_h]$. The associated ring $B_\sigma[x_h]$ is obtained by setting $P_{hj} = 0$ for all j. Having made this remark we see that we can inductively modify the relations (a) so that at the h–th step we have an algebra A_h^n with the same type of relations but $P_{ij} = 0$ for all $i > n - h$ and all j. Since A_h^n and A_{h-1}^n are of type $B_{\sigma,D}[x]$ and $B_\sigma[x]$ respectively we have constructed the required sequence of simple degenerations. □

§6. Representation theory of twisted derivation algebras.

6.1. We want to analyze some interesting cases of the previous constructions for which the resulting algebras are finite modules over their centers and thus we can develop for them the notion of degree and a good representation theory.

Let us first consider a finite dimensional semisimple algebra A over an algebraically closed field $\mathbb{F}$, let $\bigoplus_i \mathbb{F}e_i$ be the fixed points of the center of A under σ where the e_i are central idempotents. We have $D(e_i) = D(e_i^2) = 2D(e_i)e_i$ hence $D(e_i) = 0$ and, if $x = xe_i$, then $D(x) = D(x)e_i$. It follows that, decomposing $A = \bigoplus_i Ae_i$, each component Ae_i is stable under σ and D and thus we have

$$A_{\sigma,D}[x] = \bigoplus_i (Ae_i)_{\sigma,D}[x].$$

This allows us to restrict our analysis to the case in which 1 is the only fixed central idempotent.

The second special case is described by the following:

Lemma. *Consider the algebra $A = \mathbb{F}^{\oplus k}$ with σ the cyclic permutation of the summands, and let D be a twisted derivation of this algebra relative to σ. Then D is an inner twisted derivation.*

Proof. Compute D on the idempotents: $D(e_i) = D(e_i^2) = D(e_i)(e_i + e_{i+1})$. Hence we must have $D(e_i) = a_i e_i - b_i e_{i+1}$ and from $0 = D(e_i e_{i+1}) = D(e_i)e_{i+1} + e_{i+1}D(e_{i+1})$ we deduce $b_i = a_{i+1}$. Let now $a = (a_1, a_2, \dots, a_k)$; an easy computation shows that $D = ad_\sigma a$. □

Proposition. *Let σ be the cyclic permutation of the summands of the algebra $\mathbb{F}^{\oplus k}$. Then*

(a) $\mathbb{F}_\sigma^{\oplus k}[x, x^{-1}]$ is an Azumaya algebra of degree k over its center $\mathbb{F}[x^k, x^{-k}]$.

(b) $R := \mathbb{F}_\sigma^{\oplus k}[x, x^{-1}] \otimes_{\mathbb{F}[x^k, x^{-k}]} \mathbb{F}[x, x^{-1}]$ is the algebra of $k \times k$ matrices over $\mathbb{F}[x, x^{-1}]$.

Proof. It is enough to prove (b). Let $u := x \otimes x^{-1}$, $e_i := e_i \otimes 1$; we have $u^k = x^k \otimes x^{-k} = 1$ and $ue_i = e_{i+1}u$. From these formulas it easily follows that the elements $e_i u^j$ ($i, j = 1, \dots, k$) span a subalgebra A and that there exists an isomorphism $A \xrightarrow{\sim} M_k(\mathbb{F})$ mapping

$\mathbb{F}^{\oplus k}$ to the diagonal matrices and u to the matrix of the cyclic permutation. Then $R = A \otimes_{\mathbb{F}} \mathbb{F}[x, x^{-1}]$. □

6.2 Assume now that A is semisimple and that σ induces a cyclic permutation of the central idempotents.

Lemma. *(a)* $A = M_d(\mathbb{F})^{\oplus k}$.

(b) Let D be a twisted derivation of A relative to σ. Then the pair (σ, D) is equivalent to the pair $(\sigma', 0)$ where

$$\sigma'(a_1, a_2, \ldots, a_k) = (a_k, a_1, a_2, \ldots, a_{k-1}). \tag{6.2.1}$$

Proof. Since σ permutes transitively the simple blocks they must all have the same degree d so that $A = M_d(F)^{\oplus k}$. Furthermore we can arrange the identifications of the simple blocks with matrices so that:

$$\sigma(a_1, a_2, \ldots, a_k) = (\tau(a_k), a_1, a_2, \ldots, a_{k-1}),$$

where τ is an automorphism of $M_d(\mathbb{F})$. Any such automorphism is inner, hence after composing σ with an inner automorphism, we may assume in the previous formula that $\tau = 1$. Then we think of A as $M_d(\mathbb{F}) \otimes \mathbb{F}^{\oplus k}$, the new automorphism being of the form $1 \otimes \sigma'$ where $\sigma' : F^{\oplus k} \to F^{\oplus k}$ is given by (6.2.1).

We also have that $M_d(\mathbb{F}) = A^\sigma$ and $\mathbb{F}^{\oplus k}$ is the centralizer of A^σ. Next observe that D restricted to A^σ is a derivation of $M_d(\mathbb{F})$ with values in $\oplus_{i=1}^k M_d(\mathbb{F})$, i.e., $D(a) = (D_1(a), D_2(a), \ldots, D_k(a))$ where each D_i is a derivation of $M_d(\mathbb{F})$. Since for $M_d(\mathbb{F})$ all derivations are inner we can find an element $u \in A$ such that $D(a) = [u, a]$ for all $a \in M_d(\mathbb{F})$. So $(D - ad_\sigma u)(a) = [u, a] - (ua - \sigma(a)u) = 0$ for $a \in A^\sigma$. Thus, changing D by adding $-ad_\sigma u$ we may assume that $D = 0$ on $M_d(\mathbb{F})$.

Now consider $b \in \mathbb{F}^{\oplus k}$ and $a \in M_d(\mathbb{F})$; we have $D(b)a = D(ba) = D(ab) = aD(b)$. Since $\mathbb{F}^{\oplus k}$ is the centralizer of $M_d(\mathbb{F})$ we have $D(b) \in \mathbb{F}^{\oplus k}$ and D induces a twisted derivation of $\mathbb{F}^{\oplus k}$. By Lemma 6.1 this last derivation is inner and the claim is proved. □

Summarizing we have

Proposition. *Let A be a finite-dimensional semisimple algebra over an algebraically closed field $\mathbb{F}$. Let σ be an automorphism of A which induces a cyclic permutation of order k of the central idempotents of A. Let D be a twisted derivation of A relative to σ. Then:*

$$A_{\sigma,D}[x] \cong M_d(\mathbb{F}) \otimes \mathbb{F}_\sigma^{\oplus k}[x],$$
$$A_{\sigma,D}[x, x^{-1}] \cong M_d(\mathbb{F}) \otimes \mathbb{F}_\sigma^{\oplus k}[x, x^{-1}].$$

This last algebra is Azumaya of degree dk.

6.3 We can now globalize the previous construction. Let A be a prime algebra (i.e. $aAb = 0$, $a, b \in A$, implies that $a = 0$ or $b = 0$) over a field $\mathbb{F}$ and let Z be the center of A. Then Z is a domain and A is a torsion free module over Z. Assume that A is a finite module over Z. Then A embeds in a finite-dimensional central simple algebra $Q(A) = A \otimes_Z Q(Z)$, where $Q(Z)$ is the ring of fractions of Z. If $\overline{Q(Z)}$ denotes the algebraic closure of $Q(Z)$ we have that $A \otimes_Z \overline{Q(Z)}$ is the full ring $M_d(\overline{Q(Z)})$ of $d \times d$ matrices over $\overline{Q(Z)}$. Then d is called the *degree* of A.

Let σ be an automorphism of the algebra A and let D be a twisted derivation of A relative to σ. Assume that

(a) There is a subalgebra Z_0 of Z, such that Z is finite over Z_0.

(b) D vanishes on Z_0 and σ restricted to Z_0 is the identity.

These assumptions imply that σ restricted to Z is an automorphism of finite order. Let d be the degree of A and let k be the order of σ on the center Z.

Definition. *If A is an order in a finite dimensional central simple algebra and σ, D satisfy the previous conditions we shall say that the triple A, σ, D is finite.*

Assume that the field $\mathbb{F}$ has characteristic 0. The main result of this section is:

Theorem. *Under the above assumptions the twisted polynomial algebra $A_{\sigma,D}[x]$ is an order in a central simple algebra of degree kd.*

Proof. Let Z^σ be the fixed points in Z of σ. By the definition, it is clear that D restricted to Z^σ is a derivation. Since it vanishes on a subalgebra over which it is finite hence algebraic and since we are in characteristic zero it follows that D vanishes on Z^σ. Let us embed Z^σ in an algebraically closed field $\mathbb{F}$ and let us consider the algebra $A \otimes_{Z^\sigma} \mathbb{F}$. This algebra of course equals $A \otimes_Z Z \otimes_{Z^\sigma} F$, but clearly $Z \otimes_{Z^\sigma} \mathbb{F} = \mathbb{F}^{\oplus k}$ and $A \otimes_Z \mathbb{F} = M_d(\mathbb{F})$. Thus we get that $A \otimes_{Z^\sigma} \mathbb{F} = \oplus_{i=1}^k M_d(\mathbb{F})$. The pair σ, D extends to $A \otimes_{Z^\sigma} \mathbb{F}$ and using the same notations we have that $(A \otimes_{Z^\sigma} \mathbb{F})_{\sigma,D}[x] = (A_{\sigma,D}[x]) \otimes_{Z^\sigma} \mathbb{F}$. We are now in the situation of a semisimple algebra which we have already studied and the claim follows. □

Corollary. *Under the above assumptions, $A_{\sigma,D}[x]$ and $A_\sigma[x]$ have the same degree.*

Recall that passing from $A_{\sigma,D}[x]$ to $A_\sigma[x]$ we perform a step which we have decided to call a simple degeneration in section 5.3. In the next paragraph we will in fact encounter sequences of such degenerations.

Remark.. *The previous analysis yields in fact a stronger result.*

Consider the open set of Spec Z where A is an Azumaya algebra; it is clearly σ–stable. In it we consider the open part where σ has order exactly k. Every orbit of k elements of the group generated by σ gives a point $F(p)$ in Spec Z^σ and $A \otimes_Z Z \otimes_{Z^\sigma} F(p) = \oplus_{i=1}^k M_d(F(p))$. Thus we can apply the previous theory which allows us to describe the fiber over $F(p)$ of the spectrum of $A_{\sigma,D}[x]$.

6.4 Let A be a prime algebra over a field $\mathbb{F}$ of characteristic 0, let $x_1, \ldots, x_n$ be a set of generators of A and let Z_0 be a central subalgebra of A. For each $i = 1, \ldots, k$, denote by A^i the subalgebra of A generated by $x_1, \ldots, x_i$, and let $Z_0^i = Z_0 \cap A^i$. We assume, strengthening the discussion preceding theorem 5.3 that the following three conditions hold for each $i = 1, \ldots, k$:

(a) $x_i x_j = b_{ij} x_j x_i + P_{ij}$ if $i > j$, where $b_{ij} \in \mathbb{F}$, $P_{ij} \in A^{i-1}$.

(b) A^i is a finite module over Z_0^i.

(c) Formulas $\sigma_i(x_j) = b_{ij} x_j$ for $j < i$ define an automorphism of A^{i-1} which is the identity on Z_0^{i-1}.

Note that letting $D_i(x_j) = P_{ij}$ for $j < i$, we obtain $A^i = A^{i-1}_{\sigma_i, D_i}[x_i]$, so that A is an iterated twisted polynomial algebra. Note also that each triple (A^{i-1}, σ_i, D_i) satisfies assumptions 6.4(a) and (b).

We may consider the twisted polynomial algebras $\overline{A}^i$ with zero derivations, so that the relations are $x_i x_j = b_{ij} x_j x_i$ for $j < i$. We call this the ***associated quasipolynomial algebra*** (as in [DK1]).

We can prove now the main theorem of this section.

Theorem. *Under the above assumptions, the degree of A is equal to the degree of the associated quasipolynomial algebra $\overline{A}$.*

Proof. By Theorem 5.3 $\overline{A}$ is obtained from A with a sequence of simple degenerations, hence by Corollary 6.3, it follows that they have the same degree. □

6.5 We want to discuss now some criteria under which, by degeneration arguments, we can deduce that an algebra is a maximal order.

One can find in [MC-R] a general approach.

The setting that we choose is suggested by the work on quantum groups.

We assume to have an algebra R (over some field F) with a commutative subalgebra A and elements $x_1, \ldots, x_k$ satisfying certain special conditions which we will presently explain.

Let us first introduce some notations. For an integral vector $\underline{n} := (n_1, \ldots, n_k)$, $n_i \in \mathbb{N}$ we set $deg\underline{n} := n_1 + \ldots + n_k$, and $x^{\underline{n}} := x_1^{n_1} \ldots x_k^{n_k}$ we call such an element a *monomial*. Furthermore we define on the set of integral vectors the degree-lexicographic ordering, i.e. set $\underline{n} < \underline{m}$ if either $deg\underline{n} < deg\underline{m}$ or $deg\underline{n} = deg\underline{m}$ but $\underline{n}$ is less than $\underline{m}$ in the usual lexicographic order, in this way $\mathbb{N}^k$ becomes an ordered monoid.

We now impose first:

1) The monomials $x^{\underline{n}}$ are a basis of R as a left A module.

Let us now denote by

$$R_{\underline{n}} := \sum_{\underline{m} \leq \underline{n}} A x^{\underline{m}}.$$

We impose next:

2) The subspaces $R_{\underline{n}}$ give a structure of filtered algebra with respect to the ordered monoid $\mathbb{N}^k$.

Furthermore we restrict the commutation relations among the elements x_i and A.

3) $x_i x_j = a_{ij} x_j x_i + b_{ij}$ with $a_{ij} \in F^*$ and b_{ij} lower than $x_i x_j$ in the filtration.

4) $x_i a = \sigma_i(a) x_i +$ *lower term*, with σ_i an automorphism of A.

Notice that:

i) $x^{\underline{n}} x^{\underline{m}} = \lambda x^{\underline{n}+\underline{m}} +$ *lower term*, with $\lambda \in F^*$.

ii) The associated graded algebra $\overline{R}$ is a twisted polynomial ring over A. In fact the classes $\overline{x}_i$ of the x_i satisfy

3') $\overline{x}_i \overline{x}_j = a_{ij} \overline{x}_j \overline{x}_i$

4') $\overline{x}_i a = \sigma_i(a) \overline{x}_i$.

We wish to make 2 further restrictions:

5) A is integrally closed.

6) For every vector $\underline{n}$ there exists a monomial $a := x^{\underline{m}}$ such that $\underline{n} \leq \underline{m}$ and its class $\overline{a}$ is in the center of $\overline{R}$.

Let us say for such an $\underline{m}$ that $x^{\underline{m}}$ is an *almost central* monomial. Such monomials have simple special commutation rules:

$a x^{\underline{m}} = x^{\underline{m}} a +$ *lower term*, $\forall a \in A$

$x^{\underline{m}} x^{\underline{s}} = x^{\underline{m}+\underline{s}} +$ *lower term*, $\forall \underline{s}$.

We can finally state our result:

Theorem. *Assume that R satisfies hypotheses 1–6 then R is a maximal order.*

Proof. Let $z = b x^{\underline{r}} +$ *lower term*, $b \in A$ be an element in the center of R and B be an algebra with $R \subset B \subset z^{-1} R$. We must show that $B = R$. Let us then take any element $u \in B$ and let $y := zu \in R$.

We develop $y := a x^{\underline{s}} +$ *lower term*, $a \in A$, , we need to show that $u \in R$ by induction on $\underline{s}$. In order to do this we first want to prove that b divides a. Using hypotheses 6 and 4 (in the form ii) we deduce that there is a monomial $v \in R$ such that $yv = z(uv)$ has the form $a x^{\underline{m}} +$ *lower term* with $x^{\underline{m}}$ an almost central monomial.

Next write $(yv)^k = z^{k-1}z(uv)^k$ and remark that $z(uv)^k \in R$. Now $(yv)^k = a^k x^{k\underline{m}} +$ *lower term*; furthermore we claim that $z^h = \lambda_h b^h x^{h\underline{r}} +$ *lower term*, for all h, $\lambda_h \in F^*$;. This is easily proved, since z is central, by induction, remarking that

$$z^h = z^{h-1}(bx^{\underline{r}} + \text{lower term}) = bz^{h-1}x^{\underline{r}} + \text{lower term}$$

We deduce that

$$(6.5.1)\quad a^k x^{k\underline{m}} + l.\ t. = (yv)^k = z^{k-1}z(uv)^k = (\lambda_{k-1}b^{k-1}x^{(k-1)\underline{r}} + \text{lower term})z(uv)^k.$$

This relation implies that $\forall k$, b^{k-1} divides a^k in A, in other words $(a/b)^k \in b^{-1}A$. Since A is integrally closed we deduce that b divides a in A as requested.

Next we claim that $\underline{r}$ divides $\underline{s}$. In fact from $y^k = z^{k-1}u^k$ as for the identity 6.5.1 we also deduce that in the monoid $\mathbb{N}^k$ the vector $(k-1)\underline{r}$ divides $k\underline{s}$ for all k. By an obvious integral closure of the monoid $\mathbb{N}^k$ follows that $\underline{r}$ divides $\underline{s}$ so we can find an element w in R so that $x^{\underline{r}} = x^{\underline{s}}w +$ *lower term*.

We can now finish our argument by induction. Assume by contradiction that there is an element $u \in B$ and not in R we may choose it in such a way that the degree of $y := zu \in R$ is minimal. By the previous argument we know that $a = bf$, $f \in A$. Then $fzw = zfw$ has the same leading term as y and $u - fw \in B$. By induction $u - fw \in R$ which gives a contradiction.

§7. Representation theory of twisted polynomial rings.

7.1 Let $\mathbb{F}$ be a field and $q \in \mathbb{F}^\times$ a given element. Given an $n \times n$ skew–symmetric matrix $H = (h_{ij})$ over $\mathbb{Z}$, we construct the *twisted polynomial algebra* $\mathbb{F}_H[x_1, \dots, x_n]$. This is the algebra on generators $x_1, \dots, x_n$ and the following defining relations:

$$(7.1.1)\qquad x_i x_j = q^{h_{ij}} x_j x_i \quad (i,j = 1, \dots, n).$$

It can be viewed as an iterated twisted polynomial algebra with respect to any ordering of the indeterminates x_i. Similarly, we can define the twisted Laurent polynomial algebra $\mathbb{F}_H[x_1, x_1^{-1}, \dots, x_n, x_n^{-1}]$. Both algebras have no zero divisors.

To study its spectrum we start with a simple general lemma.

Lemma. *If M is an irreducible module over $A_\sigma[x]$, then there are two possibilities:*
(i) $x = 0$, hence M is actually an A–module,
(ii) x is invertible, hence M is actually an $A_\sigma[x, x^{-1}]$-module.

Proof. It is clear that $\text{Im}(x)$ and $\text{Ker}(x)$ are submodules of M. □

Corollary. *In any irreducible $\mathbb{F}_H[x_1, \dots, x_n]$–module each element x_i is either 0 or invertible.* □

Given $a = (a_1, \dots, a_n) \in \mathbb{Z}^n$, we shall write $x^a = x_1^{a_1} \dots x_n^{a_n}$. The torus $\mathbb{F}^{\times n}$ acts by automorphisms of the algebra $\mathbb{F}_H[x_1, \dots, x_n]$ and $\mathbb{F}_H[x_1, x_1^{-1}, \dots, x_n, x_n^{-1}]$ in the usual way, the monomial x^a being a weight vector of weight a. Consider the group G of inner automorphisms of the Laurent polynomials generated by conjugation by the variables x_i. Clearly G induces a group of automorphisms of the twisted polynomial algebra which are, by (7.1.1) in this torus of automorphisms. In fact one can formalize this as follows:

Let $\Gamma := \{\alpha x^a | \alpha \in \mathbb{F}^\times\}$ be the set of non zero monomials. Then Γ is a group, $\mathbb{F}^\times$ is a central subgroup and $\Gamma/\mathbb{F}^\times$ is free abelian, the homomorphism $\Gamma \to (\mathbb{F}^\times)^n$ given by considering the associated inner automorphisms has as kernel the monomials in the center.

Let ε be a primitive ℓ–th root of 1 in $\mathbb{F}$ and take now $q = \varepsilon$. We consider the matrix H as a matrix of a homomorphism $H: \mathbb{Z}^n \to (\mathbb{Z}/\ell\mathbb{Z})^n$, and we denote by K the kernel of H and by h the cardinality of the image of H.

Proposition. *(a) The elements x^a with $a \in K \cap \mathbb{Z}_+^n$ (resp. $a \in K$) form a basis of the center of $\mathbb{F}_H[x_1, \ldots, x_n]$ (resp. $\mathbb{F}_H[x_1, x_1^{-1}, \ldots, x_n, x_n^{-1}]$).*

(b) Let $a^{(1)}, \ldots, a^{(h)}$ be a set of representatives of $\mathbb{Z}^n$ mod K. Then the monomials $x^{a^{(1)}}, \ldots, x^{a^{(h)}}$ form a basis of the algebra $\mathbb{F}_H[x_1, x_1^{-1}, \ldots, x_n, x_n^{-1}]$ over its center.

(c) degree $\mathbb{F}_H[x_1, \ldots, x_n] =$ degree $\mathbb{F}_H[x_1, x_1^{-1}, \ldots, x_n, x_n^{-1}] = \sqrt{h}$.

Proof. Define a skewsymmetric bilinear form on $\mathbb{Z}^n$ by letting for $a = (a_1, \ldots, a_n)$, $b = (b_1, \ldots, b_n) \in \mathbb{Z}^n : \langle a|b\rangle = \sum_{i,j=1}^n h_{ij} a_i b_j$. Then we have

(7.1.2)
$$x^a x^b = \varepsilon^{\langle a|b\rangle} x^b x^a.$$

Since the center is invariant with respect to the action of $\mathbb{F}^{\times n}$, it must have a basis of elements of the form x^a. This together with (7.1.2) implies (a).

(b) follows from (a) and the fact that

(7.1.3)
$$x^a x^b = \varepsilon^{c(a,b)} x^{a+b}, \text{ where } c(a,b) = \sum_{i>j} h_{ij} a_i b_j.$$

(c) follows from (b). □

The center of a twisted polynomial algebra is the ring of invariants of a torus acting on a polynomial ring hence it is integrally closed, moreover the algebra is finite over its center hence these algebras are closed under trace and in fact from (7.1.3) one can easily deduce a formula for the trace

(7.1.4)
$$tr(x^a) = 0 \text{ if } x^a \text{ is not in the center.}$$

Proposition. *A twisted polynomial ring is a maximal order.*

Proof. All the hypotheses of Theorem 6.5 are satisfied.

7.2. We start from:

Proposition. *Let $\mathbb{F}$ be an algebraically closed field. Then any Laurent quasipolynomial algebra $\mathbb{F}_H[x_1, x_1^{-1}, \ldots, x_n, x_n^{-1}]$ is an Azumaya algebra over its center. In particular, all irreducible representations of the algebra $\mathbb{F}_H[x_1, \ldots, x_n]$ for which all $x_i \neq 0$ have dimension $\sqrt{h}$.*

Proof. Let $Z_0 = \{x^a | a \in (\ell\mathbb{Z})^n\}$. This is a finitely generated central subalgebra over which the algebra $A := \mathbb{F}_H[x_1, x_1^{-1}, \ldots, x_n, x_n^{-1}]$ is finitely generated. Recall that we have the surjective map $\chi_0 :$ Spec $A \longrightarrow$ Spec Z_0 and that the set $\Omega_A^0 = \{a \in \text{Spec } Z_0 | \chi_0^{-1}(a)$ consists of representations of maximal dimension$\}$ is a dense open subset of Spec Z_0 (Theorem 4.5 e). But the group $\mathbb{F}^{\times n}$ of automorphisms of A acts transitively on Spec Z_0, hence $\Omega_A^0 =$ Spec Z_0, proving the proposition. □

Remark. *We can make a monomial change of variables in a twisted polynomial ring and bring the skew form in canonical form.*

A canonical form for a skew symmetric matrix over $\mathbb{Z}$ is as direct sum of 2×2 blocks, which reduces to tensor products of algebras of twisted Laurent polynomials in 2 variables with commutation relations $xy = \varepsilon yx$.

7.3 Let $A := \mathbb{F}_H[x_1, \dots, x_n]$ be as in §7.2 (of degree $\sqrt{h}$). Recall that the torus $T := \mathbb{F}^n$ acts by automorphisms of A and hence of its center, so that the representation picture looks like a non-commutative version of affine torus embeddings. First of all remark that, by Lemma 7.2, the vanishing of the central element x_i^l in an irreducible representation implies the vanishing of x_i. Thus it is natural to stratify the *Spec* A according to the set S of indices i for which $x_i^l \neq 0$ and remark that this stratification is just the stratification by orbits under T. Let A_S denote the twisted Laurent polynomial algebra in the variables x_i, $i \in S$. From §7.1 we have that A_S is an Azumaya algebra whose degree d_S is computed as in Proposition 7.2 by restricting the homomorphism H to the subgroup of $\mathbb{Z}^n$ formed by the vectors with zero coordinates in the indices not in S, i.e. by analyzing the skew submatrix H_S of H which defines A_S. The spectrum of its center is isomorphic to a quotient T_S of the torus T.

On the other hand we can pass from A to A_S as follows. First we can invert in A the elements x_i, $i \in S$, to get an algebra which we may call A'_S. In A'_S we have the ideal I_S generated by the variables x_i, $i \notin S$, and we clearly have that $A_S = A'_S/I_S$. The center of A'_S is the center of A localized at the elements x_i^l and its points parametrize equivalence classes of semisimple representations of degree $\sqrt{h}$ where the central character is non-zero in the x_i^l, $i \in S$. The algebra A_S inherits from A'_S a trace map tr with values in the quotient Z'_S of the center of A'_S by the ideal generated by the elements x_i, $i \in S$. It is not hard to see that the picture is the one predicted by proposition 4.5.

In A_S we have the center Z_S and its subring Z'_S over which Z_S is finite. The spectrum of Z'_S is also isomorphic to a quotient T'_S of the torus T and T'_S is a quotient T_S/Γ by a finite subgroup Γ. In particular each fiber of the map π : *Spec* $Z_S \rightarrow$ *Spec* Z'_S is reduced and consists of a coset of the finite group Γ. We have several trace maps: the reduced trace tr_{A_S/Z_S} to the center, the trace of the finite map tr_{Z_S/Z'_S} and the composition tr_{A_S/Z'_S}. From 4.5) and the torus description it follows that there exists a positive integer d such that $tr = d\, tr_{A_S/Z'_S}$. From this and §7.1 it follows that each point of the spectrum of Z'_S corresponds to a semisimple representation which is obtained counting with multiplicity d each irreducible representations of A_S appearing in the fiber of the map π: *Spec* $Z_S \rightarrow$ *Spec* Z'_S. Of course we have: $d|\Gamma|d_S = \sqrt{h}$.

CHAPTER 3

QUANTUM GROUPS

§8. Some properties of finite root systems.

8.1. The quantum groups which will be the object of our study arise as q-analogues of the universal enveloping algebras of semisimple Lie algebras. The way in which we will introduce them, which is also the one in which they have been discovered, is to generalize the classical presentation of semisimple Lie algebras by Chevalley generators and Serre's relations. In this approach the main starting ingredient, which one might consider as a genetic code for the theory, is the Cartan matrix or its Dynkin diagram and the associated root system. We thus recall first quickly some basic notions and notations of this theory.

Let $C := (a_{ij})$ be a $n \times n$ matrix with integer entries such that $(i, j = 1, \dots, n)$:

(8.1.1) $$a_{ii} = 2, \ a_{ij} \leq 0 \text{ if } i \neq j,$$

and there exists a vector $(d_1, \dots, d_n)$ with relatively prime positive integral entries d_i such that

(8.1.2) $$(d_i a_{ij}) \text{ is a symmetric positive definite matrix.}$$

This is the definition of a **Cartan matrix.**

To the Cartan matrix C there is associated a finite reduced root system R, its weight and root lattices P and Q, the Weyl group W, a set of positive roots R^+, the set of simple roots Π, the fundamental weights $\omega_1, \dots, \omega_n$, etc. Let us recall for convenience the basic definitions.

Let P be a lattice over $\mathbb{Z}$ with basis $\omega_1, \dots, \omega_n$ P is called the *weight lattice* and the elements ω_i the *fundamental weights*. Let $Q^\vee = \mathrm{Hom}_{\mathbb{Z}}(P, \mathbb{Z})$ be the dual lattice with dual basis $\alpha_1^\vee, \dots, \alpha_n^\vee$ (called the *coroots*), i.e. $\langle \omega_i, \alpha_j^\vee \rangle = \delta_{ij}$.

One introduces the following objects:

(Dominant integral weights) $$P_+ = \sum_{i=1}^n \mathbb{Z}_+ \omega_i$$

(A special weight) $$\rho = \sum_{i=1}^n \omega_i,$$

(The simple roots) $$\alpha_j = \sum_{i=1}^n a_{ij} \omega_i \quad (j = 1, \dots, n),$$

(Root lattice) $$Q = \sum_{j=1}^{n} \mathbb{Z}\alpha_j \subset P,$$

(Positive root lattice) $$Q_+ = \sum_{j=1}^{n} \mathbb{Z}_+\alpha_j$$

Define the usual partial ordering on P (*the dominant order*) by $\lambda \geq \mu$ if $\lambda - \mu \in Q_+$. For $\beta = \sum_i k_i\alpha_i \in Q$ let $ht\beta = \sum_i k_i$.

Define reflection automorphisms s_i of P by $s_i(\omega_j) = \omega_j - \delta_{ij}\alpha_i$ $(i, j = 1, \dots, n)$. Then $s_i(\alpha_j) = \alpha_j - a_{ij}\alpha_i$. Let W (the Weyl group) be the subgroup of $GL(P)$ generated by $s_1, \dots, s_n$. Recall that W is a *Coxeter group* on generators s_i $(i = 1, \dots, n)$ and defining relations

$$s_i^2 = 1 \text{ and } (s_is_j)^{m_{ij}} = 1 \text{ when } i \neq j,$$

where $m_{ij} = 2, 3, 4$ or 6 for $a_{ij}a_{ji} = 0, 1, 2$ or 3 respectively $(i \neq j)$.

Together with the Weyl group it is useful to introduce the (generalized) braid group. It is an infinite group $\mathcal{B}$ generated by elements T_i, $i = 1, \dots, n$ and the *braid relations*: For $i \neq j$ we take the word of (even) length $(T_iT_j)^{m_{ij}}$, split it in half and impose that the first half be equal to the second written in reverse order.

Of course the Weyl group W is the quotient of $\mathcal{B}$ under the further relations $T_i^2 = 1$.

It will be convenient to use the following abbreviated notation:

$$T_{ij}^{(m)} = T_iT_jT_i \dots (m \text{ factors}).$$

For example, the braid relations read:

(8.1.3.) $$T_{ij}^{(m_{ij})} = T_{ji}^{(m_{ji})} \text{ if } i \neq j$$

Let

$$\Pi = \{\alpha_1, \dots, \alpha_n\},\ \Pi^\vee = \{\alpha_1^\vee, \dots, \alpha_n^\vee\},$$
$$R = W\Pi,\ R^+ = R \cap Q_+,\ R^\vee = W\Pi^\vee.$$

R is the set of *roots*, $R^\vee$ the *coroots*, Π the *simple roots*. The map $\alpha_i \longmapsto \alpha_i^\vee$ extends uniquely to a bijective W-equivariant map $\alpha \longmapsto \alpha^\vee$ between R and $R^\vee$. The reflection s_α defined by $s_\alpha(\lambda) = \lambda - \langle\lambda, \alpha^\vee\rangle\alpha$ lies in W for each $\alpha \in R$, so that $s_{\alpha_i} = s_i$.

Define a bilinear pairing $P \times Q \to \mathbb{Z}$ by $(\omega_i|\alpha_j) = \delta_{ij}d_j$. Then $(\alpha_i|\alpha_j) = d_ia_{ij}$, giving a symmetric $\mathbb{Z}$-valued W-invariant bilinear form on Q such that $(\alpha|\alpha) \in 2\mathbb{Z}$. We may identify $Q^\vee$ with a sublattice of the $\mathbb{Q}$-span of P (containing Q) using this form. Then:

(8.1.4) $$\alpha_i^\vee = d_i^{-1}\alpha_i,\ \alpha^\vee = 2\alpha/(\alpha|\alpha).$$

8.2 Every element of W has a length $\ell(w)$ which can be defined as the length of a shortest expression of w as a product of s_i and equals the cardinality of $R_w := \{\beta \in R^+ | w(\beta) < 0\}$; such an expression is called a *reduced expression* of w. Recall that

$$\ell(ws_i) = \ell(w) + 1 \text{ if } w(\alpha_i) > 0 \text{ and } \ell(ws_i) = \ell(w) - 1 \text{ if } w(\alpha_i) < 0.$$

In particular, since W acts transitively on the bases there exists a unique element w_0 of longest length $N = |R^+|$ such that $w_0(R^+) = -R^+$. Of course $w_0 = w_0^{-1}$, $w_0(\Pi) = -\Pi$ and $w_0(P_+) = -P_+$. For $\alpha \in R^+$ (resp. $\lambda \in P_+$) we let ${}^t\alpha = -w_0(\alpha) \in R^+$ (resp. ${}^t\lambda = -w_0(\lambda)$). For a fundamental weight ω, the weight ${}^t\omega$ is also fundamental and since $\Pi = -w_0(\Pi)$ we denote by $j \mapsto \bar{j}$ the permutation of $1, 2, \dots, n$ such that $\alpha_{\bar{j}} = -w_0(\alpha_j)$.

We have that $s_j w_0 = w_0 s_{\bar{j}}$. More precisely, writing

$$w_0 = s_j s_{i_1} s_{i_2} \dots s_{i_{N-1}} = s_{i_1} s_{i_2} \dots s_{i_{N-1}} s_{\bar{j}}$$

we deduce:

Lemma. $s_{i_1} s_{i_2} \dots s_{i_{N-1}}(\alpha_{\bar{j}}) = \alpha_j$.

Proof. $s_{i_1} s_{i_2} \dots s_{i_{N-1}}(\alpha_{\bar{j}}) = s_j w_0(\alpha_{\bar{j}}) = s_j(-\alpha_j)$. □

If $w = ab \in W$ is such that $\ell(w) = \ell(a) + \ell(b)$ we will say that this is a *reduced* decomposition.

Given an element $w \in W$ we set $\overline{w} := w_0 w w_0^{-1}$ (so that $\bar{s}_j = s_{\bar{j}}$).

Fix a reduced expression

$$w_0 = s_{i_1} s_{i_2} \dots s_{i_N}, \text{ where } N = |R^+|, \tag{8.2.1}$$

then we have the corresponding *convex* ordering of R^+:

$$\beta_1 = \alpha_{i_1},\ \beta_2 = s_{i_1}(\alpha_{i_2}) \dots,\ \beta_N = s_{i_1} \dots s_{i_{N-1}}(\alpha_{i_N}).$$

(The name "convex" refers to the property that if $i < j$ and $\beta_i + \beta_j \in R^+$, then $\beta_i + \beta_j = \beta_k$ for some k between i and j.) More generally (cf. [Pa]):

Proposition. *If*

$$w = s_{i_1} s_{i_2} \dots s_{i_k}, \text{ where } k = |R_w|, \tag{8.2.2}$$

the elements

$$\beta_1 = \alpha_{i_1},\ \beta_2 = s_{i_1}(\alpha_{i_2}) \dots,\ \beta_k = s_{i_1} \dots s_{i_{N-1}}(\alpha_{i_k})$$

produce a convex ordering of $R_{w^{-1}}$, *and conversely.*

For $R_{w^{-1}}$ the notion of convex ordering needs one further condition. If $\beta_i, \beta_k \in R_{w^{-1}}$, $\beta_k = \beta_i + \gamma$ with γ a positive root not in $R_{w^{-1}}$, then $i < k$.

A major tool of the computations to follow is a result of Matsumoto [M]:

Theorem. *If* $s_{i_1} s_{i_2} \dots s_{i_m}$ *is a reduced expression of an element* $w \in W$, *then the element* $T_w := T_{i_1} T_{i_2} \dots T_{i_m}$ *in* $\mathcal{B}$ *depends only on* w *(and not on its reduced expression).* □

In other words we can pass from a reduced expression of w to another using only the braid relations.

In particular we have defined a canonical section $w \to T_w$ of W in $\mathcal{B}$; of course this section is not multiplicative but $T_a T_b = T_{ab}$ if $\ell(ab) = \ell(a) + \ell(b)$.

§9. Quantum groups.

9.1 There are several objects, which have recently appeared and may deserve the name of quantum group. Here we introduce as quantum groups a simple variation of the construction of Drinfeld and Jimbo of the quantized enveloping algebra associated to a given Cartan matrix (cf. also the definitions of twisted derivations in §5.1). The *dual* of such Hopf algebras could also be justly considered as interesting quantum groups (cf. DC-L).

As we have already explained at the beginning of this chapter we follow the line of describing a $q-$analogue of the presentation of a semisimple Lie algebra by generators and relations. The reader should look at the standard presentation where the Chevalley generators are denoted by e_i, f_i, h_i and compare the usual Serre's relations with the ones we will present soon, he will find some mysterious differences. The explanation of these is really unclear, the discovery was made for $SL(2)$ first of all by Kulish-Reshetikhin who were trying to quantize the sin-Gordon equation.

One defines the *simply connected quantum group* U associated to the matrix (a_{ij}) as an algebra over the field $\mathbb{C}(q)$ on generators E_i, F_i $(i = 1, \ldots, n)$, K_α $(\alpha \in P)$ subject to the following relations

$$
\begin{aligned}
&K_\alpha K_\beta = K_{\alpha+\beta},\ K_0 = 1,\\
&\sigma_\alpha(E_i) = q^{(\alpha|\alpha_i)}E_i,\ \sigma_\alpha(F_i) = q^{-(\alpha|\alpha_i)}F_i,\\
&[E_i, F_j] = \delta_{ij}\frac{K_{\alpha_i} - K_{-\alpha_i}}{q^{d_i} - q^{-d_i}},\\
&(ad_{\sigma_{-\alpha_i}}E_i)^{1-a_{ij}}E_j = 0,\ (ad_{\sigma_{-\alpha_i}}F_i)^{1-a_{ij}}F_j = 0\ (i \neq j),
\end{aligned}
\tag{9.1.1}
$$

where $\sigma_\alpha = Ad\ K_\alpha$. One should think of E_i, F_i as $q-$analogues of the e_i, f_i while the K_{α_i} are $q-$analogues of $q^{d_i\alpha_i}$. It has been shown that U has a Hopf algebra structure with comultiplication Δ, antipode S and counit η defined by:

$$
\begin{aligned}
&\Delta E_i = E_i \otimes 1 + K_{\alpha_i} \otimes E_i,\ \Delta F_i = F_i \otimes K_{-\alpha_i} + 1 \otimes F_i,\ \Delta K_\alpha = K_\alpha \otimes K_\alpha,\\
&SE_i = -K_{-\alpha_i}E_i,\ SF_i = -F_iK_i,\ SK_\alpha = K_{-\alpha},\\
&\eta E_i = 0,\ \eta F_i = 0,\ \eta K_\alpha = 1.
\end{aligned}
\tag{9.1.2}
$$

In order to verify this statement one has, first of all to verify that the maps are well defined, i.e. compatible with the relations. and then the various Hopf algebras identities. To make these verifications is rather tedious (specially when $a_{ij} = -3$ but straightforward). It is usual to write $K_i := K_{\alpha_i}$. The quantum group of Drinfeld–Jimbo is the subalgebra of U_P over $\mathbb{C}(q)$ generated by the E_i, F_i, K_i, K_i^{-1} $(i = 1, \ldots, n)$. More generally, for any lattice M between P and Q one may consider the intermediate quantum group U_M generated by the E_i, F_i $(i = 1, \ldots, n)$ and the K_β with $\beta \in M$. In this paper by a *quantum group* we mean one of these algebras. We denote by U^+, U^- and U^0 the $\mathbb{C}(q)$–subalgebra of U_M generated by the E_i, the F_i and the K_β respectively. We shall sometimes add the subscript M to emphasize the dependence on M, like U_M^0, etc.

Before going into deeper facts about the structure of these algebras we want to make a remark. We wish to analyze its 1-dimensional representations.

In other words we need to substitute to the given generators numbers satisfying the same relations.

From $[E_i, F_i] = \frac{K_{\alpha_i} - K_{-\alpha_i}}{q^{d_i} - q^{-d_i}}$ we deduce $K_{\alpha_i} - K_{-\alpha_i} = 0$ or $K_{\alpha_i} = \pm 1$.

From the commutation relations between the K_i and E_i, F_i we deduce $E_i = F_i = 0$. Thus in the adjoint case we have 2^n 1-dimensional representations given assigning the values ± 1 to the K_i, in the simply connected case we identify these representations with the characters of $P/2Q$. Given any representation M and a character δ, since U is a Hopf algebra we can perform the tensor product between these representations which will be denoted by M_δ.

9.2 As usual, for $n \in \mathbb{Z}$ and $d \in \mathbb{N}$ we let

$$[n]_d = (q^{dn} - q^{-dn})/(q^d - q^{-d}),\ [n]_d! = [1]_d[2]_d \dots [n]_d,$$

$$\begin{bmatrix} n \\ j \end{bmatrix}_d = [n]_d[n-1]_d \dots [n-j+1]_d/[j]_d! \text{ for } j \in \mathbb{N},\ \begin{bmatrix} n \\ 0 \end{bmatrix}_d = 1.$$

We shall omit the subscript d when $d = 1$. We review the method which has been developed to construct quantum analogues of Poincaré Birkhoff Witt bases.

Lusztig ([L2]) has defined an action of the braid group $\mathcal{B}_W$ (associated to W), whose canonical generators one denotes by T_i, as a group of automorphisms of the algebra U by the formulas:

$$\begin{aligned} &T_iK_\alpha = K_{s_i(\alpha)},\ T_iE_i = -F_iK_{\alpha_i}, \\ (9.2.1)\qquad &T_iE_j = \frac{1}{[-a_{ij}]_{d_i}!}(ad_{\sigma_{-\alpha_i}}(-E_i))^{-a_{ij}}E_j = (-adE_i)^{a_{i,j}}(E_j), \\ &T_i\kappa = \kappa T_i, \end{aligned}$$

where if $\Delta(x) = \sum x_j \otimes y_j$, $ad(x)(y) = \sum x_j y S(y_j)$, and where κ is a conjugate-linear anti-automorphism of U, viewed as an algebra over $\mathbb{C}(q)$ with conjugation given by $\kappa q = q^{-1}$, defined by:

$$(9.2.2)\qquad \kappa E_i = F_i,\ \kappa F_i = E_i,\ \kappa K_\alpha = K_{-\alpha},\ \kappa q = q^{-1}.$$

Again one might verify directly that the maps and their inverses are well defined and satisfy the braid relations.

Notice that these automorphisms do not respect the comultiplication or the antipode so they are not Hopf algebra automorphisms.

One of the main ingredients consists in using the braid group to construct analogues of root vectors associated to non simple roots.

9.3 Let us now take a reduced expression $w_0 = s_{i_1} \cdots s_{i_N}$ for the longest element w_0 in the Weyl group. Set $\beta_t = s_{i_1} \cdots s_{i_{t-1}}(\alpha_{i_t})$ and get a total ordering on the set $\{\beta_1, \dots, \beta_N\}$ of positive roots (the convex ordering defined in 8.2).

We define the elements $E_{\beta_1}, \dots, E_{\beta_N}$ by $E_{\beta_t} = T_{i_1} \cdots T_{i_{t-1}}(E_{\alpha_{i_t}})$. Similarly the elements $F_{\beta_1}, \dots, F_{\beta_N}$ by $F_{\beta_t} = T_{i_1} \cdots T_{i_{t-1}}(F_{\alpha_{i_t}})$. These elements depend heavily on the choice of the reduced expression.

Theorem. *([L],[L-S]). i) Let U^+ (resp. U^-) be the subalgebra generated by the E_i (resp. F_i). Then $E_{\beta_t} \in U^+, \forall t = 1, \dots, N$ (resp. $F_{\beta_t} \in U^-, \forall t = 1, \dots, N$.)*

ii) The monomials $E_{\beta_1}^{k_1} \cdots E_{\beta_N}^{k_N}$ (resp. $F_{\beta_N}^{k_N} \cdots F_{\beta_1}^{k_1}$) are a $\mathbb{C}(q)$ basis of U^+ (resp. U^-).

iii) The monomials

$$E_{\beta_1}^{k_1} \cdots E_{\beta_N}^{k_N} K_\lambda F_{\beta_N}^{h_N} \cdots F_{\beta_1}^{h_1} \tag{9.3.1}$$

are a $\mathbb{C}(q)$ basis of $U_{q,\Lambda}$.

iv) [L–S] For $i < j$ one has:

$$E_{\beta_i}E_{\beta_j} - q^{(\beta_j|\beta_i)}E_{\beta_j}E_{\beta_i} = \sum_{k \in \mathbb{Z}_+^N} c_k E^k, \tag{9.3.2}$$

where $c_k \in \mathbb{Q}[q, q^{-1}]$ *and* $c_k \neq 0$ *only when* $k = (k_1, \ldots, k_N)$ *is such that* $k_s = 0$ *for* $s \leq i$ *and* $s \geq j$ *and* $E^k = E_{\beta_1}^{k_1} \cdots E_{\beta_N}^{k_N}$.

$$F_{\beta_i} F_{\beta_j} - q^{-(\beta_j|\beta_i)} F_{\beta_j} F_{\beta_i} = \sum_{k \in \mathbb{Z}_+^N} c_k F^k, \tag{9.3.3}$$

where $c_k \in \mathbb{Q}[q, q^{-1}]$ *and* $c_k \neq 0$ *only when* $k = (k_N, \ldots, k_1)$ *is such that* $k_s = 0$ *for* $s \leq i$ *and* $s \geq j$ *and* $F^k = F_{\beta_N}^{k_N} \cdots F_{\beta_1}^{k_1}$.

v) [R2] We have the tensor product decomposition (as algebras over the field $\mathbb{C}(q)$.

$$U_q = U_q^- \otimes U_q^0 \otimes U_q^+.$$

Proof. Let us discuss only part iv) leaving the rest to the literature. We proceed by induction on $l - t$. And also assume that the result is proven in the rank two case. In particular we can assume that there are no components of type G_2. Using the rank two case it is clear that is $l - t \leq 1$ there is nothing to prove. Assume now that everything is proved for $l - t = m$. It is also clear that we can always assume $t = 1$, so that $E_{\beta_t} = E_h$ for some h Set $i_l = i$, $i_{l+1} = j$ and $w = s_{i_1} \cdots s_{i_{l-1}}$. We distinguish various cases.

Case 1) $a_{ij} = 0$. We have $T_w T_i E_j = T_w E_j$, so that again by applying the relation $s_i s_j = s_j s_i$ to the reduced expression for w_0 we deduce that $E_{ws_i\alpha_j} E_h - q^{-(\alpha_h, ws_i\alpha_j)} E_h E_{ws_i\alpha_j}$ can be expressed as a linear combinations of ordered monomials involving only $E_{\beta_2}, \ldots, E_{\beta_{l-1}}$.

Case 2) $a_{ij} = (\alpha_i, \alpha_j) = -1$ and $l(ws_j) > l(w)$.

The assumption clearly implies that $l(ws_is_js_i) = l(w)+3$. It follows that for $T_w T_i T_j E_i = T_w E_j$ we can repeat the argument above and get that

$$T_w T_i T_j E_i E_h - q^{-(\alpha_h, w\alpha_j)} E_h T_w T_i T_j E_i = \Sigma$$

can be expressed as a linear combinations of ordered monomials involving only the elements $E_{\beta_2}, \ldots, E_{\beta_{l-1}}$. Also we have that $E_{ws_i\alpha_j} = T_w T_i E_j = -T_w E_i T_w E_j + q^{-1} T_w E_j T_w E_i = -E_{\beta_l} T_w E_j + q^{-1} T_w E_j E_{\beta_l}$.

Write $E_{\beta_l} E_h - q^{-(\alpha_h, \beta_l)} E_h E_{\beta_l} = \Omega$. We obtain

(9.3.1)

$$\begin{aligned}
&E_{ws_i\alpha_j} E_h = \\
&-q^{-(\alpha_h, w\alpha_j)} E_{\beta_l} E_h T_w E_j - E_{\beta_l} \Sigma + q^{-(\alpha_h, \beta_l)-1} T_w E_j E_h E_{\beta_l} + q^{-1} T_w E_j \Omega = \\
&-q^{-(\alpha_h, ws_i\alpha_j)} E_h E_{\beta_l} T_w E_j - q^{-(\alpha_h, w\alpha_j)} \Omega T_w E_j - E_{\beta_l} \Sigma \\
&+ q^{-(\alpha_h, ws_i\alpha_j)-1} E_h T_w E_j E_{\beta_l} + q^{-(\alpha_h, \beta_l)-1} \Sigma E_{\beta_l} + q^{-1} T_w E_j \Omega = \\
&q^{-(\alpha_h, ws_i\alpha_j)} E_h E_{ws_i\alpha_j} - q^{-(\alpha_h, w\alpha_j)} \Omega T_w E_j - E_{\beta_l} \Sigma + q^{-(\alpha_h, \beta_l)-1} \Sigma E_{\beta_l} + q^{-1} T_w E_j \Omega
\end{aligned}$$

Consider now $-q^{-(\alpha_h, w\alpha_j)} \Omega T_w E_j + q^{-1} T_w E_j \Omega$ Since Ω has clearly degree $\alpha_h + \beta_l$ and involves only the elements $E_{\beta_2}, \ldots, E_{\beta_{l-1}}$ it follows reasoning as above that

$$T_w E_j \Omega = q^{-(w\alpha_j, \alpha_h + \beta_l)} \Omega T_w E_j + \Omega'$$

with Ω' involving only $E_{\beta_2}, \ldots, E_{\beta_{l-1}}$. Substituting and using the fact that $(\alpha_i, \alpha_j) = (w\alpha_j, \beta_l) = -1$ we deduce that $-q^{-(\alpha_h, w\alpha_j)} \Omega T_w E_j + q^{-1} T_w E_j \Omega = q^{-1} \Omega'$. Since in (9.3.1) the terms $E_{\beta_l} \Sigma$ and $q^{-(\alpha_h, \beta_l)-1} \Sigma E_{\beta_l}$ only involve the elements $E_{\beta_2}, \ldots, E_{\beta_l}$ our claim follows.

Case 3) $a_{ij} = (\alpha_i, \alpha_j) = -1$ and $l(ws_j) < l(w)$.

In this case we can write $w = us_j$ with $u = ws_j$. Also we claim $l(us_i) > l(u)$. Indeed suppose the contrary then $l(w)+2 = l(ws_is_j) = l(us_js_is_j) = l(us_is_js_i) < l(u)+2 = l(w)+1$ which is clearly a contradiction. So applying the above considerations to the elements E_h and $T_uE_i = T_wT_iE_j$, and recalling that since by hypothesis $\alpha_h \neq w\alpha_i$ so that either $u^{-1}\alpha_h < 0$ or $u^{-1}\alpha_h = \alpha_j$, in this last case we are commuting T_uE_i, T_uE_j and the claim follows from the rank 2 case otherwise we get that

$$E_{ws_i\alpha_j}E_h - q^{-(\alpha_h, ws_i\alpha_j)}E_hE_{ws_i\alpha_j} \in U_u$$

By the definition of u we deduce that $U_u \subset U_w$ so $E_{ws_i\alpha_j}E_h - q^{-(\alpha_h, ws_i\alpha_j)}E_hE_{ws_i\alpha_j}$ can be written as a linear combination of ordered monomials involving only the elements $E_{\beta_2}, \dots, E_{\beta_{l-1}}$ and E_h. Assume that a monomial of the form $E_h^m E_{\beta_2}^{m_2} \cdots E_{\beta_{l-1}}^{m_{l-1}}$ appears with non zero coefficient then by homogeneity $\sum_{k=2}^{l-1} m_k\beta_k + (m-1)\alpha_h = ws_i\alpha_j$. Since for $w^{-1}\beta_k < 0$, and $w^{-1}\alpha_h < 0$ we deduce $0 > w^{-1}ws_i\alpha_j = \alpha_j + \alpha_i$ a contradiction.

Case 4) $a_{ij} = -1$, $(\alpha_i, \alpha_j) = -2$ and $l(ws_j) > l(w)$. Then the same argument as in case 2) applies as soon as we notice that $l(ws_is_js_is_j) = l(w)+4$, $T_wT_iT_jT_iE_j = T_wE_j$ and $E_{ws_i\alpha_j} = T_wT_iE_j = -E_{\beta_l}T_wE_j + q^{-2}T_wE_jE_{\beta_l}$.

Case 5) $a_{ij} = -2$ and $l(ws_j) > l(w)$.. In this case again $l(ws_is_js_is_j) = l(w)+4$, $T_wT_iT_jT_iE_j = T_wE_j$ so that $T_wE_jE_h - q^{-(\alpha_h, w\alpha_j)}E_hT_wE_j = \Sigma$ involves only the elements $E_{\beta_2}, \dots, E_{\beta_{l-1}}$. Furthermore one has $T_wT_iT_jE_i = -T_wE_jE_{\beta_l} + q^{-2}E_{\beta_l}T_wE_j$ so that, reasoning as above, we deduce that $T_wT_iT_jE_iE_h - q^{(\alpha_h, ws_is_j}E_hT_wT_iT_jE_i = \Omega$ can be involves only the elements $E_{\beta_2}, \dots, E_{\beta_l}$. Also we have $E_{ws_i\alpha_j} = -E_{\beta_l}T_wT_iT_jE_i + T_wT_iT_jE_iE_{\beta_l}$. So applying the same reasoning as in the proof of case b) above we get the result in this situation.

Case 6) $(\alpha_i, \alpha_j) = -2$ and $l(ws_j) > l(w)$. This case has two subcases. The first happens when also $l(ws_js_i) < l(ws_j)$. In this case one easily shows that setting $u = ws_js_i$ $l(us_i) = l(us_j) > l(u)$. And then reasons in a completely analogous way to case b). In the second on has $l(ws_js_i) > l(lws_j)$. In this case one puts $u = ws_j$ and again reasoning as in the above cases easily reaches the conclusion. We leave the details to the reader.

An immediate corollary is the following:

Let w be any element of the Weyl group. We can choose for it a reduced expression $w = s_{i_1} \dots s_{i_k}$ which we complete to a reduced expression $w_0 = s_{i_1} \dots s_{i_N}$ of the longest element of W. Consider the elements E_{β_j}, $j = 1, \dots, k$. Then we have:

Proposition. *(a) The elements E_{β_j}, $j = 1, \dots, k$, generate a subalgebra U^w which is independent of the choice of the reduced expression of w.*

(b) If $w' = ws$ with s a simple reflection and $l(w') = l(w) + 1 = k + 1$, then $U^{w'}$ is a twisted polynomial algebra of type $U^w_{\sigma,D}[E_{\beta_{k+1}}]$, where the formulas for σ and D are implicitly given in the formulas of the previous theorem.

Proof. (a) Using the fact that one can pass from one reduced expression of w to another by braid relations one reduces to the case of rank 2 where one repeats the analysis made by Lusztig ([L4]). (b) is clear also by the previous theorem. □

The elements K_α clearly normalize the algebras $\mathcal{U}^w$ and when we add them to these algebras we are performing an iterated construction of Laurent twisted polynomials. The resulting algebras will be called $\mathcal{B}^w$.

9.4 In §12 we will define a suitable form A of U defined over $\mathcal{A} := \mathbb{C}[q, q^{-1}$, this will allow us to specialize q to any number ε which in not an hth root of 1 where $h \leq max\ d_i$. We will indicate the corresponding algebras with the subscript ε. Letting $\mathbb{Q}(\varepsilon) = \mathcal{A}/(q-\varepsilon)$ we have thus $U_\varepsilon := A \otimes_{\mathcal{A}} \mathbb{Q}(\varepsilon)$, etc..

If we wish, as we do, induce the antiautomorphism κ on $\mathcal{U}_\varepsilon$ it is convenient to restrict ε to be a complex number of absolute value 1 so that κ on $\mathbb{Q}(\varepsilon)$ coincides with complex conjugation. Often we may want to extend the scalars $\mathbb{Q}(\varepsilon)$ to the complex numbers keeping the same notations if there is no ambiguity.

It is useful to introduce other forms of these algebras. We recall an important construction of Lusztig. Given $s \in \mathbb{N}$ we shall introduce (cf. [L3]) the notations:

$$E_i^{(s)}, := E_i^s/[s]_{d_i}!,\ F_i^{(s)} := F_i^s/[s]_{d_i}! \tag{9.4.1.}$$

Theorem [L]. *The monomials*

$$E_{\beta_1}^{(k_1)} \cdots E_{\beta_N}^{(k_N)} K_\lambda F_{\beta_N}^{(h_N)} \cdots F_{\beta_1}^{(h_1)} \tag{9.4.2}$$

are a basis of a $\mathcal{A}$ algebra.

Another form, useful to make the connection with Poisson groups, will be discussed in 11.8.

9.5 First we need a fact on root systems. Consider a simple root α_h and the set A_h formed by all the pairs (w, i) such that $w \in W$ and $w(\alpha_i) = \alpha_h$. We make this set into a graph by joining (w, i) with (ws_j, i) if $m_{ij} = 2$, with (ws_js_i, j) if $m_{ij} = 3$, with $(ws_js_is_j, i)$ if $m_{ij} = 4$ and with $(ws_js_is_js_is_j, i)$ if $m_{ij} = 6$. We have:

Lemma. *The graph A_h is connected.*

Proof. We connect any given element (w, i) with $(1, h)$ as follows. We consider ws_i as part of a reduced expression of w_0 and pass from this expression to one starting with s_h with a sequence of braid relations. We follow stepwise the braid relations and associate in an obvious way to each step an edge of the graph. □

Theorem. *Let $\mathcal{B}$ act on a set S and let $u_1, \dots, u_n \in S$ be elements with the properties:*

$$T_{ij}^{(2)} u_i = u_j \text{ if } m_{ij} = 3 \text{ and } T_{ij}^{(m_{ij}-1)} u_j = u_j \text{ if } m_{ij} = 2, 4, \text{ or } 6.$$

Then:

(a) For a pair (w, i) in A_h we have $T_w(u_i) = u_h$.

(b) $T_i(u_i) = T_0(u_{\bar{i}})$ and $T_0^2(u_i) = T_i^2(u_i)$.

Proof. (a) is just a restatement of the lemma using the relations of the theorem. As for (b) we have $T_0 = T_iT_w = T_wT_{\bar{i}}$ where $w(\alpha_{\bar{i}}) = \alpha_i$. Thus $T_i(u_i) = T_iT_w(u_{\bar{i}}) = T_0(u_{\bar{i}})$, so $T_i^2(u_i) = T_iT_0(u_{\bar{i}}) = T_0T_{\bar{i}}(u_{\bar{i}}) = T_0^2(u_i)$. □

In particular we can apply the theorem to the action of the braid group on a quantum group and the elements $u_i = E_i$ or $u_i = F_i$. In this case the relations of the theorem are easily verified by the definitions of the T_i. Thus, we have: if α_i, α_j are two simple roots and $w \in W$ is such that $w(\alpha_i) = \alpha_j$ then

$$T_w(E_i) = E_j.$$

We also have:

$$T_0(E_{\bar{i}}) = T_i(E_i) = -F_iK_i,\ T_0(F_{\bar{i}}) = T_i(F_i) = -K_i^{-1}E_i.$$

9.6 Recall the notations and results of §8.2 where we denoted by $j \mapsto \bar{j}$ the permutation of $1, 2, \dots, n$ such that $\alpha_{\bar{j}} = -w_0(\alpha_j)$ and hence we have that $s_jw_0 = w_0s_{\bar{j}}$. Recall the result of Matsumoto (8.3) allowing to define a canonical section $w \to T_w$ of W in $\mathcal{B}$ with $T_aT_b = T_{ab}$ if $\ell(ab) = \ell(a) + \ell(b)$. Going back to the longest element we set $T_0 := T_{w_0}$ and will often use the following:

Lemma. . *If $a \in W$ we have $T_a T_0 = T_0 T_{\overline{a}}$.*

Proof. From 8.2 there exists b such that $w_0 = ab = b\overline{a}$ are reduced decompositions. We have thus $T_0 = T_a T_b = T_b T_{\overline{a}}$ from which the claim follows. □

§10 Degenerations of quantum groups.

10.1 We want to introduce now some interesting degenerations of the quantum groups, constructed as graded algebras associated to suitable filtrations, and which can be thought as being obtained by simplifying the defining relations of quantum groups.

Consider a monomial

$$M_{k,r,\alpha} := F^k K_\alpha E^r, \tag{10.1.1}$$

where $k = (k_1, \dots, k_N)$, $r = (r_1, \dots, r_N) \in \mathbb{Z}_+^N$ and $\alpha \in P$. Define its total height by

$$d_0(M_{k,r,\alpha}) = \sum_i (k_i + r_i) \mathrm{ht}\ \beta_i,$$

and its total degree by

$$d(M_{k,r,\alpha}) = (k_N, k_{N-1}, \dots, k_1, r_1, \dots, r_N, d_0(M_{k,r,\alpha})) \in \mathbb{Z}_+^{2N+1}.$$

We shall view $\mathbb{Z}_+^{2N+1}$ as a totally ordered semigroup with the lexicographical order $<$ such that $u_1 < u_2 < \dots < u_{2N+1}$ where $u_i = (\delta_{i,1}, \dots, \delta_{i,2N+1})$.

Introduce a $\mathbb{Z}_+^{2N+1}$-filtration of the algebra U by letting U_s ($s \in \mathbb{Z}_+^{2N+1}$) be the span of the monomials $M_{k,r,\alpha}$ such that $d(M_{k,r,\alpha}) \le s$. Theorem 9.3 implies:

Proposition. *The associated graded algebra Gr U of the $\mathbb{Z}_+^{2N+1}$–filtered algebra U is an algebra over $\mathbb{C}(q)$ on generators E_α ($\alpha \in R$) and K_β ($\beta \in P$) subject to the following relations:*

$$K_\alpha K_\beta = K_{\alpha+\beta}, \ K_0 = 1; \tag{10.1.2}$$

$$K_\alpha E_\beta = q^{(\alpha|\beta)} E_\beta K_\alpha; \tag{10.1.3}$$

$$E_\alpha E_{-\beta} = E_{-\beta} E_\alpha \text{ if } \alpha, \beta \in R^+; \tag{10.1.4}$$

$$E_\alpha E_\beta = q^{(\alpha|\beta)} E_\beta E_\alpha, \ E_{-\alpha} E_{-\beta} = q^{(\alpha|\beta)} E_{-\beta} E_{-\alpha} \tag{10.1.5}$$

if $\alpha, \beta \in R^+$ and $\alpha > \beta$ in our convex ordering of R^+. □

REMARK (a) Considering the degree by total height d_0, we obtain a $\mathbb{Z}_+$–filtration of U; let $U^{(0)} = \overline{U}$ be the associated graded algebra. Letting $d_1(M_{k,r,\alpha}) = r_N$, we obtain a $\mathbb{Z}_+$–filtration of $U^{(0)}$; let $U^{(1)} = \overline{U}^{(0)}$ be the associated graded algebra. Letting $d_2(M_{k,r,\alpha}) = r_{N-1}$, we similarly obtain $U^{(2)} = \overline{U}^{(1)}$, etc. It is clear that at the last step we get the algebra Gr U defined by (10.1.2-5):

(10.1.6) $$U^{(2N)} \simeq \mathrm{Gr}\, U.$$

(b) The algebra Gr U is a twisted polynomial algebra over $\mathbb{C}(q)$ on generators $E_{\beta_1}, \dots, E_{\beta_N}$, $E_{-\beta_N}, \dots E_{-\beta_1}, K_{\omega_1}, \dots, K_{\omega_n}$, with the elements $K_{\omega_1}, \dots, K_{\omega_n}$ inverted.

(c) One could as well use two different convex orderings for the E_β and the F_β or even reverse the ordering in one or both of these sets. All the results will be essentially unchanged.

Let us remark at this moment that the first degeneration $U^{(0)} = \overline{U}$ is obtained by modifying the defining relations of quantum groups only imposing that the elements E_i, F_j should commute. In particular the subalgebras of U or $U^{(0)}$ generated only by the E_i, K_α are isomorphic.

A first simple application of this method is:

Theorem. *The algebra U has no zero divisors.*

Proof. This follows from Lemma 4.8 a) and 5.2 or 7.1.

We can apply the same ideas to the algebras U^w and $\mathcal{B}^w$ introduced in the previous section. Since the algebras U^w and $\mathcal{B}^w$ are iterated twisted polynomial rings with relations of the type 5.3 we can consider the associated quasipolynomial algebras, and we will denote them by $\overline{U}^w$ and $\overline{\mathcal{B}}^w$. Notice that the latter algebras depend on the reduced expression chosen for w. Of course the defining relations for these algebras are obtained from Proposition 10.1 since these algebras are subalgebras of the algebra $\overline{U}$. We could of course also perform the same construction with the negative roots but this is not strictly necessary since we can simply apply the anti-automorphism κ to define the analogous negative objects.

10.2. We want to study the previous algebras in the spirit of §7.1 when q is an ℓth root of 1. The reader should look at §12 where we define a quantum group A over $\mathbb{C}[q, q^{-1}]$. For this we have a P.B.W. basis and the Leverdowsky-Soibelman relations hence we can consider its subalgebras $\mathcal{U}^w$ and $\mathcal{B}^w$ which can be specialized at $q = \varepsilon$. We have anticipated some of the discussion here as an example.

As has been already remarked, the algebras $\mathcal{U}^w_\varepsilon$ and $\mathcal{B}^w_\varepsilon$ are iterated twisted polynomial algebras with relations of the type 5.3. We can apply to these algebras the theory developed in §6.4 and get:

Theorem. *The algebras U^w_ε and $\overline{U}^w_\varepsilon$ (resp. $\mathcal{B}^w_\varepsilon$ and $\overline{\mathcal{B}}^w_\varepsilon$) have the same degree.* □

Since the algebras $\overline{U}^w_\varepsilon$, $\overline{\mathcal{B}}^w_\varepsilon$ are twisted polynomials (or Laurent polynomials) where the commutation relations are of type $x_i x_j = q^{c_{ij}} x_j x_i$, in order to compute their degree d it is necessary to identify and study the corresponding matrix (c_{ij}) since, according to proposition 7.1, d^2 equals the number of elements of the image of (c_{ij}) modulo ℓ.

Let us explicit in all cases the matrix (c_{ij}). Recall the relations 10.1.2-5. For a reduced expression $w = s_{i_1} \dots s_{i_k}$ of w, let

$$\beta_1 = \alpha_{i_1},\ \beta_2 = s_{i_1}(\alpha_{i_2}), \dots, \beta_k = s_{i_1} \dots s_{i_{N-1}}(\alpha_{i_k})$$

be the corresponding convex ordering of $R_{w^{-1}}$.

Let x_m denote the class of E_{β_m} in the corresponding graded algebra. From 10.1.5 we have

$$x_i x_j = q^{(\beta_i|\beta_j)} x_j x_i, \text{ if } i < j.$$

Thus we introduce the skew symmetric matrix $(a_{ij}) = A$ with $a_{ij} = (\beta_i|\beta_j)$ if $i < j$. Of course this matrix depends both on w and its reduced expression but we will not stress this in the notations. A is associated to the twisted polynomial ring $\overline{U}^w$.

We wish to think of this matrix as the matrix of a skew form on a free module V with basis $u_1, \dots, u_k$. Identifying V with its dual V^* using the given basis, we may also think of A as a linear operator from V to itself.

For $\overline{\mathcal{B}}^w$ we add the generators L_i and use the relations 10.1.3 which then give rise to a $n \times k$ matrix

(10.2.1)
$$B = ((\omega_i|\beta_j))_{1\le i\le n, 1\le j\le k}.$$

We may think of the matrix B as a linear map from the module V with the basis $u_1, \dots, u_k$ to the module $Q^\vee \otimes_{\mathbb{Z}} \mathbb{Z}'$ with the basis $\alpha_1^\vee, \dots, \alpha_n^\vee$, by duality its transpose is a linear map from the module $P \otimes_{\mathbb{Z}} \mathbb{Z}'$ with the basis $\omega_1, \dots, \omega_n$ to the module $V^* = V$ with the basis $u_1, \dots, u_k$. Then we have:

(10.2.2)
$$A(u_i) = \sum_{h<i}(\beta_h|\beta_i)u_h - \sum_{h>i}(\beta_h|\beta_i)u_h, \quad B(u_i) = \beta_i, \ i = 1, \dots, k, \quad {}^tB(\beta) = \sum_h (\beta|\beta_h)u_h.$$

Construct the matrix S:

(10.2.3)
$$S = \begin{pmatrix} A & -{}^tB \\ B & 0 \end{pmatrix}$$

S is the matrix associated to the twisted polynomial ring $\overline{\mathcal{B}}^w$.

10.3 To study the matrices A and S we need the following:

Lemma. *Given $\omega = \sum_{i=1}^n \delta_i\omega_i$ with $\delta_i = 0$ or 1, set*
$$I_\omega(w) := \{t \in \{1, \dots, k\} | s_{i_t}(\omega) \neq \omega\}.$$
Then $\omega - w(\omega) = \sum_{t\in I_\omega} \beta_t$.

Proof. Let us drop the w and write simply I_ω. By induction on the length of w, the hypothesis made implies $s_i(\omega) = \omega$ or $\omega - \alpha_i$. Write $w = w's_{i_k}$. If $k \notin I_\omega$ then $w(\omega) = w'(\omega)$ and we are done by induction. Otherwise $w(\omega) = w'(\omega - \alpha_{i_k}) = w'(\omega) - \beta_k$ and again we are done by induction. □

Note that $\{1, 2, \dots, k\} = \coprod_{i=1}^n I_{\omega_i}$.

10.4 Consider the operators: $M = (A \ -{}^tB)$ and $N = (B \ 0)$ so that $S = M \oplus N$.

Lemma. *(a) The operator M is surjective.*

(b) The vectors $v_\omega := (\sum_{t\in I_\omega} u_t) - \omega - w(\omega)$, as ω runs through the fundamental weights, form a basis of the kernel of M.

(c) $N(v_\omega) = \omega - w(\omega) = \sum_{t\in I_\omega} \beta_t$.

Proof. (a) We have by a straightforward computation:

$$S(u_i + \beta_i) = A(u_i) + B(u_i) -{}^t B(\beta_i) = \sum_{j<i}(\beta_j|\beta_i)u_j - \sum_{j>i}(\beta_j|\beta_i)u_j + \beta_i - \sum_j(\beta_i|\beta_j)u_j =$$
$$= -(\beta_i|\beta_i)u_i - 2\sum_{j>i}(\beta_i|\beta_j)u_j + \beta_i,$$

and

$$M(u_i+\beta_i) = -(\beta_i|\beta_i)u_i - 2\sum_{j>i}(\beta_i|\beta_j)u_j.$$

Since $(\beta_i|\beta_i)$ is invertible in $\mathbb{Z}'$ the claim follows.

(b) Since the n vectors v_ω are part of a basis and, by (a), the kernel of M is a direct summand of rank n, it is enough to show that these vectors lie in the kernel. To check that $M(v_{\omega_i}) = 0$ is equivalent to seeing that v_{ω_i} lies in the kernel of the corresponding skew–symmetric form, i.e. $\langle u_j|v_{\omega_i}\rangle = 0$ for all $j = 1,\ldots,k$: Recall that:

$$\langle u_i|u_j\rangle = (\beta_i|\beta_j) \text{ if } i<j, \langle u_j|\beta\rangle = -(\beta_j|\beta),\ \beta \in P.$$

Using Lemma 10.3, $\omega = w(\omega) + \sum_{t\in I_\omega}\beta_t$ so $v_\omega := \sum_{t\in I_\omega}(u_t-\beta_t) - 2w(\omega)$ we have

$$\langle u_j|v_{\omega_i}\rangle = \sum_{t\in I_{\omega_i}}\langle u_j|u_t-\beta_t\rangle - 2\langle u_j|w(\omega_i)\rangle$$

$$(10.4.1) \qquad = 2\sum_{t>j}(\beta_j|\beta_t) + 2(\beta_j|w(\omega_i)) + a_j,$$

where $a_j = 0$ if $j \notin I_{\omega_i}$ and $a_j = (\beta_j|\beta_j)$ otherwise.

We proceed by induction on $k = l(w)$. Let us write $v_{\omega_i}(w)$ to stress the dependence on w. For $k = 0$ there is nothing to prove. Let $w = w's_{i_k}$ with $l(w') = l(w) - 1$. We distinguish two cases according to whether $i = i_k$ or not.

Case 1) $i \neq i_k$, i.e. $k \notin I_{\omega_i}$.

We have $I_{\omega_i}(w) = I_{\omega_i}(w')$ and $w(\omega_i) = w'(\omega_i)$ so that $v_{\omega_i}(w) = v_{\omega_i}(w')$ hence the claim follows by induction if $j < k$. For $j = k$ we obtain from (10.4.1):

$$\langle u_k|v_{\omega_i}\rangle = 2(\beta_k|w(\omega_i)) = 2(w'(\alpha_{i_k})|w'(\omega_i)) = 2(\alpha_{i_k}|\omega_i) = 0.$$

Case 2) $i_k = i$ so that $w = w's_i$ and $w(\omega_i) = w's_i(\omega_i) = w'(\omega_i) - w'(\alpha_i) = w'(\omega_i) - \beta_k$. Then $v_{\omega_i}(w) = v_{\omega_i}(w') + u_k + \beta_k$. For $j < k$ by induction we get:

$$\langle u_j|v_{\omega_i}\rangle = \langle u_j|u_k\rangle + \langle u_j|\beta_k\rangle = (\beta_j|\beta_k) - (\beta_j|\beta_k) = 0.$$

Finally if $j = k$ we have:

$$2(\beta_k|w(\omega_i)) + (\beta_k|\beta_k) = 2(w'\alpha_i|w'(\omega_i - \alpha_i)) + (\alpha_i|\alpha_i) = 2(\alpha_i|\omega_i) - (\alpha_i|\alpha_i) = 0.$$

(c) Using (10.2.2), we have: $N(v_\omega) = \sum_{t\in I_\omega}\beta_t = \omega - w(\omega)$, from Lemma 10.3. □

10.5 Since S is the direct sum of M, N its kernel is the intersection of the 2 kernels of these operators. We have computed the kernel of M in 10.4; thus the kernel of S equals the kernel of N restricted to the submodule spanned by the vectors v_{ω_i}.

Let us identify this module with the weight lattice P by identifying v_ω with ω. By 10.4 (c) $N(v_\omega) = \omega - w(\omega)$ and we see that N is identified with the map $1 - w : P \to Q$. At this point we need the following fact:

Lemma. *Let $\theta = \sum_{i=1}^{n} a_i\alpha_i$ be the highest root of the root system R.*

Let $\mathbb{Z}'' = \mathbb{Z}''[a_1^{-1}, \dots, a_n^{-1}]$, and let $M' = M \otimes_{\mathbb{Z}} \mathbb{Z}'$, $M'' = M \otimes_{\mathbb{Z}} \mathbb{Z}''$ for $M = P$ or Q. Then for any $w \in W$, the $\mathbb{Z}''$-submodule $(1-w)P''$ of Q'' is a direct summand.

Proof. Recall that one can represent w in the form $w = s_{\gamma_1} \dots s_{\gamma_m}$ where $\gamma_1, \dots, \gamma_m$ is a linearly independent set of roots (see e.g. [C]). Since in the decomposition $\gamma^\vee = \sum_i r_i\alpha_i^\vee$ one of the r_i is 1 or 2, it follows that $(1-s_\gamma)P' = \mathbb{Z}'\gamma$. Since:

$$1 - w = (1 - s_{\gamma_1} \dots s_{\gamma_{m-1}})s_{\gamma_m} + (1 - s_{\gamma_m}),$$

we deduce by induction that

$$(1-w)P' = \sum_{i=1}^{m} \mathbb{Z}'\gamma_m \tag{10.5.1}$$

Recall now that any sublattice of Q spanned over $\mathbb{Z}$ by some roots is a $\mathbb{Z}$-span of a set of roots obtained from Π by iterating the following procedure: add a highest root to the set of simple roots, then remove several other roots from this set. The index of the lattice M thus obtained in $M \otimes_{\mathbb{Z}} \mathbb{Q} \cap Q$ is equal to the product of coefficients of removed roots in the added highest root. Hence it follows from (10.5.1) that

$$((1-w)P'') \otimes_{\mathbb{Z}} \mathbb{Q} \cap Q'' = (1-w)P'',$$

proving the claim. □

We call $\ell > 1$ a *good* integer if it is relatively prime to d and to all the a_i.

Theorem. *If ℓ is a good integer, then*

$$\deg \mathcal{B}_\varepsilon^w = \deg \overline{\mathcal{B}}_\varepsilon^w = \ell^{\frac{1}{2}(\ell(w)+\mathrm{rank}\ (1-w))}.$$

Proof. From the above discussion we see that $\deg \overline{\mathcal{B}}_\varepsilon^w = \ell^s$, where $s = (\ell(w)+n) - (n - \mathrm{rank}(1-w))$, which together with Corollary 6.1 proves the claim. □

10.6 We pass now to U_ε^w. For this we need to compute the image of the matrix A. The operator A can be thought as M restricted to the subspace V. So $Ker A = V \cap Ker M$ is identified with the set of linear combinations $\sum_i c_i v_{\omega_i}$ for which $\sum_i c_i(\omega_i + w(\omega_i)) = 0$ i.e. $\sum_i c_i\omega_i \in \ker(1+w)$. To understand this module requires a case by case analysis. A simple case is when $w_0 = -1$, so that $Ker(1+w) = Ker(1 - w_0 w)$ and one reduces to the previous case for $w_0 w$. Thus we get

Proposition. *If $w_0 = -1$ (i.e. for types different from A_n, D_{2n+1} and E_6) and if ℓ is a good integer, we have:*

$$\deg U_\varepsilon^w = \deg \overline{\mathcal{U}}_\varepsilon^w = \ell^{\frac{1}{2}(\ell(w)+\mathrm{rank}(1+w)-n)}.$$

Let us note the special case $w = w_0$. Remark that defining ${}^t\omega := -w_0(\omega)$ we have an involution $\omega \longrightarrow {}^t\omega$ on the set of fundamental weights. Let us denote by s the number of orbits of this involution.

Theorem. *If ε is a primitive ℓ-th root of 1, where ℓ is an integer greater than 1 and relatively prime to d, then the algebras $U_\varepsilon^{w_0}$ and $\mathcal{B}_\varepsilon^{w_0}$ have degrees $\ell^{\frac{N-s}{2}}$ and $\ell^{\frac{N+s}{2}}$ respectively.*

Proof. In this case $l(w_0) = N$ and the maps $\omega \longrightarrow \omega + w_0(\omega)$ and $\omega \longrightarrow \omega - w_0(\omega)$ are $\omega \longrightarrow \omega - {}^t\omega$ and $\omega \longrightarrow \omega + {}^t\omega$ and so their ranks are clearly $n - s$ and s respectively. □

10.7 In this section we may consider finally the matrix associated to the full degeneration of the quantum group $\overline{U}$ which is a twisted polynomial ring (localized) in the generators $E_{\beta_i}, E_{-\beta_j}, L_j = K_{\omega_k}$ where $i, j = 1, \ldots, N$ and $k = 1, \ldots n$. We think again of its matrix as the matrix of a skew symmetric form on a module. Since we will eventually reduce modulo ℓ an odd integer we start inverting 2.

Thus consider the free $\mathbb{Z}\left[\frac{1}{2}\right]$–modules V_+ with basis $u_1, \ldots, u_N$, V_- with basis $u'_1, \ldots, u'_N$ and V_0 with basis $\omega_1, \ldots, \omega_n$. On the $\mathbb{Z}\left[\frac{1}{2}\right]$–module $V = V_+ \oplus V_- \oplus V_0$ define a skew symmetric bilinear form $\langle . | . \rangle$ by the following formulas:

$$(10.7.1)\qquad \begin{cases} \langle V_- | V_+ \rangle = 0, \ \langle V_0 | V_0 \rangle = 0, \\ \langle u_i | u_j \rangle = -\langle u'_i | u'_j \rangle = (\beta_i | \beta_j) \text{ if } i < j, \\ \langle \omega_i | u_j \rangle = -\langle \omega_i | u'_j \rangle = (\omega_i | \beta_j). \end{cases}$$

With the previously defined matrices (cf. 10.2):

$$A = (\langle u_i | u_j \rangle)_{1 \le i,j \le N}, \ B = ((\omega_i | \beta_j))_{1 \le i \le n, 1 \le j \le N},$$

the matrix T of the skew symmetric form $\langle . | . \rangle$ in the above basis has the form:

$$(10.7.2)\qquad T = \begin{pmatrix} A & 0 & -{}^tB \\ 0 & -A & {}^tB \\ B & -B & 0 \end{pmatrix}$$

Proposition. *All non-zero elementary divisors of the matrix T over the ring $\mathbb{Z}\left[\frac{1}{2}\right]$ are $\frac{1}{2}(\beta_i | \beta_i)$, $i = 1, \ldots, N$, each repeated twice.*

Proof. Consider the matrix

$$C = \begin{pmatrix} I_N & I_N & 0 \\ -\frac{1}{2} I_N & \frac{1}{2} I_N & 0 \\ 0 & 0 & I_n \end{pmatrix} \in GL\left(2N + n, \mathbb{Z}[\tfrac{1}{2}]\right).$$

Let $T' = CT\,{}^tC$. We have:

$$T' = \begin{pmatrix} 0 & -A & 0 \\ -A & 0 & {}^tB \\ 0 & -B & 0 \end{pmatrix}.$$

Since $T' = T_1 - {}^tT_1$, where $T_1 = \begin{pmatrix} 0 & 0 & 0 \\ -A & 0 & {}^tB \\ 0 & 0 & 0 \end{pmatrix}$, we see that $T_1(V) \subset V_-$, ${}^tT_1(V) \subset V_+ \oplus V_0$. Hence it suffices to show that all non-zero elementary divisors of the matrix T_1 over $\mathbb{Z}\left[\frac{1}{2}\right]$ are $\frac{1}{2}(\beta_i | \beta_i)$, $i = 1, \ldots, N$. Note that

$$T_1(u_i) = \sum_j \operatorname{sign}(j-i)(\beta_i|\beta_j)u'_j,$$

$$T_1(\beta_i) = -\sum_j (\beta_i|\beta_j)u'_j.$$

Hence we have:

$$T_1(u_i+\beta_i) = -(\beta_i|\beta_i)u'_i - 2\sum_{j<i}(\beta_i|\beta_j)u'_j.$$

Since $(\beta_i|\beta_j)$ is divisible by $\frac{1}{2}(\beta_i|\beta_i)$, this completes the proof. □

We can think of $T' = T_1 - {}^tT_1$ as a direct sum decomposition with $T_1 = (-A\ {}^tB)$. Thus in order to compute its image it is enough to analyze the operator $(-A\ {}^tB)$. This is just the operator we called M for $w = w_0$ thus from Lemma 10.4 it is surjective and so we deduce:

Theorem. *(a) The image of T in $\mathbb{Z}/(\ell)$ has ℓ^{2N} elements.*
(b) The algebra $\overline{U}$ has degree ℓ^N.

Onc of our goals will be to show that also U has the same degree. For this we cannot apply a simple degeneration argument since U is not an iterated twisted derivation algebra. Thus we will give a more complicated argument via the theory of Verma modules and Poisson structures in §.

10.8 We describe the intersection of U_ε^+ with the center of U_ε

Proposition. *The intersection of U_ε^+ with the center of U_ε is the span of the monomials in the elements E_α^ℓ.*

Proof. Since these monomials lie in the center, it is enough to show the same statement for the associated graded $\overline{U}_\varepsilon^+$. In this case it is enough to compute the elements in $\overline{U}_\varepsilon^+$ which lie in the center of $\overline{B}_\varepsilon^+$. These, by the usual argument, are spanned by the monomials with exponents a vector in the kernel of the associated matrix

$$S = \begin{pmatrix} A & -{}^tB \\ B & 0 \end{pmatrix}$$

since the monomial is in $\overline{U}_\varepsilon^+$ the vector is of the form: $\begin{pmatrix} X \\ 0 \end{pmatrix}$. Hence the claim follows from Lemma 10.4 a) which implies that the transpose of the operator M is injective, hence X is 0 modulo ℓ.

§11 Poisson structures.

11.1 Let us recall the notion of a Poisson structure over a commutative k–algebra A:

Definition. *A Poisson structure on A is a Lie product*

$$\{-,-\} : A \otimes_k A \longrightarrow A$$

satisfying the (Leibniz rule):

$$\{f_1 f_2, f_3\} = f_1\{f_2,f_3\} + f_2\{f_1,f_3\}. \tag{11.1.1}$$

If A, B are 2 Poisson algebras we have a canonical Poisson structure on $A \otimes_k B$ for which the 2 factors Poisson commute and which extends the given 2 structures on the factors. It is given clearly by:

$$\{a \otimes b, c \otimes d\} := \{a, c\} \otimes bd + ac \otimes \{b, d\}.$$

Note that 11.1.1 implies that the map:

$$g \to \{f, g\}$$

is a derivation.

In a more geometric language, if A is the coordinate ring of a smooth affine variety, f gives rise to a vector field X_f, such that:

$$X_f(g) = \{f, g\}.$$

X_f is called an *Hamiltonian vector field* and f is its Hamiltonian function.

The assumptions imply that the map $f \to X_f$ is a Lie algebra homomorphism (where the Lie product of vector fields is the usual commutator as operators). It follows that, if P is a point and g a function which vanishes in P to order ≥ 2 (i.e. $g(P) = 0, dg(P) = 0$) then $\{f, g\}(P) = 0$. This implies that a Poisson structure on the coordinate ring of a smooth algebraic variety X can be interpreted as a particular type of bivector u (a section of $\wedge^2 T(X)$). Given a point P and two cotangent vectors α, β we have:

(11.1.2) $$u_P(\alpha, \beta) := \{f, g\}(P), \text{ where } df(P) = \alpha, \ dg(P) = \beta.$$

The Jacobi identity can be interpreted as a cocycle identity on u (cf. L-W). In general this bivector gives a skew symmetric form on the cotangent spaces which may be degenerate. In fact it is clear by the definitions that, if N is the kernel of this form in the space $T^*_X(P)$, its orthogonal in the tangent space $T_X(P)$ is the span of the values in P of the Hamiltonian vector fields. Let us call $H(P)$ this subspace, of course it can (and it will) be a subspace of variable dimension.

One should remark first of all that the definition can be globalized to any algebraic variety and also that it can be formulated as well in the analytic or differentiable categories.

Going back to the spaces $H(P)$ we have the following:

Definition. *A connected submanifold Y of X is called a symplectic leaf if for each $P \in Y$ the space $H(P)$ is the tangent space to Y in P and if Y is maximal with this property. The theorem of Frobenius implies the existence of a decomposition of X in symplectic leaves.*

A special case is when u is always non degenerate and so, if the manifold is connected, we have a unique leaf. In this case we also say that X is a *symplectic manifold*. Thus the leaves have an induced structure of symplectic manifolds. Given a Hamiltonian vector field H_f since it is tangent in any point to the leaf through that point it follows that if we integrate it locally to a germ of a 1-parameter subgroup, this subgroup of local diffeomorphisms preserves the leaves.

11.2 Assume now that $A = k[H]$ is the coordinate ring of an affine algebraic group H. We know that $k[H]$ is a Hopf algebra with comultiplication

$$\Delta : k[H] \to k[H] \otimes k[H] = k[H \times H]$$

given by $(\Delta f)(h_1, h_2) = f(h_1 h_2), \forall h_1, h_2 \in H$;
antipode

$$S : k[H] \to k[H]$$

given by $(Sf)(h) = f(h^{-1}), \forall h \in H$;
counit

$$\varepsilon : k[H] \to k$$

given by $\varepsilon(f) = f(e)$, where e $\in H$ is the identity element.

Definition. *H is a Poisson algebraic group if $k[H]$ has a Poisson structure compatible with the Hopf algebra structures:*

$$\Delta\{f_1, f_2\} = \{\Delta f_1, \Delta f_2\}, \quad (11.2.1)$$

$$(11.2.2) \quad S(\{f_1, f_2\}) = \{Sf_1, Sf_2\}$$

$$(11.2.3) \quad \varepsilon\{f_1, f_2\} = 0.$$

11.3 It is useful to translate the axioms of a Poisson group in a more explicit form. As we have remarked a Poisson structure is a special type of bivector, i.e. a skew form on the tangent bundle. For a group H it is convenient of course to trivialize the tangent bundle. Given that $\mathfrak{h} = Lie(H) = T_1(H)$ one identifies $T_1(H)$ with $T_a(H)$ using the differential at 1 of the map $l_a : x \to ax$. With this trivialization we think of a bivector as a family of skew forms ψ_a on $\mathfrak{h}$ and we want to traslate the Poisson group axioms in this language.

First remark how the basic formulas for the group translate.

Consider the multiplication map $\mu : H \times H \to H$ and its differential in a point (a, b), we use the identifications $T_{(a,b)}H \times H = \mathfrak{h} \oplus \mathfrak{h}$, $T_{ab}(H) = \mathfrak{h}$ and call $\mu_{a,b} : \mathfrak{h} \oplus \mathfrak{h} \to \mathfrak{h}$ the resulting map, and $\mu^*_{a,b} : \mathfrak{h}^* \to \mathfrak{h}^* \oplus \mathfrak{h}^*$ its dual map. Since $(ab)^{-1}(axby) = (b^{-1}xby)$ we deduce that:

$$(11.3.1) \quad \mu_{a,b}(u, v) = b^{-1}u + v, \text{ resp. } \mu^*_{a,b}(\phi) = (b\phi, \phi)$$

where $b^{-1}u$, $b\phi$ refer respectively to adjoint and coadjoint action.

Similarly consider the inverse map $i : x \to x^{-1}$ its differental induces the family of maps:

$$u \to -au, \ \phi \to -a^{-1}\phi.$$

Now let us translate the Poisson axioms in terms of the forms:

The compatibility with multiplication means that the inclusion maps $\mu^*_{a,b}$ are compatible with the corresponding forms ψ_{ab}, $\psi_a \oplus \psi_b$, from the formula 11.3.1 this means:

$$(11.3.2) \quad \psi_{ab} = b^{-1}\psi_a + \psi_b.$$

Where we have used the action of H on skew forms on $\mathfrak{h}$ induced by the coadjoint action.

The compatibility with the inverse means that the inverse map takes the Poisson form to minus itself and hence that:

$$(11.3.3) \quad \psi_{a^{-1}} = -a\psi_a.$$

It is clear that from 11.3.2 it follows that $\psi_1 = 0$ and also 11.3.3.

Take now the differential at 1 of the map $\psi : H \to \wedge^2 \mathfrak{h}$

it defines a copruduct $\gamma : \mathfrak{h} \to \wedge^2\mathfrak{h}$ and dually an antisymmetric product on $\mathfrak{h}^*$.

One can easily verify from the Jacobi identity for the Poisson bracket that this is a Lie algebra product.

In fact one can easily deduce the same Lie algebra structure as follows:

The maximal ideal M of 1 is a Lie ideal as well as M^2 (under Poisson bracket) hence $h^* = T^*_H(1)$ is a Lie algebra. This LIe structure is the same as the one previously described.

Let us change for a while notations. Set $L := Lie(H) = T_H(1)$. We have a Lie algebra structure on L and on L^*, we have moreover an action (the coadjoint one) of L on L^* since L is a Lie algebra, and similarly an action of L^* on L. Define thus a multiplication on $L \oplus L^*$ setting:

$$(11.2.2) \quad [a + \varphi, b + \psi] := [a, b] + a.\psi - \psi.a + \varphi.b - b.\varphi + [\varphi, \psi]$$

where the first and last term refer to the Lie algebra structures on L, L^* respectively, while the middle terms refer to the 2 coadjoint actions, we have then:

Theorem. *The given multiplication on $L \oplus L^*$ gives it a Lie algebra structure.*
The quadratic form $\langle(a,\varphi)|(b,\psi)\rangle := \psi(a) + \varphi(b)$ is associative (invariant) with respect to the Lie product.

We leave the proof of this theorem which can be verified from the various identities.
Notice that L, L^* are maximal isotropic with respect to the given form. It is then natural to define:

Definition. *A triple of Lie algebras $(\mathfrak{g}, \mathfrak{h}, \mathfrak{k})$ is called a Manin triple if it satisfies:*
i) $\mathfrak{h}, \mathfrak{k} \subset \mathfrak{g}$ as Lie subalgebras, $\mathfrak{g} = \mathfrak{h} \oplus \mathfrak{k}$ as vector spaces.
ii) There is a non degenerate invariant symmetric bilinear form $(\; , \;)$ on $\mathfrak{g}$ with respect to which both $\mathfrak{h}$ and $\mathfrak{k}$ arc maximal isotropic subspaces.

We aim to show that, as a Lie algebra is the infinitesimal datum of a Lie group, so is a Manin triple one for a Poisson Lie group.

We go back to the identity 11.3.2 applied to $a = exp(tu), b = exp(su)$ 1-parameter subgroups. Set

$$A_u(t) := \psi_{exp(tu)}$$

recall by definition that $\gamma(u) = A'_u(0)$ and differentiate getting:

$$A'_u(t) = -u\, A(t) + \gamma(u).$$

This equation shows that the Manin triple determines completely the Poisson structure. We explicit thus the solution of this differential equation from a given Manin triple reconstruct the structure.
Let G be an algebraic group, $H, K \subset G$ two closed subgroups. Set $\mathfrak{g}=LieG$ (resp. $\mathfrak{h}=LieH$, $\mathfrak{k}=LieK$).

Definition. *The triple (G, H, K) is called an algebraic Manin triple if the triple of Lie algebras $(\mathfrak{g}, \mathfrak{h}, \mathfrak{k})$ is a Manin triple.*

Let us see how we can associate to an algebraic Manin triple, a structure of Poisson algebraic group on H (and symmetrically on K).
Let us now denote by $\pi : \mathfrak{g} \to \mathfrak{h}$ the projection with kernel $\mathfrak{k}$.
First we notice that the definition immediately implies that there is a natural isomorphism between $\mathfrak{h}^*$ and $\mathfrak{k}$. Under this isomorphism the coadjoint action of $\mathfrak{h}^*$ on $\mathfrak{k}$ is clearly read from formula 11.2.2:

$$a.\phi = (1-\pi)[a, \phi].$$

For any $h \in H$, we set $\pi^h = adh^{-1} \circ \pi \circ adh$. Notice that also π^h maps $\mathfrak{g}$ to $\mathfrak{h}$. We can now define the bilinear form $\psi_h :=< \; , \; >_h$ on $\mathfrak{k} \cong \mathfrak{h}^*$ by

$$< x, y >_h := (\pi^h x, y), \forall x, y \in \mathfrak{k}. \tag{11.3.4}$$

Since the bilinear form $(\; , \;)$ is invariant and the two subalgebras are maximal isotropic, we get

$$< x, y >_h = (\pi^h x, y) = (x, adh^{-1} \circ (1-\pi) \circ adhy) =$$
$$-(x, adh^{-1} \circ \pi \circ adhy) = -< y, x >_h .$$

Thus $< \; , \; >_h$ is antisymmetric $\forall h \in H$.
As before consider the tangent and cotangent bundles on H as trivialized as $H \times \mathfrak{h}$, $H \times \mathfrak{h}^* = H \times \mathfrak{k}$ respectively. Their direct sum is the trivial bundle $H \times \mathfrak{g}$ in which we consider the flat connection making $\mathfrak{g}$ a space of horizontal sections for which the canonical scalar product is constant.

In this bundle we consider the constant projection π of $\mathfrak{g}$ to $\mathfrak{h}$ with kernel $\mathfrak{h}^*$ and the variable projection Π whose value in a point h is π^h. Notice that the restriction of π^h to $\mathfrak{h}^*$ is exactly the skew form ψ_h defined at the beginning of the section by 11.3.1.

Thus in compact form working in this larger bundle we have the formula for the Poisson bracket of two functions $f_1, f_2 \in k[H]$:

$$\{f_1, f_2\} = (\Pi df_1, df_2) \tag{11.3.5}$$

Proposition. *The bracket { } defined above, provides H with the structure of a Poisson algebraic group with given infinitesimal Manin triple.*

Proof. From the definition we only have to verify the Jacobi identity and the compatibility with with the Hopf structure.

A simple direct computation in the Lie algebra of sections of the bundle $H \times \mathfrak{g}$ shows that

$$d\{f,g\} = (1-\pi)[(1-\Pi)df, (1-\Pi)dg]. \tag{11.3.6}$$

. In fact given X a vector field we have:

$$\begin{aligned}(d\{f,g\}, X) =&([\Pi, ad(X)]df, dg) = \\ &(\Pi[X, df], dg) - ([X, \Pi df, dg) = \\ &([X, df], (1-\Pi)dg) - (X, [\Pi df, dg]) = \\ &(X, [df, (1-\Pi)dg] - [\Pi df, dg]) = \\ &(X, (1-\pi)[(1-|Pi)df, (1-\Pi)dg]).\end{aligned}$$

From this we immediately get that

$$d\{f_1, \{f_2, f_3\}\} = (1-\pi)[(1-\Pi)df_1, [(1-\Pi)df_2, (1-\Pi)df_1]],$$

so that the Jacobi identity for $\mathfrak{g}$ clearly implies that

$$\{f_1, \{f_2, f_3\}\} + \{f_2, \{f_3, f_1\}\} + \{f_3, \{f_1, f_2\}\}$$

is a constant function $\forall f_1, f_2, f_3 \in k[H]$. Since clearly $\{f_1, f_2\}(e) = 0, \forall f_1, f_2 \in k[H]$, one gets that

$$\{f_1, \{f_2, f_3\}\} + \{f_2, \{f_3, f_1\}\} + \{f_3, \{f_1, f_2\}\} = 0$$

It remains to prove that ψ_h (which is the restriction of π^h to $\mathfrak{h}^*$ satisfies the relation 11.3.2. We work in the bundle $H \times \mathfrak{g}$.

$$Ad(b)^{-1}\Pi_a(1-\pi)Ad(b)(1-\pi) + \psi_b =$$

$$Ad(b)^{-1}\Pi_a Ad(b)(1-\pi) - Ad(b)^{-1}\pi Ad(b)(1-\pi) + \psi_b = \psi_{ab}.$$

There is an alternative description of the Poisson structure which can be described as follows:

Given an affine algebraic manifold X, denote by $\mathbb{C}[X]$ (resp. Vect X, resp. $\mathcal{D}X$) the space of regular functions (resp. vector fields, resp. differential 1–forms) on X. To define a Poisson bracket $\{,\}$ on $\mathbb{C}[X]$ is equivalent to defining a homomorphism of $\mathbb{C}[X]$–modules $\tau : \mathcal{D}X \to$ Vect X (satisfying the appropriate conditions), so that $\{f,g\} = <\tau(df), dg>$.

In our case $X = H$. We define a map $\tau : \mathcal{D}H \to$ Vect H as follows. We identify (Lie H)* with the space of left–invariant 1–forms on H. On the other hand, consider the unramified map $\tilde{\pi} : H \to K/G$. The right action of K on K/G gives an embedding Lie $K \subset$ Vect K/G. This gives a linear map $\tau_0 :$ (Lie H)$^* \to$ Vect K/G. Since $\tilde{\pi}$ is unramified we have a map $\tilde{\pi}^*$: Vect $K/G \to$ Vect H. Then τ is the homomorphism of $\mathbb{C}[H]$–modules defined by the linear map $\tilde{\pi}^* \circ \tau_0$. One can easily verify that this definition agrees with the one we previously described.

Let us briefly discuss Poisson subgroups and quotients:

Consider a quotient Poisson group S of H, that is S is a quotient group of H and the ring $\mathbb{C}[S] \subset \mathbb{C}[H]$ is a Poisson subalgebra. Let U be the kernel of the quotient homomorphism $\varphi: H \to S$, let $\mathfrak{s} = Lie\ S$, $\mathfrak{u} = Lie\ U$ and $d\varphi : \mathfrak{h} \to \mathfrak{s}$ the Lie algebra quotient map. Then $\mathfrak{u}$ is an ideal in $\mathfrak{h}$ and we identify $\mathfrak{s}^*$ with a subspace of $\mathfrak{h}^* = \mathfrak{k}$ by taking $\mathfrak{u}^\perp \subset \mathfrak{g}$ under the invariant form and intersecting it with $\mathfrak{k}$. Then for $p \in S$ the linear map: $\overline{\gamma}_p : \mathfrak{s}^* \to \mathfrak{s}$ giving rise to the Poisson structure is given by:

$$\overline{\gamma}_p = (d\varphi) \cdot (\gamma_{\tilde{p}}|_{\mathfrak{s}^*})$$

where $\tilde{p} \in H$ is any representative of p ($\overline{\gamma}_p$ is independent of the choice of $\tilde{p}$).

The construction of the Manin triple corresponding to the Poisson manifold S is obtained from the following simple fact:

Lemma. *Let* $(\mathfrak{g}, \mathfrak{h}, \mathfrak{k})$ *be a Manin triple of Lie algebras, and let* $\mathfrak{u} \subset \mathfrak{h}$ *be an ideal such that* $\mathfrak{u}^\perp$ *(in* $\mathfrak{g}$*) intersected with* $\mathfrak{k}$ *is a subalgebra of the Lie algebra* $\mathfrak{k}$*. Then*

(a) $\mathfrak{u}^\perp$ *is a subalgebra of* $\mathfrak{g}$ *and* $\mathfrak{u}$ *is an ideal of* $\mathfrak{u}^\perp$.

(b) $(\mathfrak{u}^\perp/\mathfrak{u},\ \mathfrak{h}/\mathfrak{u},\ \mathfrak{k} \cap \mathfrak{u}^\perp)$ *is a Manin triple, where the bilinear form on* $\mathfrak{u}^\perp$ *is induced by that on* $\mathfrak{g}$.

Proof. is straightforward. □

Notice that in this construction we have at the same time a quotient Poisson group of H and a Poisson subgroup of K, we are in a way in a *self dual* picture.

11.4 The construction that we have just described allows us to give various examples of Poisson algebraic groups.

Examples

(1) The first such example is given by the so called Kostant-Kirillov structure on the coadjoint representation of a group K. In this case $H = \mathfrak{k}^*$ with the trivial Lie algebra structure and G is the semidirect product $K \times \mathfrak{k}^*$, K acting on $\mathfrak{k}^*$ via the coadjoint action. So as a vector space $\mathfrak{g} = \mathfrak{k} \oplus \mathfrak{k}^*$ and has a canonical symmetric bilinear form. We leave to the reader the easy verification that the triple (G, H, K) is an algebraic Manin triple and that the corresponding Poisson structure on H is the usual Kostant-Kirillov structure i.e. $\forall f_1, f_2 \in k[H], h \in H$,

$$\{f_1, f_2\}(h) = [df_1(h), df_2(h)](h).$$

It is also clear that dually the Poisson structure one gets on K is trivial.

(2) The second class of examples are the ones we shall be mostly interested in these notes. Let K be a semisimple group. Fix a maximal torus $T \in K$ and a Borel subgroup $B^+ \supset T$, let B^- be the unique Borel subgroup such that $T = B \cap B^-$. Set $G = K \times K$. Denote by $\mu_\pm : B^\pm \to T$ the canonical projection homomorphisms, and consider the homomorphism $\phi : B^- \times B^+ \to T$ defined by $\phi(b_-, b+) = \mu_-(b_-)\mu_+(b_+)$. Then $H = Ker\phi$, while K is embedded in G as the diagonal subgroup. Thus we have a triple (G, H, K) and in order to see that it is an algebraic Manin triple we need to define a non degenerate symmetric invariant bilinear form on $\mathfrak{g}$ satisfying the various properties in the definition. We first rescale the Killing form so that on the Cartan subalgebra induces the normalized form of 8.1 giving square length 2 to the short roots. Then take the difference of the scaled Killing form on the second and first factor of $\mathfrak{g} = \mathfrak{k} \oplus \mathfrak{k}$, i.e. as the unique form on $\mathfrak{g}$ coinciding with minus the scaled Killing form (resp. the scaled Killing form) on $\mathfrak{k} \oplus \{0\}$ (resp. $\{0\} \oplus \mathfrak{k}$), and such that $\mathfrak{k} \oplus \{0\}$ and $\{0\} \oplus \mathfrak{k}$) are mutually orthogonal. It is then trivial to verify all the required properties.

This example of course appears in the Beliavin Drinfeld [B-D] classification of classical r-matrices, together with some of its Poisson subgroups it will be carefully studied in the next chapter 4.

11.5 A very interesting feature of algebraic Manin triples is the fact that the so called symplectic leaves in H are open sets in algebraic subvarieties and can be quite explicitly described.

Let us now consider an algebraic Manin triple (G, H, K). Let $p : H \to K\backslash G$ denote the restriction to H of the quotient map from G to $K\backslash G$. Then

Proposition. *The symplectic leaves in H coincide with the connected components of the preimages under p of the K orbits in $K\backslash G$.*

Proof. First notice that the map p is etale onto its image so that it induces an isomorphism between the tangent space TH_h and the tangent space $TK\backslash G_{p(h)}$ for each $h \in H$. Furthermore the right action of K on $K\backslash G$ induces a map of $\mathfrak{k}$ to $TK\backslash G_{p(h)}$ whose image clearly coincides with the tangent space to the K orbit through $p(h)$. Hence, composing with the above isomorphism we obtain a map $\phi : \mathfrak{k} \to TH_h$ whose image coincides with the tangent space to our candidate to be the symplectic leaf through h. If we now identify TH_h with $\mathfrak{h}$ using left translation we finally obtain a map $\phi_h : \mathfrak{k} \to \mathfrak{h}$ and a simple computation which we leave to the reader shows that

$$\phi_h = \pi^h. \tag{11.4.1}$$

Thus the vector fields $\{f, -\}, f \in k[H]$ span at h the tangent space to $p^{-1}(p(h)K)$ and our claim follows.

The above proposition allows us to determine the symplectic leaves in each of our examples. In the case of the Kostant-Kirillov structure on $\mathfrak{k}^*$ they are of course the coadjoint orbits, for the second example see the next chapter.

11.6 Let R be a commutative algebra and $h \in R$ an element. Suppose we have an R algebra A with the property that h is a non zero divisor in A and $A/(h)$ is commutative. Given an element $a \in A$ we shall denote by $\overline{a}$ its image modulo h.

Given $x, y \in A/(h)$, choose a and b such that $x = \overline{a}$ and $y = \overline{b}$. Since $A/(h)$ is commutative and h is a non zero divisor in A, the element $[a, b]/h \in A$ is well defined. Set:

$$\{x, y\} = \overline{[a, b]/h}, \tag{11.6.1}$$

It is immediate to see that $\{x, y\}$ is independent from the choice of the representatives a and b and in this way one defines a Poisson bracket on $A/(h)$.

Borrowing the language from Physics one says:

Definition. *A is a* **quantization** *or a quantum deformation of the Poisson algebra $A/(h)$.*

In practice one may be given a Poisson algebra and ask the problem of giving a suitable quantization of it. This problem does not a priori have a canonical solution, so that the procedures of quantization are quite variable.

If we assume that A is a Hopf algebra over R then $A/(h)$ is also a Hopf algebra and we easily see that its Hopf and Poisson structures are compatible, i.e. $A/(h)$ is a Poisson Hopf algebra.

Thus given a Poisson algebraic group H we may conversely ask for the existence of quantum deformations of H as Poisson group.

Definition. *A Hopf algebra A over R is called a quantum deformation of $k[H]$ if*

i) A is flat over R.

ii) there is a isomorphism $j : A/(h) \to k[H]$ of Hopf algebras such that

$$j(\{x, y\} = \{j(x), j(y)\},$$

the Poisson brackct on the commutative algebra $A/(h)$ being the one defined above.

Notice that one can give a more local definition of a quantum deformation using as base algebra the algebra $k[[h]]$ instead of the algebra R. In this case Reshetikin [Re],extending work of Drinfeld has recently shown that any Lie bialgebra admits a quantum deformation . We do not know whether one can obtain similar results in this more global situation.

We end this section showing that in our first example, the Kirillov-Kostant structure on $\mathfrak{k}^*$, a quantum deformation is indeed the universal enveloping algebra $U(\mathfrak{k})$. Indeed if one sets A equal to the R algebra generated by $\mathfrak{k}$ with relations

$$xy - yx = (h)[x,y], \forall x, y \in \mathfrak{k},$$

with comultiplication

$$\Delta(x) = x \otimes 1 + 1 \otimes x, \forall x \mathfrak{k},$$

antipode

$$S(x) = -x, \forall x \mathfrak{k},$$

counit

$$\varepsilon(x) = 0, \forall x \mathfrak{k},$$

one easily sees that A is a quantum deformation of the coordinatc ring of $\mathfrak{k}^*$ with the Kirillov-Kostant structurc.

11.7 We want to enrich our description of Poisson structures.

Let us start as in 11.6 with R be a commutative algebra, $h \in R$ an element, an R algebra A with the property that h is a non zero divisor in A but we do not assume anymore that $A/(h)$ is commutative. Consider an element $a \in A$ such that its image $\overline{a}$ modulo h is in the center of $A/(h)$.

Given $y \in A/(h)$, $y = \overline{b}$. Since $\overline{a}$ is in the center $\mathcal{Z}$ of $A/(h)$ and h is a non zero divisor in A, the element $[a,b]/h \in A$ is well defined. Set:

$$D_a[y] = \overline{[a,b]/h}, \tag{11.7.1}$$

It is immediate to see that:

Theorem.

i) D_a is a derivation.

ii) If a' is a different representative of $\overline{a}$ then $D_a - D_{a'}$ is an inner derivation.

iii) If φ is an automorphism of A(inducing one on $A/(h)$) we have:

$$\varphi \circ D_a \circ \varphi^{-1} = D_{\varphi(a)}.$$

In particular we get:

Corollary. *i) The center $\mathcal{Z}$ of $A/(h)$ has a natural Poisson structure induced by:*

$$\{x,y\} := D_a(y), \text{ where } x = \overline{a}.$$

ii) The group of automorphisms of A induces a group of Poisson automorphisms of $\mathcal{Z}$.

It may be worth to remark that we have several Lie algebras of Derivations:

a) $[D_a, D_b] = D_{[a,b]/h}$ thus the derivations $D_a, \{a \in A | \underline{a} \in \mathcal{Z}\}$, form a Lie algebra $\mathcal{L}$.

b) The Poisson structure induces a Lie algebra of Derivations $\mathcal{L}'$ of $\mathcal{Z}$ and we have an exact sequence:

$$0 \to \mathcal{L}^0 \to \mathcal{L} \to \mathcal{L}' \to 0$$

where $\mathcal{L}^0$ is the algebra of inner derivations of $A/(h)$.

If A is a Hopf algebra and furthermore we assume that also $\Delta(\overline{a})$ is in the center of $A/(h) \otimes A/(h)$ then we also have $D_{\Delta(a)}(\Delta(y)) = \Delta(D_a(y))$. We want to deduce a useful corollary:

Assume we are in the previous setting A is a Hopf algebra and that we are given a subhopfalgebra S of the center $\mathcal{Z}$ of $A/(h)$. Let T be the minimal subalgebra of $\mathcal{Z}$ containing S and closed under Poisson bracket, then:

Proposition. *T is a sub-Hopf algebra and compatible with the Poisson structure.*

Proof. From the previous analysis let $U := \{t \in T | \Delta(t) \in T \otimes T\}$. Clearly $S \subset U$ and U is a subalgebra. But now we have that U is also closed under Poisson bracket since, from the last remarks we have for $x, y \in U$ that $\Delta(\{x, y\}) = \{\Delta(x), \Delta(y)\} \in T \otimes T$.

11.8 Before we revert to our examples we want to make a general remark which applies to the setting we have just discussed of an algebra $A/(h)$ which is a free module over a Poisson subalgebra Z^0 of its center, we have then the derivations extending the Poisson structure.

Let us use the differentiable language rather than that of algebras. Assume that we have a manifold M and a vector bundle V of algebras with 1 (i.e. 1 and the multiplication map are smooth sections). We identify the functions on M with the sections of V which are multiples of 1. Let D be a derivation of V, i.e. a derivation of the algebra of sections which maps the algebra of functions on M into itself, and let X be the corresponding vector field on M.

Proposition. *For each point $p \in M$ there exists a neighborhood U_p and a map φ_t defined for $|t|$ sufficiently small on $V|U_p$ which is a morphism of vector bundles covering the germ of the 1–parameter group generated by X and is also an isomorphism of algebras.*

Proof. The hypotheses on D imply that it is a vector field on V linear on the fibers, hence we have the existence of a local lift of the 1–parameter group as a morphism of vector bundles. The condition of being a derivation implies that the lift preserves the multiplication section i.e. it is a morphism of algebras. □

We will have to consider a variation of the previous proposition:

Suppose M is a Poisson manifold. Assume furthermore that the Poisson structure lifts to V, i.e. each (local) function f induces a derivation on sections (as in 11.7) extending the given Poisson bracket. This means that we have a lift of the Hamiltonian vector fields as in the previous proposition and deduce:

Corollary. *Under the previous hypotheses, the fibers of V over points of a given symplectic leaf of M are all isomorphic as algebras.*

Proof. The proposition implies that in a neighborhood of a point in a leaf the algebras are isomorphic but since the notion of isomorphism is transitive this implies the claim. □

CHAPTER 4

The Poisson group H.

§12 The quantum group A.

12.1 For $R = k[q, q^{-1}], h = q - 1$ we want now to define a remarkable R subalgebra A of $U_{q,\Lambda}$ which will be the quantum deformation of of the Poisson algebraic group of example 2 in section 11.4.

We use again the action of the generalized Braid group $\mathcal{B}$ introduced in 9.2.

We now define A to be the smallest $\mathcal{B}$ stable R subalgebra of $U_{q,\Lambda}$ containing the elements:

$$\overline{E}_i := (q^{(\alpha_i,\alpha_i)/2} - q^{-(\alpha_i,\alpha_i)/2})E_i; \ \overline{F}_i := (q^{(\alpha_i,\alpha_i)/2} - q^{-(\alpha_i,\alpha_i)/2})F_i$$

for $i = 1, \ldots, r$ and the K_λ's.

Notice that, if we had chosen these new elements as generators of the quantum group U, we would have the same defining relations except for

$$[\overline{E}_i, \overline{F}_j] = \delta_{ij}(q^{d_i} - q^{-d_i})(K_{\alpha_i} - K_{-\alpha_i}).$$

Let H be the Poisson algebraic group associated to the algebraic Manin triple of example 2 in section 1, then:

Theorem. *The algebra A is a quantum deformation of $k[H]$.*

The proof of this theorem will be broken in a series of steps. We give the proof of the theorem in the rank 1 and rank 2 cases in the appendix and prove it here assuming these cases.

With the notations of 9.3 set $\overline{E}_{\beta_t} = (q^{(\beta_t,\beta_t)/2} - q^{-(\beta_t,\beta_t)/2})E_{\beta_t}, \ \overline{F}_{\beta_t} = (q^{(\beta_t,\beta_t)/2} - q^{-(\beta_t,\beta_t)/2})F_{\beta_t}$.

Step 1. *The monomials*

$$\overline{E}^k K_\lambda \overline{F}^t \tag{11.8.1}$$

with $\overline{E}^k = \overline{E}_{\beta_1}^{k_1} \cdots \overline{E}_{\beta_N}^{k_N}, \ \overline{F}^t = \overline{F}_{\beta_N}^{t_N} \cdots \overline{F}_{\beta_1}^{t_1}$ are a R basis of A.

Proof. The fact that the monomials (11.8.1) lie in A and that are linearly independent, follows immediately from the definitions and Theorem 9.3.

We have to see that the product of two monomials is a R linear combination of monomials and that that their R span is $\mathcal{B}$ stable.

We set A' equal to the R span of the monomials (11.8.1), A^+ equal to the span of those among these monomials of the form $\overline{E}_{\beta_1}^{k_1} \cdots \overline{E}_{\beta_N}^{k_N}$, A^- the span of those of the form $\overline{F}_{\beta_N}^{t_N} \cdots \overline{F}_{\beta_1}^{t_1}$, A^0 equal to the span of the K_λ's.

Let us first show that A^+ and A^- are independent from the choice of the reduced expression for w_0. For this we shall use the fact that one can pass from one reduced expression to another by applying a finite sequence of Braid relations (Theorem 9.2).

We are reduced to show that the R span of the monomials $\overline{E}^k$ (resp.$\overline{F}^t$) for a given reduced expression $w_0 = s_{i_1} \cdots s_{i_N}$ coincides with the span of the monomials constructed from a reduced expression obtained from the given one applying one of the braid relations.

The step is to pass from a reduced decomposition $w_0 = AUB$ to another $w_0 = AU'B$ by a braid relation $U = U'$ of type 8.1.3. These are rank two relations, by the very definition of the elements $\overline{E}_\beta$ it is enough to verify the statement in the rank 2 case, this is done in the appendix.

We next show that A^+ is closed under product. It is clear that we are reduced to see that, if $h < k$, then $\overline{E}_{\beta_k}\overline{E}_{\beta_h}$ can be written as R linear combination of the monomials (11.8.1). Of course this amounts to say that if we write the Leverdowski-Soibelmann relations in terms of the $\overline{E}$'s their coefficients still lie in R.

Set $w = s_{i_1} \cdots s_{i_{h-1}}$ and $T_w = T_{i_1} \cdots T_{i_{t-1}}$, complete the given expression of w to a reduced expression of w_0, let $\{\gamma_1, \dots, \gamma_N\}$ be the corresponding ordering of the set of positive roots, and $\overline{E}'_{\gamma_i}$ be the corresponding elements in A^+.

Since it is clear that we obtain the Leverdowski-Soibelmann relation for $\overline{E}_{\beta_k}\overline{E}_{\beta_h}$ applying T_w to the analogous relation for $\overline{E}'_{\gamma_{k-h+1}}\overline{E}'_{\gamma_1}$, we can assume that $h = 1$ and hence $\overline{E}_{\beta_h} = \overline{E}_i$ for some i.

Thus in order to show our claim we need to see that A^+ is close under left and right multiplication by the $\overline{E}_i$'s.

Fix i and take a reduced expression $w_0 = s_{i_1} \cdots s_{i_N}$, with $i_1 = i$, by construction the set of monomials $\overline{E}^k$ for this reduced expression is stable under left multiplication by $\overline{E}_i$ and A^+ is stable under left multiplication by $\overline{E}_i$.

Now complete $s_{i_2} \cdots s_{i_N}$ to a reduced expression for $w_0 = s_{i_2} \cdots s_{i_N} s_j$. Since $\alpha_i = s_{i_2} \cdots s_{i_N}\alpha_j$, we have that $\overline{E}_i = T_{i_2} \cdots T_{i_N} E_j$ so the set of monomials $\overline{E}^k$ for this reduced expression is stable under right multiplication by $\overline{E}_i$ and A^+ is stable under right multiplication by $\overline{E}_i$.

Similarly one has that A^- is closed under product.

We are now in the position to see that A' is stable under the action of $\mathcal{B}$.

Fix $i = 1, \dots, r$ and take the two reduced expressions $w_0 = s_i s_{i_2} \cdots s_{i_N}$, and $w_0 = s_{i_2} \cdots s_{i_N} s_j$ considered above. Take

$$\overline{E}_{\beta_1}^{k_1} \cdots \overline{E}_{\beta_{N-1}}^{k_{N-1}} \overline{E}_i^{k} K_\lambda \overline{F}_i^{h} \overline{F}_{\beta_{N-1}}^{h_{N-1}} \cdots \overline{F}_{\beta_1}^{h_1}$$

to be a monomial (11.8.1) relative to the second reduced expression.

Then $T_i(\overline{E}_{\beta_1}^{k_1} \cdots \overline{E}_{\beta_{N-1}}^{k_{N-1}}) \in A^+$ and $T_i(\overline{F}_{\beta_{N-1}}^{h_{N-1}} \cdots \overline{F}_{\beta_1}^{h_1}) \in A^-$, so we only need to show that $T_i(\overline{E}_i^k K_\lambda \overline{F}_i^h) \in A^+A^0A^-$ to see that the image of our monomial under T_i lies in A'. This last claim follows immediately from the definition of the action of $\mathcal{B}$ and the defining relations (9.1.1). The proof of stability under T_i^{-1} is completely analogous.

We finally show that A' is closed under product completing the proof of our step 1. From the stability under $\mathcal{B}$ it suffices to show that A' is stable under right and left multiplication by the $\overline{E}_i$'s, $\overline{F}_i$'s and K_λ's. The only non trivial facts are to show invariance under right multiplication by $\overline{E}_i$'s, and left multiplication by $\overline{F}_i$'s.

Let us show it for the $\overline{E}_i$'s (the proof for the $\overline{F}_i$'s is analogous).

Letting $a \in A'$, write $a = T_i^{-1}(b)$, so $a\overline{E}_i = T_i^{-1}(-bK_{\alpha_i}^{-1}\overline{F}_i)$ and everything follows from the invariance under right multiplication by K_{α_i} and $\overline{F}_i$).

12.2 Before we proceed we notice that both A^+ and A^- are naturally Q^+-graded, Q^+ being the monoid of positive linear combinations of the α_i's. This is of course obtained by giving $\overline{E}_\beta$ and $\overline{F}_\beta$ degree β. For $\delta \in Q^+$ we shall denote by A_δ^+ and A_δ^- the components of degree δ in A^+ and A^-.

Before proceeding we need

Lemma. *$adE_i(A^+) \in A^+$; $adF_i(A^-) \in A^-$. for all $n \geq 0$.*

Proof. Let us see that adE_α, α simple, preserves A^+. Let $x \in A^+$. We can assume $x \in A^+_\delta$ for some δ. Also if $x = yz$ with $y \in A^+_\gamma$, $z \in A^+_{\delta-\gamma}$, we have

$$adE_\alpha(x) = adE_\alpha(y)z - q^{(\alpha,\gamma)}yadE_\alpha(z),$$

so we can assume $x = \overline{E}_\beta$ for some root β.

If $x = \overline{E}_i$ we are in the rank two case that will be treated in the appendix, we shall then proceed by induction on $ht(\beta)$ assuming β is not simple. Notice that by our induction and the fact that the R module $\frac{A^+_\beta}{\sum A^+_\delta A^+_{\beta-\delta}}$ has rank one, it suffices to show our claim for a arbitrary choice of $\overline{E}_\beta$. We know that there is a $w \in W$ and two simple roots α_i and α_j such that $w^{-1}\alpha = \alpha_i$ and $w^{-1}\beta = \gamma$ lies in the rank two system spanned by α_i and α_j. Take $w = s_{i_1} \cdots s_{i_k}$ a reduced expression for w, set $T_w = T_{i_1} \cdots T_{i_k}$. Then for $T_w E_i = E_\alpha$ and for $\overline{E}_\beta$ we can take $T_w \overline{E}_\gamma$. We then have $adE_\alpha(\overline{E}_\beta) = T_w(adE_i(\overline{E}_\gamma))$, and we are reduced to the rank two case. The proof in the case of F_i's is completely analogous.

Step 2. *The algebra $B := A/(q-1)$ is commutative.*

Proof. Since the action of $\mathcal{B}$ on A induces an action on $A/(q-1)$, it suffices to see that the classes of the elements $\overline{E}_i$ and K_λ modulo $q-1$ lie in the center. For the K_λ this is clear from the definitions. As for the $\overline{E}_i$ notice that since $\overline{E}_i = (q^{(\alpha_i,\alpha_i)/2} - q^{-(\alpha_i,\alpha_i)/2})E_i$, by the preceding lemma $ad\overline{E}_i(x) \in (q-1)A^+$ for all $x \in A^+$. Since

$$ad\overline{E}_i(x) = \overline{E}_i x - K_{\alpha_i} x K_{\alpha_i}^{-1} \overline{E}_i,$$

we immediately get from the fact that the elements K_{α_i} lie in the center modulo $q-1$, that $\overline{E}_i$ commutes with A^+ modulo $q-1$.

It remains to show that $\overline{E}_i$ commutes with A^- modulo $q-1$. Reasoning as in the previous Lemma we are reduced to see that $\overline{E}_i$ commutes, modulo $q-1$ with $\overline{F}_\beta$ for an arbitrary choice of $\overline{F}_\beta$. To see this fix a positive root β and choose a reduced expression $w_0 = s_{i_1} \cdots s_{i_N}$ in such a way that $\alpha_i = s_{i_1} \cdots s_{i_{N-1}}(\alpha_{i_N})$. Then $\beta = s_{i_1} \cdots s_{i_{t-1}}(\alpha_{i_t})$ for some $t < N$. Set $w = s_{i_1} \cdots s_{i_{t-1}}$, $T_w = T_{i_1} \cdots T_{i_{t-1}}$. A choice for $\overline{F}_\beta$ is given by $T_w \overline{F}_{i_t}$. Also $\overline{E}_i = T_w T_{i_t} \cdots T_{i_{N-1}} \overline{E}_{i_N}$. So in order to show that $\overline{F}_\beta$ and $\overline{E}_i$ commute modulo $q-1$ it suffices to see that $T_{i_t}^{-1} T_w^{-1} \overline{F}_\beta = -\overline{E}_{i_t} K_{\alpha_i}^{-1}$ and $T_{i_{t+1}} \cdots T_{i_{N-1}} \overline{E}_{i_N}$ do. Since $T_{i_{t+1}} \cdots T_{i_{N-1}} \overline{E}_{i_N} \in A^+$ this follows from our preceding discussion.

12.3

Step 3. $\Delta(A) \subset A \otimes A$, $S(A) \subset A$, $\varepsilon(A) \subset R$ *so that A is a Hopf R algebra.*

Proof. We need to show that $\Delta(x) \in A \otimes A$, $S(x) \in A$, $\varepsilon(x) \in R$, for $x = K_\lambda, \overline{E}_\beta, \overline{F}_\beta$. All these statements are clear for the K_λ's ,the $\overline{E}_i$'s and the $\overline{F}_i$'s. Let us show them for $\overline{E}_\beta$ and $\overline{F}_\beta$ with β not simple. Notice that using induction on $ht(\beta)$ and reasoning as above we reduce to prove our claims for a arbitrary choice of $\overline{E}_\beta$, and similarly for $\overline{F}_\beta$.

Let us start to show that $\Delta\overline{E}_\beta \in A \otimes A$. We shall assume here that there are no components of type G_2. This case will be treated in the appendix. Since β is positive there exist a simple root α such that $(\alpha, \beta) > 0$ and since β is not simple $s_\alpha\beta$ is also positive and of smaller height. As in 12.2 there exists $w \in W$ such that both $w^{-1}\alpha = \alpha_i$

and (since $(\alpha,\beta)>0$) $w^{-1}\beta=\alpha_j+m\alpha_i$, with α_i and α_j simple roots and $m=-a_{i,j}=1$ or 2. If $m=1$ a choice for $\overline{E}_\beta$ is given (since α is simple) by:

$$\overline{E}_\beta=T_wT_i(\overline{E}_j)=T_w(-adE_i(\overline{E}_j))=-adE_\alpha(T_w(\overline{E}_j)).$$

Recall that $T_w(\overline{E}_j)$ is a choice for $\overline{E}_{\beta-\alpha}$, so we have $\Delta((T_w(\overline{E}_j))\in A\otimes A$. If $m=2$, $\beta=ws_js_i\alpha_j$, and a choice for $\overline{E}_\beta$ is given by:

$$T_wT_jT_i(\overline{E}_j)=T_w(adE_i(T_{\overline{E}_i}))=adE_\alpha(T_wT_j\overline{E}_i)$$

(see the appendix, formulas A6). Now we have that $T_wT_jE_i$ is a choice for $E_{\beta-\alpha}$, so that again by induction on the height $\Delta(T_wT_jE_i)\in A\otimes A$. Thus our claim will follow from:

Lemma. *Let $x\in A^+$ be such that $\Delta(x)\in A\otimes A$. For any simple root α,*

$$\Delta(adE_\alpha(x))\in A\otimes A.$$

Proof. We can assume $x\in A^+_\delta$. Then it follows easily from the definition of Δ that

$$\Delta(x)=\sum_{\gamma_1+\gamma_2=\delta}K_{\gamma_2}x_{\gamma_1}\otimes y_{\gamma_2},x_{\gamma_1},y_{\gamma_2}\in A^+.$$

Also $adE_\alpha(x)=E_\alpha x-q^{(\delta,\alpha)}xE_\alpha$. Thus

$$\Delta(adE_\alpha(x))=\sum_{\gamma_1+\gamma_2=\delta}(E_\alpha K_{\gamma_2}x_{\gamma_1}\otimes y_{\gamma_2}+K_{\gamma_2+\alpha_i}x_{\gamma_1}\otimes E_\alpha y_{\gamma_2}-$$
$$q^{(\delta,\alpha_i)}K_{\gamma_2}x_{\gamma_1}E_\alpha\otimes y_{\gamma_2}-q^{(\delta,\alpha_i)}K_{\gamma_2}x_{\gamma_1}K_{\alpha_i}\otimes y_{\gamma_2}E_\alpha)$$

Now

$$E_\alpha K_{\gamma_2}x_{\gamma_1}\otimes y_{\gamma_2}-q^{(\delta,\alpha_i)}K_{\gamma_2}x_{\gamma_1}E_\alpha\otimes y_{\gamma_2}=q^{-(\gamma_2,\alpha_i)}K_{\gamma_2}E_\alpha x_{\gamma_1}\otimes y_{\gamma_2}$$
$$-q^{(\delta,\alpha_i)}K_{\gamma_2}x_{\gamma_1}E_\alpha\otimes y_{\gamma_2}$$
$$=(q^{-(\gamma_2,\alpha_i)}-q^{(\gamma_2,\alpha_i)})K_{\gamma_2}E_\alpha x_{\gamma_1}\otimes y_{\gamma_2}+q^{(\gamma_2,\alpha_i)}K_{\gamma_2}adE_\alpha(x_{\gamma_1})\otimes y_{\gamma_2}.$$

Since (γ_2,α_i) is a multiple of $(\alpha_i,\alpha_i)/2$, we get using Lemma 3.5 that

$$E_\alpha K_{\gamma_2}x_{\gamma_1}\otimes y_{\gamma_2}-q^{(\delta,\alpha_i)}K_{\gamma_2}x_{\gamma_1}E_\alpha\otimes y_{\gamma_2}\in A\otimes A.$$

On the other hand

$$K_{\gamma_2+\alpha_i}x_{\gamma_1}\otimes E_\alpha y_{\gamma_2}-q^{(\delta,\alpha_i)}K_{\gamma_2}x_{\gamma_1}K_{\alpha_i}\otimes y_{\gamma_2}E_\alpha=K_{\gamma_2+\alpha_i}x_{\gamma_1}\otimes adE_\alpha(y_{\gamma_2}),$$

which lies in $A\otimes A$ by Lemma 12.2. The proof for $\overline{F}_\beta$ is similar.

Let us now pass to the antipode. Reasoning exactly as before, we reduce to see that, if $x\in A^+$ (resp. $x\in A^-$ and $S(x)\in A$, then $S(ad(E_i(x))\in A$ (resp. $S(ad(F_i(x))\in A)$. Let us show it for the E's, the case of the F's being similar. We can assume $x\in A_\delta$ for some δ, it the follows from the definitions that $S(x)=K_{-\delta}y$ with $y\in A_\delta$ Then

$$S(ad(E_i(x))=-K_{-\delta}yK_{-\alpha_i}E_i+q^{(\delta,\alpha_i)}K_{-\alpha_i}E_iK_{-\delta}y=K_{-\delta-\alpha_i}(q^{2(\delta,\alpha_i)}E_iy-q^{(\delta,\alpha_i)}yE_i)$$
$$=K_{-\delta-\alpha_i}((q^{2(\delta,\alpha_i)}-1)E_iy+adE_i(y))$$

From the fact that $(\alpha_i, \alpha_i)/2$ divides (δ, α_i) and Lemma 12.2 we get that $S(adE_i(x)) \in A$.
Finally, since we have $\varepsilon(Tx) = \varepsilon(x)$, $\forall T \in \mathcal{B}$, $x \in U_{q,\Lambda}$, we get $\varepsilon(\overline{E}_\beta) = \varepsilon(\overline{F}_\beta) = 0$.

12.4 It is convenient, for the formulas which will follow to change our notations.

$$z_\beta := K_\beta \ (\beta \in M),\ z_i = z_{\alpha_i},\ y_i := \overline{F}_i,\ x_i := T_0(y_{\bar{i}})\ (i = 1, \dots, n)$$

As classes in $B = A/(q-1)$.
Following the definitions of 9.3:

$$E_{\beta_i} = T_{i_1} \cdots T_{i_{t-1}}(E_{\alpha_{i_t}}),\ F_{\beta_i} = T_{i_1} \cdots T_{i_{t-1}}(F_{\alpha_{i_t}})$$

set

$$z_{\beta_t} := T_{i_1} \cdots T_{i_{t-1}}(z_{\alpha_{i_t}}),\ y_{\beta_t} := T_{i_1} \cdots T_{i_{t-1}}(y_{\alpha_{i_t}}) \in B.$$

From Theorem 9.5 we have $T_i(y_i) = T_0(y_{\bar{i}})$, which implies:

$$x_i = -\overline{E}_i z_i^{-1}.$$

It follows from what we have proved up to now that the k algebra $B = A/(q-1)$ is a finitely generated Hopf algebra with a Poisson structure satisfying the properties of Definition 1.1, so $SpecB$ has the structure of a Poisson algebraic group. Also A being free over R is clearly flat. So A is a quantum deformation of the Poisson group $SpecB$.
Before we continue our discussion let us remark that by the definitions of the Hopf algebra structure on U we get

$$\begin{aligned} &\Delta y_i = y_i \otimes z_i^{-1} + 1 \otimes y_i,\ \Delta x_i = x_i \otimes z_i^{-1} + 1 \otimes x_i,\ \Delta z_\alpha = z_\alpha \otimes z_\alpha, \\ (12.4.1) \qquad &S x_i = -z_i x_i,\ S y_i = -y_i z_i,\ S z_\alpha = z_{-\alpha}, \\ &\eta x_i = 0,\ \eta y_i = 0,\ \eta z_\alpha = 1. \end{aligned}$$

As an immediate consequence we have:

Corollary. *The algebra $U := k[x_i, y_i, z_\alpha]$ is a sub-Hopfalgebra.*

Step 4. *There is a isomorphism of Poisson groups between SpecB and the Poisson group H.*

We notice that both B and the coordinate ring $k[H]$ of H are polynomial rings on $h = dimH$ generators with r elements inverted. So they are isomorphic as algebras. We need to give a isomorphism which is also compatible with the Hopf and Poisson structures.
For this purpose we will have to explicit a sufficient number of explicit formulas both for B that for the coordinate ring of H.
In view of this we postpone the proof of our Step 4 to §14, §15. In the next section we shall instead try to understand in a deeper way the action of the Braid group on B which will then be interpreted as an action on H.

§13. A universal construction associated to the braid group.

13.1 We need to remark several formulas for the algebra B which can be verified quite trivially from the basic defining relations of the quantum group and formulas A1-9 of the appendix setting $q = 1$.
The following formulas are immediate:
1) $T_i(x_i) = z_i^2 y_i$, $T_i(y_i) = x_i$, $T_i(z_\beta) = z_{s_i\beta}$.
In order to write down further relations, introduce one more notation (cf. 8.1.3):

$$y_{ij}^{(m)} = T_{ij}^{(m)} y_j \text{ if } m \text{ is odd},\ y_{ij}^{(m)} = T_{ij}^{(m)} y_i \text{ if } m \text{ is even}.$$

We have the following formulas ($i \neq j$):

2) $y_{ij}^{(m_{ij}-1)} = y_j$.

3) $y_{ij}^{(m_{ij}-2)} + y_{ji}^{(1)} = y_i y_j$ if $m_{ij} \geq 3$ and $a_{ij} = -1$.

4) $y_{ij}^{(\frac{1}{2}m_{ij}-1)} - y_{ji}^{(\frac{1}{2}m_{ij})} = y_i^2 y_j - 2y_{ji}^{(1)} y_i$ if $m_{ij} \geq 4$ and $a_{ji} = -1$.

5) $y_{ij}^{(1)} + y_{ji}^{(4)} = y_i^3 y_j - 3y_i^2 y_{ji}^{(1)} + 3y_i y_{ji}^{(3)}$, $y_{ij}^{(3)} - y_{ji}^{(2)} = y_i^3 y_j^2 - 3y_i^2 y_j y_{ji}^{(1)} + 3y_i y_{ji}^{(1)2} - 3y_{ji}^{(1)} y_{ji}^{(3)}$ $+ y_j y_{ji}^{(3)}$ if $m_{ij} = 6$ and $a_{ji} = -1$.

Formula 2) follows from 9.5. The remaining formulas are deduced rewriting the relations in the appendix and setting $q = 1$.

Recall that we also have an involution κ (9.2.2), defined by $\kappa E_i = F_i$, $\kappa K_\beta = K_{-\beta}$. It commutes with the braid group and we have $\kappa(z_\beta) = z_{-\beta}$ and $\kappa(x_i) = z_i y_i$, we claim:

Proposition. *κ is an antihomomorphism of the Poisson structure.*

Proof. We know that κ is an automorphism and $\kappa(q) = q^{-1}$ hence

$$\kappa(\{\overline{a}, \overline{b}\}) = \kappa \frac{\overline{[a, b]}}{(q - q^{-1})} = \frac{\overline{[\kappa(a), \kappa(b)]}}{(q^{-1} - q)} = \{\kappa(\overline{b}), \kappa(\overline{a})\}.$$

In view of the previous identities we start the following construction.

13.2 Denote by Z_0^0 the algebra with basis z_β ($\beta \in M$) and multiplication $z_\alpha z_\beta = z_{\alpha+\beta}$. The Weyl group and hence the braid group, acts on Z_0^0 by $s_i(z_\beta) = z_{s_i\beta}$. Consider the polynomial algebra over Z_0^0 in the indeterminates Ty_i, as $i = 1, 2, \ldots, n$ and $T \in \mathcal{B}$; we write $y_i := 1y_i$. We extend the action of $\mathcal{B}$ from Z_0^0 to this polynomial algebra by

$$T_1(T_2 y_i) := (T_1 T_2) y_i.$$

We define a quotient algebra of this polynomial algebra by imposing the minimal ideal of relations, stable under $\mathcal{B}$ and containing the following ones:

1) $T_i^2 y_i = z_i^2 y_i$,

2) $y_{ij}^{(m_{ij}-1)} = y_j$ if $i \neq j$,

3) relations 3)–5) from 13.1.

Denote by Z_0 the resulting algebra. Since we have defined Z_0 by extracting some formal properties of the algebra B it is clear that we have a $\mathcal{B}$ equivariant homomorphism $\pi : Z_0 \to B$. We will prove in fact in Theorem 13.5 that π is an isomorphism and so the 2 algebras coincide.

Remark that relations 1) and 3) are not homogeneous and we can also consider the associated homogeneous relations:

1′) $T_i^2(y_i) = y_i$,

3′) the left-hand sides of 3)–5) are zero.

In this case we can ignore the variables z_i and consider the algebra D over $\mathbb{C}$ generated by the Ty_i ($T \in \mathcal{B}$, $i = 1, \ldots, n$) and the relations 1′), 2) and 3′).

13.3 Next given an element $w \in W$ we need to consider the set of all possible reduced expressions J of w. If $J : \quad w = s_{i_1} s_{i_2} \ldots s_{i_t}$ is such a reduced expression we define $J^{-1} : w^{-1} = s_{i_t} s_{i_{t-1}} \ldots s_{i_1}$ a reduced expression of w^{-1}. For $k = 1, \ldots, t$ we also set:

$$w_k^J := s_{i_1} s_{i_2} \ldots s_{i_{k-1}}, \; \beta_k^J := s_{i_1} s_{i_2} \ldots s_{i_{k-1}}(\alpha_{i_k}) = w_{k-1}^J(\alpha_{i_k}).$$

From Proposition 8.2 the β_k^J give a total ordering of the set $R_{w^{-1}}$. Moreover we have $w^{-1}(\beta_k^J) = s_{i_t}s_{i_{t-1}}\dots s_{i_1}s_{i_1}s_{i_2}\dots s_{i_{k-1}}(\alpha_{i_k}) = -\beta_{t-k+1}^{J^{-1}}$.

We can apply this analysis to the element w_0 and the set of all positive roots. In this case for a given reduced expression J we can construct another reduced expression $\overline{J}: w_0 = s_{\overline{i}_1}s_{\overline{i}_2}\dots s_{\overline{i}_N}$. It is clear that $\beta_k^{\overline{J}} = -w_0(\beta_k^J)$.

We now define in $\mathcal{B}$ and in Z_0 the elements:

$$T_{k-1}^J := T_{i_1}T_{i_2}\dots T_{i_{k-1}}, \quad y_k^J := T_{i_1}T_{i_2}\dots T_{i_{k-1}}(y_{i_k}) = T_{k-1}^J(y_{i_k}).$$

Let us choose a reduced expressions $J: w = s_{i_1}s_{i_2}\dots s_{i_t}$ of an element w. Denote by Z^J the Z_0^0-subalgebra of Z_0 generated by the elements y_k^J.

Proposition. *Z^J is independent of the choice of the reduced expression J of w.*

Proof. It suffices to show that, given two expressions J and J', we have $y_k^J \in Z^{J'}$. Since one can pass from one expression to another by the use of braid relations we may by induction restrict to the case where J' is obtained from J by a single braid relation. We give a proof in the cases $m_{ij} = 2$ and $m_{ij} = 3$. In the remaining cases the proof is similar:

1) $J': w = as_is_jb = as_js_ib; \; s_is_j = s_js_i \; (m_{ij} = 2)$

2) $J': w = as_is_js_ib = as_js_is_jb; \; s_is_js_i = s_js_is_j \; (m_{ij} = 3)$.

Let $k = \ell(a)$. In case 1) we remark that $y_h^J = y_h^{J'}$ if $h \neq k, k+1$, $y_k^J = T_a(y_i) = T_aT_j(y_i) = y_{k+1}^{J'}$ and similarly $y_{k+1}^J = y_k^{J'}$.

In case 2) we get $y_h^J = y_h^{J'}$ if $h \neq k, k+1, k+2$, and $y_k^J = T_a(y_i) = T_aT_jT_i(y_j) = y_{k+2}^{J'}$ and similarly $y_k^{J'} = y_{k+2}^J$.

As for y_{k+1}^J we have:

$$y_{k+1}^J = T_aT_i(y_j) = T_a(-T_j(y_i) - y_iy_j) = -y_{k+1}^{J'} - y_k^Jy_k^{J'} = -y_{k+1}^{J'} - y_{k+2}^{J'}y_k^{J'} \in Z^{J'}. \quad \square$$

13.4 The previous proof and Theorem 9.5 have an important corollary:

Corollary. *(a) If for a given J and k, $\beta_k^J = \alpha_i$ is a simple root then $y_k^J = y_i$.*

(b) $T_i(x_i) = z_i^2y_i$.

(c) If $\beta_k^J = \beta_h^{J'}$, then $y_k^J = \pm y_h^{J'} + P$ where P is a polynomial in the elements $y_i^{J'}$ involving only the indices i for which $\beta_i^{J'}$ has height strictly less than that of $\beta_h^{J'}$.

Proof. (a) This is a special case of Theorem 9.5a.

(b) This follows directly from Theorem 9.5 and the defining relations 13.2.

(c) We can complete any reduced expression of any element w to a reduced expression of the longest element w_0, then (c) follows from the proof of the proposition. $\square$

13.5 We will now denote $Z_w^+ = Z^J$ and $Z_w^- := T_0(Z_w^+)$ and let $x_k^{\overline{J}} := T_0(y_k^J)$. If $w = w_0$ is the longest element we will simply write $Z_0^{\pm}$.

Theorem. *The algebra Z_0 is the tensor product $Z_0^+ \otimes Z_0^0 \otimes Z_0^-$ and, given a reduced expression J of w_0, we have:*

$Z_0 = Z_0^0[y_k^J, x_k^J;\ k = 1, \dots, N]$ is a polynomial ring.

The mapping $\pi: Z_0 \to B$ (B being the one introduced in §12.) is an isomorphism.

Proof. First of all we will show that the subring $S := Z_0^0[y_k^J, x_k^J;\ k = 1, \dots, N]$ is stable under the action of $\mathcal{B}$ which will prove that it is the entire algebra, then we will see that the $2N$ given elements are algebraically independent by using the map π, this will prove simultaneously all statements.

Let us begin with the first claim. We need to show that, for a given j, $T_j(S) \subset S$. From the independence of S from J we can choose

$$J: w_0 = s_{i_1} s_{i_2} \dots s_{i_N}, \quad J': w_0 = s_j s_{i_1} s_{i_2} \dots s_{i_{N-1}}.$$

Then we have $T_j(y_k^J) = y_{k+1}^{J'}$ unless $k = N$; in this case $T_j(y_N^J) = T_0(y_{i_N}) = x_{\bar{i}_N}$. Now $T_j(x_k^J) = T_j T_0(y_k^J) = T_0 T_{\bar{j}}(y_k^J)$. Again $T_{\bar{j}}(y_k^J)$ is in Z_0^+ and so $T_j(x_k^J)$ is in Z_0^- unless $y_k^J = y_{\bar{j}}$, in which case we use 13.2.1 and get $T_j(x_{\bar{j}}) = z_j^2 y_j$ as desired.

Consider the mapping $\pi : Z_0 \to B$, which by definition is $\mathcal{B}$ equivariant, we have seen in §12.4, as a consequence of the existence and form of a PBW basis, that the monomials in the elements y_k^J, x_k^J; $k = 1, \dots, N$ are linearly independent over Z_0^0, this completes the proof of all the remaining statements. □

Remark. *i) The construction and statements of Sections 13.2–13.4 as well as the first statement of Theorem 13.5 below hold over* $\mathbb{Z}$.

ii) The same proof, applied to the algebra D defined in 13.2 with the homogenized relations gives a more precise result:

(c') If $\beta_k^J = \beta_h^{J'}$ *, then in D;* $y_k^J = \pm y_h^{J'}$.

13.6 We have remarked, in various points, that we can consider an algebra D defined in 13.2 by generators and relations as a $\mathcal{B}$ algebra. We wish to show now how this algebra appears naturally in the classical Lie theory.

Consider the simple Lie algebra $\mathfrak{g}$ corresponding to the Cartan matrix (a_{ij}), let $\mathfrak{h}$ be a Cartan subalgebra and let $\mathfrak{g} = \mathfrak{h} \oplus (\oplus_{\alpha \in R} \mathfrak{g}_\alpha) := \mathfrak{h} \oplus \mathfrak{r}$ be the root space decomposition. We denote by e_i, f_i, h_i the usual Chevalley generators. Let G be a connected algebraic group corresponding to this Lie algebra (and to a given lattice M, for instance we can take the simply connected group). We set:

$$t_i := \exp(f_i)\exp(-e_i)\exp(f_i) \in G.$$

Notice that, in the $SL(2)$ case, t_i is the matrix:

$$\begin{vmatrix} 1 & 0 \\ 1 & 1 \end{vmatrix} \begin{vmatrix} 1 & -1 \\ 0 & 1 \end{vmatrix} \begin{vmatrix} 1 & 0 \\ 1 & 1 \end{vmatrix} = \begin{vmatrix} 0 & -1 \\ 1 & 0 \end{vmatrix}.$$

One knows the following facts (see e.g. [K–P]):

1) The mapping $T_i \mapsto t_i$ extends to a homomorphism of $\mathcal{B}$ to G, hence we have an action of $\mathcal{B}$ on $\mathfrak{g}$ and also on G itself by conjugation.

2) If J is a reduced expression, setting $f_k^J = T_{k-1}^J(f_{i_k})$ we have that f_k^J is a root vector relative to the negative root $-\beta_k^J$.

3) $T_0(f_i) = -e_{\bar{i}}$.

Let us now consider this action of $\mathcal{B}$ on $\mathfrak{g}$. From the previous remarks $\mathcal{B}$ acts separately on $\mathfrak{h}$ (as the Weyl group) and on $\mathfrak{r}$, moreover the $\mathcal{B}$ translates of the elements f_i span $\mathfrak{r}$ (and in fact give a basis up to signs).

We verify directly the following relations

1') $T_i^2(f_i) = f_i$.

2') $T_{ij}^{(m_{ij}-1)}(f_j) = f_j$ if $m_{ij} = 2, 4$ or 6.

3′) $T_jT_i(f_j) = f_i$, $T_i(f_j) + T_j(f_i) = 0$ if $m_{ij} = 3$.
4′) $T_i(f_j) = T_jT_i(f_j)$ if $m_{ij} = 4$ and $a_{ji} = -1$
5′) $T_{ij}^{(m_{ij}-2)}(f_i) + T_j(f_i) = 0$ if $m_{ij} \geq 4$ and $a_{ji} = -1$,
6′) $T_{ij}^{(2)}(f_i) = T_{ji}^{(3)}(f_i)$, $T_i(f_j) + T_{ji}^{(4)}(f_j) = 0$, $T_{ij}^{(3)}(f_j) = T_{ji}^{(2)}(f_j)$ if $m_{ij} = 6$ and $a_{ji} = 1$.

Hence we get

Proposition. *The algebra $S(\mathfrak{r})$ is $\mathcal{B}$ isomorphic to the algebra D defined in 13.2.*

Proof. The $\mathcal{B}$-equivariant mapping sending y_i to f_i is well-defined since the relations defining D hold in $S(\mathfrak{r})$. D is generated by the elements y_k^J and x_k^J which map to the corresponding vectors f_k^J and e_k^J. These vectors are linearly independent (in $\mathfrak{g}$) and this finishes the proof. □

§14 *B* as functions on a Poisson group.

14.1 We now start the proof of Step 4 of §12:

Theorem. *B with its Poisson structure is isomorphic to the coordinate ring of the Poisson group H.*

By the definitions it is clear that we can assume that the root system is irreducible (or the Cartan matrix indecomposable). This isolates G_2 as a special case to be treated independently.

Recall that the Poisson group H coming from example 2 of 11.4 is constructed as follows: Let K be the simply connected semisimple group associated to the given Cartan matrix. Fix a maximal torus $T \in K$ and a Borel subgroup $B^+ \supset T$, let B^- be the unique Borel subgroup such that $T = B^+ \cap B^-$. Set $G = K \times K$. Denote by $\mu_\pm : B^\pm \to T$ the canonical projection homomorphisms, and consider the homomorphism $\phi : B^- \times B^+ \to T$ defined by $\phi(b_-, b+) = \mu_-(b_-)\mu_+(b_+)$. Then set $H = Ker\phi$. The elements of H are pairs $(t^{-1}u_-, tu_+)$ where $t \in T$, $u_- \in U_-$, $u_+ \in U_+$. Identifying the previous element with the triple (u_-, t, u_+) we decompose H as a variety (and as a semidirect product) as $U^- \times T \times U^+$. We have seen in 11.4 how this group acquires a Poisson group structure.

The proof of the theorem will take several steps and will not be completed until §15.

In order to prove the theorem we need first to construct an explicit isomorphism between the two algebras or equivalently between the algebraic varieties Ω and H.

Let U^+ and U^- denote the unipotent subgroups of G corresponding to positive and negative roots and let $G^0 = U^-TU^+$ be the *big cell* of G.

We have a mapping $\sigma : B^- \times B^+ \to G$ defined by $\sigma(b_-, b_+) = b_-^{-1}b_+$, this has as image the big cell G^0:

Proposition. *σ restricted to H is an unramified covering of degree 2^n of G^0.*

Proof. In fact if we write an element of H in the unique form u_-tu_+ we have $\sigma(u_-tu_+) = u_-^{-1}t^2u_+$.

14.2 We return now to the problem of constructing the isomorphism requested by Theorem 14.1. Recall that, under the identifications of §13, the algebra $B = Z_0^- \otimes Z_0^0 \otimes Z_0^+$ and let $\Omega = \text{Spec } B$ be the algebraic variety of its $\mathbb{C}$-valued points. Thus Ω is a product of the N-dimensional affine space Spec Z_0^-, the complex torus $T =$ Spec Z_0^0 and the N-dimensional affine space Spec Z_0^+. The group $\mathcal{B}$ acts on Ω algebraically.

We have seen in 13.6 that we can also act with $\mathcal{B}$ on G by inner conjugation by the elements $t_i : T_i(g) := t_igt_i^{-1}$.

In order to express in a more coherent way our future computations it is useful to introduce the set $\mathcal{F}$ of (regular) maps $f: \Omega \to G$ and act on $\mathcal{F}$ with $\mathcal{B}$ in the obvious way:

$$(T_i f)(p) := T_i(f(T_i^{-1}(p))). \tag{14.2.1}$$

We consider $\mathcal{F}$ as a (infinite dimensional) group by pointwise multiplication of the values of maps and clearly $\mathcal{B}$ acts as a group of automorphisms. We construct now some special maps $\Omega \to G$:

$$Y_k := \exp(y_k f_k) \tag{14.2.2}$$

(or more explicitly $Y_k(p) := \exp(y_k(p) f_k)$, $p \in \Omega$). For a reduced expression J:

$$Y_k^J := \exp(y_k^J f_k^J) \tag{14.2.3}$$

and finally the map Z which is the projection on the middle factor composed with the standard isomorphism of $\mathrm{Spec} Z_0{}^0$ with T, notice that the character λ on T corresponds to the function z_λ in our notations.

Since $y_k^J = T_{k-1}^J(y_{i_k})$ and $f_k^J = t_{k-1}^J(f_{i_k})$ we get from the definition of the Braid group action on $\mathcal{F}$ that

$$Y_k^J = T_{k-1}^J(Y_{i_k}). \tag{14.2.4}$$

From the definitions and the relations 13.1 2) (satisfied by the elements y_i and f_i) it follows that the set $\mathcal{F}$ and the elements Y_i satisfy all the hypotheses of Theorem 9.5 hence we deduce:

Lemma. *If $\alpha_i = w(\alpha_j)$, (α_i, α_j simple roots) then $T_w(Y_j) = Y_i$.*
If $\beta_k^J = \alpha_i$ is a simple root we have $Y_k^J = Y_i$. □

We now come to an interesting consequence of the braid relations, all the formulas are to be interpreted inside the group $\mathcal{F}$.

Proposition. *1) $Y_i T_i(Y_j) = Y_j T_j(Y_i)$ if $m_{ij} = 2$.*
2) $Y_i T_j(Y_i) Y_j = Y_j T_i(Y_j) Y_i$ if $m_{ij} = 3$.
3) $Y_i T_j T_i(Y_j) T_j(Y_i) Y_j = Y_j T_i T_j(Y_i) T_i(Y_j) Y_i$ if $m_{ij} = 4$.
4) $Y_i(T_{ji}^{(4)} Y_j)(T_{ji}^{(3)} Y_i)(T_{ji}^{(2)} Y_j)(T_j Y_i) Y_j = Y_j(T_{ij}^{(4)} Y_i)(T_{ij}^{(3)} Y_j)(T_{ij}^{(2)} Y_i)(T_i Y_j) Y_i$ if $m_{ij} = 6$.

Proof. By direct calculation. 1) is clear and as for 2) we can compute in SL_3. With the usual notation of elementary matrices we have:

$$Y_1 = 1 + y_1 e_{21},\ Y_2 = 1 + y_2 e_{32} \text{ and } t_1 = -e_{12} + e_{21} + e_{33},\ t_2 = e_{11} - e_{23} + e_{32}.$$

Hence we get

$$T_1(Y_2) = t_1(1 + T_1(y_2) e_{32}) t_1^{-1} = 1 - T_1(y_2) e_{31}$$

and

$$T_2(Y_1) = t_2(1 + T_2(y_1) e_{21}) t_2^{-1} = 1 + T_2(y_1) e_{31},$$

so

$$Y_1 T_2(Y_1) Y_2 = 1 + y_1 e_{21} + T_2(y_1) e_{31} + y_2 e_{32},$$

while

$$Y_2 T_1(Y_2) Y_1 = 1 + y_1 e_{21} - T_1(y_2) e_{31} + y_2 e_{32} + y_1 y_2 e_{31};$$

Thus the relation follows from 3.1.3. The proof of 3) and 4) is similar although lengthier since we have to compute in B_2 or G_2 which we can explicitly represent also by matrices (cf. appendix). □

14.3 We define now, for $J: w = s_{i_1} s_{i_2} \dots s_{i_t}$ the maps

$$Y^J := Y_t^J Y_{t-1}^J \dots Y_1^J,\ X^J := T_0(Y^J). \tag{14.3.1}$$

Proposition. *1) Y^J and X^J are independent of J; we denote them: Y_w and $X_{\overline{w}}$.*
2) If $w = ab$ is a reduced decomposition, then

$$Y_w = T_a(Y_b)Y_a \quad \text{and} \quad X_w = T_a(X_b)X_a. \tag{14.3.2}$$

3) Y_w can be thought of as an algebraic isomorphism between the affine space with coordinates Y_k^J and the unipotent group $w^{-1}(U^+) \cap U^-$.

Proof. 1) It suffices to do it for Y^J. Again it is enough to do it for two reduced expressions which differ by a single braid relation $aub = au'b$. In this case we see that the factors in the two products coincide except for two or three, etc. consecutive ones relative to the positions of the reduced expressions of u, u' respectively, these terms are obtained applying T_a to the factors appearing in the analogous expressions for $u = u'$. Then Proposition 14.2 finishes the proof.

2) This is clear from the definitions for the Y_w and follows from Lemma 9.6 for the X_w.

3) The unipotent group $w^{-1}(U^+) \cap U^-$ is the product (as variety) of the root subgroups relative to the roots $-\beta_k^J$ and these are the negative roots which w maps to positive roots. The map Y_w expresses an element of G as such a product of elements in these given root subgroups. □

Remark. *It will be in fact shown that the isomorphism of 3) is a group isomorphism.*

We shall refer to Y and X instead of Y_{w_0} and X_{w_0}.

Before we continue let us make some comments on the maps X_w. For a simple reflection s_i we denote:

$$X_i := X_{s_i}. \tag{14.3.3}$$

By its very definition and from 13.6 we have:

$$X_i := T_0(Y_{\overline{i}}) = t_0 \exp(T_0(y_{\overline{i}})f_{\overline{i}})t_0^{-1} = \exp(-x_i e_i) = T_i(Y_i). \tag{14.3.4}$$

We have started, somewhat arbitrarily, from the maps Y_i but we could also have started from the X_i. These elements of $\mathcal{F}$ satisfy indeed the conditions of Theorem 9.5. In fact let us consider two simple roots $\alpha_i = w(\alpha_j)$, we claim that $X_i = T_w(X_j)$. In fact $X_i = T_0(Y_{\overline{i}})$ while $T_w(X_j) = T_w T_0(Y_{\overline{j}})$ and applying Lemma 9.6 $T_w T_0 = T_0 T_{\overline{w}}$ so $T_w T_0(Y_{\overline{j}}) = T_0 T_{\overline{w}}(Y_{\overline{j}}) = T_0(Y_{\overline{i}}) = X_i$.

14.4

Proposition. $T_i(Y) = T_i(Y_i)YY_i^{-1}$, $T_i(X) = T_i(X_i)XX_i^{-1}$ and $T_i(Z) = Z$.

Proof. Let $w_0 = s_i w = w s_{\overline{i}}$ so that, from 2) of Proposition 14.3 we have $Y = T_w(Y_{\overline{i}})Y_w = T_i(Y_w)Y_i$. But, since $w(\alpha_{\overline{i}}) = \alpha_i$, from Lemma 14.2 we have $T_w(Y_{\overline{i}}) = Y_i$ and so:

$$Y = Y_i Y_w, \text{ and } YY_i^{-1} = T_i(Y_w).$$

Apply T_i to the first equality and substitute:

$$T_i(Y) = T_i(Y_i)T_i(Y_w) = T_i(Y_i)YY_i^{-1}.$$

Furthermore,

$$T_i(X) = T_i T_0(Y) = T_0 T_{\overline{i}}(Y) = T_0(T_{\overline{i}}(Y_{\overline{i}})YY_{\overline{i}}^{-1}) =$$
$$T_0(T_{\overline{i}}(Y_{\overline{i}}))XT_0(Y_{\overline{i}}^{-1}) = T_i(T_0(Y_{\overline{i}}))XT_0(Y_{\overline{i}}^{-1}) = T_i(X_i)XX_i^{-1}.$$

Finally, the statement for Z is clear from the definitions since it expresses that the map Z is equivariant. □

Theorem. $T_i(YZ^2X) = X_iYZ^2XX_i^{-1}$.

Proof. From Proposition 14.4 we have $T_i(YZ^2X) = X_iYY_i^{-1}Z^2T_i(X_i)XX_i^{-1}$, so we need to show that $Z^2 = Y_i^{-1}Z^2T_i(X_i)$ or $Z^{-2}Y_iZ^2 = T_i(X_i)$. We write:

$$T_i(X_i) = \exp(T_i(x_i)t_i(e_i)) = \exp(z_i^2y_if_i).$$

But by the definition of the map Z we have that $Z^{-2}\exp(y_if_i)Z^2 = \exp(z_i^2y_if_i)$, since z_i corresponds to the simple root associated to e_i. □

We consider then the 3 isomorphisms Y : Spec $Z_0^- \mapsto U^-$, Z : Spec $Z_0^0 \mapsto T$ and X : Spec $Z_0^+ \mapsto U^+$. We consider next $H = U^-TU^+$, our fundamental construction is now the map $Y^{-1}ZX$:

(14.4.4) $$Y^{-1}ZX : \Omega \to H.$$

This map from Ω to H composed with σ gives YZ^2X , by all the previous remarks this is an isomorphism of varieties and allows us to consider the elements y_β, z_i, x_β as coordinate functions on H.

Before we proceed we should remark about the meaning of this last theorem for algebraic groups. In order to do this we need to make some remarks which will also be useful later. Let us make them for the subgroup U^+ but similar remarks apply to U^-.

If we consider a root α we have the root subgroup $U_\alpha := exp(\lambda e_\alpha)$. The product of all these root subgroups, relative to non simple roots, is a normal subgroup $\overline{U}^+$ of U^+ such that $U^+/\overline{U}^+$ is a vector group, direct product of the root groups associated to simple roots. In particular there is a canonical homomorphism:

$$\xi_i : U^+ \to U_{\alpha_i}$$

with the property that, however we choose an ordering of the positive roots, if we write an element $u \in U^+$ as a product of elements from the various root subgroups, the factor corresponding to the simple root α_i is independent of the choice and equals $\xi_i(u)$. We may extend this map to the entire big cell composing with the obvious projection of G^0 to U^+ induced by the decomposition $G^0 = U^-TU^+$.

Having made these remarks we observe the curious consequences of the previous theorem, let us indicate for notational convenience by $\mu := \sigma(Y^{-1}ZX) = YZ^2X$ (which is a map from Ω to G^0), remark that, by its very definition μ is an etale covering of degree 2^n onto the big cell $U^- \times T \times U^+$:

Corollary. *a) The map $\overline{T}_i : G^0 \to G$ defined by*

$$\overline{T}_i(a) := t_i^{-1}\xi_i(a)a\xi_i(a)^{-1}t_i$$

maps G^0 isomorphically to G^0 and gives rise of an action of $\mathcal{B}$ on G^0 for which the map μ is equivariant.

b) Functions on G^0, invariant under conjugation, are $\mathcal{B}$ invariant.

Proof. From the statement of the theorem and the definitions it follows that, for $P \in \Omega$:

$$T_i(\mu(P)) = t_i(\mu(P))t_i^{-1} = X_i(P)\mu(P)X_i(P)^{-1}.$$

It is clear from the definition of the map μ that $X_i(P) = \xi_i(\mu(P))$. Hence to conclude we only have to remark that, since μ is a covering, the only way in which T_i can descend to a map $\overline{T}_i$ of G^0 is that this is an isomorphism of varieties.

The curious aspect of this corollary is that G^0 is not stable under conjugation by $\xi(a)$ nor by t_i but only a compensation between the 2 generates the braid action we have described.

14.5 Before we proceed let us remark some further consequences of Theorem 14.4.

Corollary. *For $w \in W$ we have $T_w(YZ^2X) = X_w YZ^2 X X_w^{-1}$.*

Proof. This is an immediate consequence of the theorem and 2) of Proposition 14.5 using induction on the length of w. □

In particular $T_0(YZ^2X) = XYZ^2$ and, since $X = T_0(Y)$ we have $T_0(X) = Z^{-1}YZ$.

One can connect the formulas between the classical Cartan involution ω in G and the transformation κ of 9.2.2. Recall that ω may be defined as automorphism of the Lie algebra $\mathfrak{g}$ giving its value on the Chevalley generators:

$$\omega(e_i) = -f_i, \omega(f_i) = -e_i, \omega(h_i) = -h_i.$$

By construction ω induces isomorphism between B^+ and B^-, U^+ and U^- and induces on T the automorphism $t \to t^{-1}$.

In fact:

Remark. $\omega Y_k^J \kappa = \exp(z_{\beta_k^J} x_k^J e_k^J)$ *so* $\omega Y \kappa = ZXZ^{-1}$.

We have arrived now at last at the main construction:

Theorem. *The map $Y^{-1}ZX$ is the required isomorphism of Poisson groups.*

It will be reduced to the verification of a few relations due to the following:

Lemma. *Let R, R' be the coordinate rings of two Poisson groups and $\psi : R \to R'$ a ring isomorphism. Assume that are given elements $r_i \in R$ which generate R as a Poisson algebra and such that:*
1) For all $r \in R$ we have $\psi(\{r_i, r\}) = \{\psi(r_i), \psi(r)\}$.
2) $\psi \otimes \psi(\Delta(r_i)) = \Delta(\psi(r_i))$.
3) $\psi(S(r_i)) = S(\psi(r_i))$.
Then ψ is an isomorphism of Poisson groups.

Proof. Let $U := \{s \in R | \psi(\{s, r\}) = \{\psi(s), \psi(r)\}$ for all $r \in R$. The Jacobi identity and Leibnitz rule imply that U is a Poisson subalgebra.

From hypothesis 1 $r_i \in U$, since the r_i generate R as Poisson algebra it follows that $U = R$ and ψ is an isomorphism of Poisson algebras.

Similarly let $V := \{v \in R | \psi \otimes \psi(\Delta(v)) = \Delta(\psi(v))\}$. From the compatibility of the coproduct with the Poisson structure it follows that T is a Poisson subalgebra. From 2) we have $r_i \in V$ hence $V = R$ and the compatibility of ψ with Δ. Similarly for the antipode S. □

We cannot yet prove this theorem which will be completed in 15.6 since we have not yet developed sufficient information on the Poisson and Hopf algebras structures.

But we can remark about the strategy. Consider the subalgebra of B generated by the elements x_i, y_i, z_i. It is a sub-Hopfalgebra (cor. 12.4) and it corresponds to the quotient of H under the normal subgroup generated by all the root subgroups in $U^- \times U^+$ generated by non simple roots. It is clear, by the formulas 12.4.1 and a simple $Sl(2)$ matrix computation, that the given map induces on these elements a Hopf algebra isomorphism, we will use the same notations in B and in the ring of functions on H, so in order to apply Lemma 14.5 it suffices to show that these elements generate one of the 2 algebras as Poisson algebras and that the isomorphism is compatible with Poisson bracket under each of these elements.

The fact that these elements generate the coordinate ring $\mathbb{C}[H]$ of H as Poisson algebra, is essentially the similar statement that the Chevalley generators generate the semisimple Lie algebra, as we now show.

Note that $\mathbb{C}[H] = \mathbb{C}[U_-] \otimes \mathbb{C}[T] \otimes \mathbb{C}[U_+]$, where $U_\pm$ and T are embedded in H in a natural way as Poisson algebraic subgroups. Hence it suffices to show that the x_i (resp. y_i) generate $\mathbb{C}[U_+]$ (resp. $\mathbb{C}[U_-]$) as a Poisson subalgebra. For this note that $\mathbb{C}[U_+]$ is $\mathbb{Z}_+$-graded by $\deg x_i = 1$. Let $\mathfrak{m}_+$ be the augmentation ideal. In order to show that the x_i generate $\mathbb{C}[U_+]$ as a Poisson subalgebra, it suffices to show that $\mathfrak{m}_+/\mathfrak{m}_+^2$ is generated as a Lie algebra by the $\overline{x}_i$. But this is clear since $\mathfrak{m}_+/\mathfrak{m}_+^2$ is the subalgebra $\mathfrak{n}_+$ in the Lie algebra $\mathfrak{m}/\mathfrak{m}_+^2 \simeq \mathfrak{g}$, the $\overline{x}_i$ corresponding to the e_i. This completes the proof.

14.6 We can now compute the Hamiltonian fields on H given by the functions x_i, y_i, z_i, let us denote by $\overline{x}_i, \overline{z}_i, \overline{y}_i$ the differential of x, z_i, y_i computed in 1.

Using the decomposition of $H = U_- \times T \times U_+$ we decompose its Lie algebra $\mathfrak{h}$ has $\mathfrak{u}_- \oplus \mathfrak{t} \oplus \mathfrak{u}_+$. Accordingly we have a basis of $\mathfrak{h}$ given by the vectors f_β, h_i, e_β (the ones associated to non simple roots are defined up to sign). The parametrization given in 14.4.1 and the formulas 14.2.2 and 14.3.4 show that $\overline{x}_i$ is orthogonal to all the basis vectors except for $\overline{x}_i(e_i) = -1$, similarly $\overline{y}_i, f_i = -1$. For $\overline{z}_i$ we also have that it is orthogonal to the elements e_β, f_β while on the h_j takes value a_{ij} by the definitions i.e.:

$$\overline{z}_i(h) = \alpha_i(h). \tag{14.6.1}$$

Lemma. *Let A be en algebraic group R its coordinate ring as Hopf algebra. Let $a \in R$ and $\Delta(a) = \sum_i a_i \otimes b_i$. The left invariant differential form coinciding with da in 1 is $\sum_i S(a_i)db_i$.*

Proof. Let us compute the translate in a point g of da in 1. We multiply on the left by g^{-1} and see that this translate equals the value of $da(g^{-1}x)$ computed for $x = g$. By definition $a(yx) = \sum_i a_i(y)b_i(x)$ and $a(g^{-1}) = (Sa)(g)$ thus $a(g^{-1}x) = \sum_i (Sa_i)(g)b_i(x)$ thus $da(g^{-1}x)$ computed for $x = g$ equals $\sum_i S(a_i)(g)db_i(g)$ as desired.

We want to apply this to $x_i \in \mathbb{C}[H]$ and get, since $\Delta(x_i) = 1 \otimes x_i + x_i \otimes z_i^{-1}$ and $S(x_i) = -z_i x_i$ that $u_i := dx_i + z_i^{-1} x_i dz_i$ is left invariant coinciding in 1 with dx_i. Clearly $u_i = z_i^{-1} d(z_i x_i)$ or $z_i u_i = d(z_i x_i)$. Thus according to 11.3 the Hamiltonian vector field $H_{z_i x_i}$ generated by $z_i x_i$ is $z_i \tau(u_i)$ where τ identifies the cotangent space of H at 1 with the Lie algebra $\mathfrak{g}$ of G (which in the Manin triple appear as the diagonal subalgebra). It only remains to verify that $\tau(u_i) = -d_i f_i$, this will prove the main lemma for the elements $z_i x_i$, by applying the Cartan involution we will get it for a set of generators. Recall that we have introduced the rescaled Killing form for which $(\alpha, \alpha) = 2$ for the short roots, by the standard theory of semisimple Lie algebras we have in the Cartan subalgebra $\mathfrak{t}$ elements t_i such that the value (t_i, h) of our form ($h \in \mathfrak{t}$) is $\alpha_i(h)$ and $h_i = \frac{t_i}{d_i}$, $(h_i, h_i) = \frac{2}{d_i}$, $(e_i, f_i) = \frac{1}{d_i}$.

Now by the normalization of the invariant form defining the Manin triple $((f_i, f_i)|(0, e_i)) = d_i^{-1}$. A similar consideration holds for y_i and:

$$H_{z_i x_i} = -d_i z_i f_i, \; H_{z_i y_i} = d_i z_i e_i. \tag{14.6.2}$$

For z_i we have the following remarks. The Cartan subalgebra $\mathfrak{t}$ is identified to the pairs (h, h) in $\mathfrak{g}$ resp. $(-h, h)$ in $\mathfrak{h}$ so that the element which is paired (in the chosen form) with $(-h, h)$ gives $\alpha_i(h)$ is $t_i/2 = \frac{d_i h_i}{2}$. From the previous Lemma and the computation of $\overline{z}_i$ we finally get:

$$H_{z_i} = d_i z_i h_i/2. \tag{14.6.3}$$

Using Leibnitz formula we can explicit the Hamiltonian fields of the x_i, z_i, y_i.

Thus in order to prove the main Theorem stated in this section it suffices to prove that the same formulas are valid for the Poisson group $Spec(B)$, verifying the remaining conditions of Lemma 14.5 for the map $Y^{-1}ZX$.

The next section 15 is thus entirely dedicated to prove the analogous formulas for B. This will conclude the proof of Theorem 14.5.

§15. Some Hamiltonian fields on $Spec(B)$.

15.1 We start to discuss the Poisson structure in B. We recall the results of §11. There, for an R algebra A with the property that $h \in R$ is a non zero divisor in A, we have defined, for every element $a \in A$ such that its image $\overline{a}$ modulo h is in the center of $A/(h)$ a derivation D_a.

Given $y \in A/(h)$, $y = \overline{b}$ we have set (12.1.1)

$$D_a[y] = \overline{[a,b]/h}$$

These derivations, restricted to the center of $A/(h)$ define a Poisson structure. Notice that this structure depends (in a simple way) on h and not just on the ideal (h).

As in §12 we want to apply these ideas to B. We consider $B = A/(h)$ where we choose $h := (q - q^{-1})$. With this choice of generator we proceed to construct the induced derivations and Poisson structures. Since we have several compatibility conditions working for us it will turn out that, in order to perform the computations, we will need only a limited amount of formulas for the action of special elements.

In particular observe that the element E_i gives a derivation:

$$\underline{e}_i(a) := [E_i, \tilde{a}]_{q=1}$$

(Where $\tilde{a}$ lifts a). Since $\frac{q^d - q^{-d}}{q - q^{-1}}\Big|_{q=1} = d$ and $x_i = -\overline{E}_i z_i^{-1}$ we have:

$$\underline{e}_i(y) = -\{\frac{x_i z_i}{d_i}, y\}. \tag{15.1.1}$$

We proceed now to give the main formulas for $\underline{e}_i$ which can be derived by direct computation. First note the general:

$$\underline{e}_i(z_\alpha) = \frac{(\alpha|\alpha_i)}{(\alpha_i|\alpha_i)} z_{\alpha+\alpha_i} x_i;$$
$$\underline{e}_i(x_i) = -z_i x_i^2;$$
$$\underline{e}_i(y_j) = \delta_{ij}(z_i - z_i^{-1})$$

and also:

$$\underline{e}_i(x_i) = \underline{e}_i(x_i) = 0 \text{ if } a_{ij} = 0.$$

The remaining formulas involve only 2 indices and can be reduced to the rank 2 case.

Consider an irreducible Cartan matrix of rank 2, i.e.:

$$A_2 := \begin{vmatrix} 2 & -1 \\ -1 & 2 \end{vmatrix}, \quad B_2 := \begin{vmatrix} 2 & -2 \\ -1 & 2 \end{vmatrix}, \quad G_2 := \begin{vmatrix} 2 & -3 \\ -1 & 2 \end{vmatrix}$$

Case A_2, $a_{12} = a_{21} = -1, m_{12} = 3$

$$\begin{aligned} &\underline{e}_1(x_2) = z_1 x_{21} \\ &\underline{e}_1(T_1(y_2)) = z_1 y_2; \\ &\underline{e}_1(T_2(x_1)) = 0. \end{aligned}$$

Case B_2, $a_{12} = -2, a_{21} = -1, m_{12} = 4$

$$\begin{aligned} &\underline{e}_1(x_2) &&= 2z_1 x_{21} &&\underline{e}_2(x_1) &&= z_2 x_{12}^{(2)} \\ &\underline{e}_1(T_1(y_2)) &&= 2z_1 T_1 T_2(y_1) &&\underline{e}_2(T_2(y_1)) &&= z_2 y_1 \\ &\underline{e}_1(T_2(x_1)) &&= z_1 T_2 T_1(x_2) &&\underline{e}_2(T_1(x_2)) &&= z_2 (T_1 T_2(x_1))^2 \\ &\underline{e}_1(T_2 T_1(x_2)) &&= z_1 T_2(x_1) &&\underline{e}_2(T_1 T_2(x_1)) &&= 0. \end{aligned}$$

Case G_2, $a_{12} = -3, a_{21} = -1, m_{12} = 6$

$$\begin{aligned} &\underline{e}_1(x_2) &&= 3z_1 T_2(x_1) &&\underline{e}_2(x_1) &&= z_2 x_{12}^{(4)} \\ &\underline{e}_1(T_2 x_1) &&= 2z_1 x_{21}^{(3)} &&\underline{e}_2(T_1 x_2) &&= z_2 (3x_{12}^{(4)} x_{12}^{(2)} - x_{12}^{(3)}) \\ &\underline{e}_1(x_{21}^{(2)}) &&= 3z_1 {x_{21}^{(3)}}^2 &&\underline{e}_2(x_{12}^{(2)}) &&= z_2 x_{12}^2 \\ &\underline{e}_1(x_{21}^{(3)}) &&= z_1 x_{21}^{(4)} &&\underline{e}_2(x_{12}^{(3)}) &&= z_2 {x_{12}^{(4)}}^3 \\ &\underline{e}_1(x_{21}^{(4)}) &&= 0 &&\underline{e}_2(x_{12}^{(4)}) &&= 0 \\ &\underline{e}_1(T_2 y_1) &&= 3z_1 T_1 T_2(y_1) &&\underline{e}_2(T_1 y_2) &&= z_2 y_1. \end{aligned}$$

In more compact form the same list reads:

$$\underline{e}_i(z_\alpha) = \frac{(\alpha|\alpha_i)}{(\alpha_i|\alpha_i)} z_{\alpha+\alpha_i} x_i;$$
$$\underline{e}_i(x_i) = -z_i x_i^2;$$
$$\underline{e}_i(x_j) = z_i x_{ji}^{(m_{ji}-2)} \text{ if } i \neq j \text{ and } |a_{ij}| \leq |a_{ji}|;$$
$$\underline{e}_i(x_j) = -a_{ij} z_i T_j(x_i) \text{ if } a_{ji} = -1;$$
$$\underline{e}_i(T_j(x_i)) = 0 \text{ if } m_{ij} = 3,$$
$$\underline{e}_i(T_j(x_i)) = z_i T_j T_i(x_j),\ \underline{e}_i(T_j T_i(x_j)) = z_i T_j(x_i) \text{ if } m_{ij} = 4 \text{ and } a_{ji} = -1;$$
$$\underline{e}_i(T_j(x_i)) = z_i (T_j T_i(x_j))^2,\ \underline{e}_i(T_j T_i(x_j)) = 0 \text{ if } m_{ij} = 4 \text{ and } a_{ij} = -1;$$
$$\underline{e}_i(T_j(x_i)) = 2 z_i x_{ji}^{(3)},\ \underline{e}_i(x_{ji}^{(3)}) = z_i x_{ji}^{(4)},$$
$$\underline{e}_i(x_{ji}^{(2)}) = 3 z_i x_{ji}^{(3)2} \text{ if } m_{ij} = 6 \text{ and } a_{ji} = -1;$$
$$\underline{e}_i(x_j^{(4)}) = 0 \text{ if } m_{ij} = 6,$$
$$\underline{e}_i(x_{ji}^{(3)}) = z_i x_{ji}^{(4)3},\ \underline{e}_i(x_{ji}^{(2)}) = z_i x_{ji}^2$$
$$\underline{e}_i(T_j(x_i)) = 3 z_i x_{ji}^{(4)} x_{ji}^{(2)} - z_i x_{ji}^{(3)} \text{ if } m_{ij} = 6 \text{ and } a_{ij} = -1;$$
$$\underline{e}_i(y_j) = \delta_{ij}(z_i - z_i^{-1});$$
$$e_i(T_i(y_j)) = z_i y_j \text{ if } a_{ij} = -1;$$
$$e_i(T_i(y_j)) = -a_{ij} z_i T_i T_j(y_i) \text{ if } a_{ji} = -1;$$
$$e_i(T_i T_j(y_i)) = z_i y_j^2 \text{ if } m_{ij} = 4 \text{ and } a_{ij} = -1;$$
$$\underline{e}_i(y_{ij}^{(3)}) = 3 z_i y_{ij}^{(4)2},\ \underline{e}_i(y_{ij}^{(2)}) = 2 z_i y_{ij}^{(4)},$$
$$\underline{e}_i(y_{ij}^{(4)}) = z_i y_j \text{ if } m_{ij} = 6 \text{ and } a_{ji} = -1;$$
$$\underline{e}_i(y_{ij}^{(3)}) = z_i y_j^2,\ e_i(y_{ij}^{(4)}) = z_i y_j^3,$$
$$\underline{e}_i(y_{ij}^{(2)}) = -z_i y_{ij}^{(4)} + 3 z_i y_j y_{ij}^{(3)} \text{ if } m_{ij} = 6 \text{ and } a_{ij} = -1.$$

In order to find the other rules of the Poisson structure we can use Theorem 12.1 iii) and its corollary, and apply the braid group which acts as a group of Poisson automorphisms or recall the formula $\phi D_b \phi^{-1}(a) = D_{\phi(b)}(a)$ when ϕ is any automorphism of U.

An important case is when α_i, α_j are two simple roots and $w \in W$ is such that $w(\alpha_i) = \alpha_j$ so that $T_w(E_i) = E_j$. Thus we have, taking $\phi = T_w$:

$$T_w(\underline{e}_i(a)) = \underline{e}_j(T_w(a)). \tag{15.1.2}$$

According to the discussion at the end §14 and the formula 15.1.1 we need to show that the vector field $\mu_*(\underline{e}_i)$ (transported by the isomorphism $\mu = Y^{-1}ZX$ from $\Omega = SpecB$ to H), equals $z_i f_i$. Here we recall how the Lie algebra $\mathfrak{g}$ has been identified to a space of vector fields on H (in view of the discussion given in 11.3 of the Poisson structure of H). The Lie algebra $\mathfrak{g}$ is a space of vector fields on G by adjoint action, the unramified covering $\sigma : H \to G^0$ allows us to pull back the elements of $\mathfrak{g}$ to vector fields on H. We consider thus f_i as a vector field on H and take the function z_i, we want to achieve one of our main results:

Theorem. $\mu_*(\underline{e}_i) = z_i f_i$.

In order to perform the computation of the vector field, it is enough to do it on functions coming from G and so we will apply it to the coordinates of the map $\sigma\mu = YZ^2X$.

The idea of the proof is that (as we have always been doing) we will perform this computation directly in the rank 2 case and then reduce to this case.

The general computational scheme comes from the following fact.

Consider a linear representation ϱ of G and the induced one on its Lie algebra $\mathfrak{g}$. In a given basis the entries of the matrix $\varrho(g)$ are functions on G.

Remark. *If $a \in \mathfrak{g}$ the entries of $[a, \varrho(g)]$ are the derivatives of the entries of $\varrho(g)$ according to the vector field a.*

We can now continue.

15.2 For SL_2 we have a fundamental weight ω and a simple root $\alpha_1 = \alpha = 2\omega$ setting $z = z_\alpha$ we have then, with the notations of 13.1 and 14.3 (dropping the index 1) the elements:

$$Y = \begin{vmatrix} 1 & 0 \\ y & 1 \end{vmatrix}, \quad Z = \begin{vmatrix} z_\omega & 0 \\ 0 & z_\omega^{-1} \end{vmatrix}, \quad Z^2 = \begin{vmatrix} z & 0 \\ 0 & z^{-1} \end{vmatrix}, \quad X = \begin{vmatrix} 1 & -x \\ 0 & 1 \end{vmatrix}$$

and their product YZ^2X:

$$A := \begin{vmatrix} z & -zx \\ zy & -zxy + z^{-1} \end{vmatrix}.$$

We apply the operator $\underline{e}$ to A and get by the formulas in 15.1:

$$\underline{e}(A) = \begin{vmatrix} z^2x & 0 \\ z^2xy + z^2 - 1 & -z^2x \end{vmatrix} = [ze_{21}, A].$$

(We have dropped the subscript 1 in all formulas.)

15.3 Next, we show that for $GL = SL_3$, again if $A = YZ^2X$, then

$$\underline{e}_1(A) = [z_1e_{21}, A].$$

We have by 14.5.2: $Y = \overline{Y}Y_1$, where $\overline{Y} = T_1(Y_{s_2s_1})$, and $X = T_{s_2s_1}(X_2)X_{s_2s_1} = X_1\overline{X}$, where $\overline{X} = X_{s_2s_1}$, since the longest element of W is $s_1s_2s_1 = s_2s_1s_2$. We also decompose $Z^2 = Z_1\overline{Z}$, where $\overline{Z}$ is the kernel of α_1 in T. Explicitly:

$$\begin{aligned}
&Y_1 = 1 + y_1e_{21}, \quad Y_2 = 1 + y_2e_{32};\\
&t_1 = -e_{12} + e_{21} + e_{33}, \quad t_2 = e_{11} - e_{23} + e_{32}, \quad t_0 := t_1t_2t_1 = e_{31} - e_{22} + e_{13};\\
&Y_{s_2s_1} = T_2(Y_1)Y_2 = 1 + y_2e_{32} + T_2(y_1)e_{31};\\
&\overline{Y} = 1 - T_1(y_2)e_{31} + y_2e_{32};\\
&Z_1 = \operatorname{diag}(z_{\omega_1}^2, z_{\omega_1}^{-2}, 1), \quad \overline{Z} = \operatorname{diag}(z_{\omega_2}^2, z_{\omega_2}^2, z_{\omega_2}^{-4});\\
&X_1 = t_0Y_2t_0 = 1 - x_1e_{12}, \quad X_2 = t_0Y_1t_0 = 1 - x_2e_{23};\\
&\overline{X} = T_2(X_1)X_2 = 1 + T_2(x_1)e_{13} + x_2e_{23}.
\end{aligned}$$

Using 15.1, we obtain:

$$\begin{aligned}
&\underline{e}_1(\overline{Y}) = -z_1y_2e_{31} = [z_1e_{21}, \overline{Y}];\\
&\underline{e}_1(\overline{X}) = z_1T_2(x_1)e_{23} = [z_1e_{21}, \overline{X}].\\
&\underline{e}_1(\overline{Z}) = 0 = [z_1e_{21}, \overline{Z}].
\end{aligned}$$

Finally since $A = \overline{Y}(Y_1 Z_1 X_1)\overline{ZX}$ and the derivations $\underline{e}_1$ and $z_1 ade_{21}$ coincide on all four factors by the above formulas and 15.2, we deduce that they coincide on A.

At this point, in order to complete the preliminary computations one should explicit the computations for the other rank 2 cases. We may use as models Sp_4 in its standard representation as 4×4 matrices, G_2 in its 7-dimensional representation and $SL(2) \times SL(2)$. Similar but lenghtier computations show that the formula holds for all these groups as well.

15.4 We can go back now to the proof of Theorem 15.1, i.e. the proof of the formula:

$$\underline{e}_i = z_i f_i$$

Proof. We assume now not to be in rank 2 and will reduce the general case to the rank 2 case discussed in the previous paragraph.

We need some preparatory steps.

1. Consider an element $z = z_\alpha \in Z_0^0$, for a derivation D we have $D(z^2) = 2zD(z)$ and so two derivations coinciding on the value z^2 coincide on Z_0^0. Therefore $\underline{e}_i$ and $z_i f_i$ coincide on Z_0^0 by the rank 2 calculations.

2. For each $\alpha \in R^+$ pick $w_\alpha \in W$ such that $\alpha = w_\alpha(\alpha_j)$, $\alpha_j \in \Pi$, and let

$$y_\alpha = T_{w_\alpha} y_j, \; x_\alpha = T_{w_\alpha} x_j.$$

By Corollary 13.4c, the elements x_α and y_α generate B over Z_0^0. It suffices thus to show that:

$$\underline{e}_i(y_\alpha) = z_i f_i(y_\alpha), \; \underline{e}_i(x_\alpha) = z_i f_i(x_\alpha), \; \alpha \in R^+. \tag{15.4.1}$$

We choose thus an α and propose to prove 15.4.1.

3. Given two non-proportional roots α and β, we denote by $R_{\alpha,\beta}$ the intersection of the $\mathbb{Z}$-span of α and β with R and let $R^+_{\alpha,\beta} = R_{\alpha,\beta} \cap R^+$. Then $R_{\alpha,\beta}$ is a rank 2 root system with $R^+_{\alpha,\beta}$ being a subset of positive roots. Notice that this system cannot be of type G_2 since it is contained in a larger irreducible root system. We apply these notations to the 2 roots α, α_i.

Lemma. *There exist two simple roots, say,* $\alpha_1, \alpha_2 \in \Pi$ *and* $w \in W$ *such that*

$$wR^+_{\alpha_1,\alpha_2} = R^+_{\alpha_i,\alpha} \text{ and } w\alpha_1 = \alpha_i. \tag{15.4.2}$$

Proof. Take the basis of simple roots in $R^+_{\alpha_i,\alpha}$ (where α_i is simple root and α is positive) and complete it to a basis of R, this is conjugate under W to the standard basis Π and in this way we perform the construction.

Consider the subalgebra $Z_0^{1,2}$ of B generated over Z_0^0 by all x_γ and y_γ with $\gamma \in R^+_{\alpha_1,\alpha_2}$. We want to prove the following formula

$$(z_i f_i)(T_w(a)) = T_w(z_1 f_1(a)) \text{ for } a \in Z_0^{1,2}. \tag{15.4.3}$$

Claim. *The previous formula (15.4.3) implies (15.4.1) (which proves the main formula).*

Proof. Indeed from Theorem 12.1 iii) we have, since $\alpha_i = w\alpha_1$:

$$\underline{e}_i(T_w(a)) = T_w \underline{e}_1(a).$$

Assume now that $a \in Z_0^{1,2}$. Using the calculation in the rank 2 case we have $\underline{e}_1(a) = z_1 f_1(a)$ and from 15.4.3:

$$T_w(\underline{e}_1(a)) = T_w(z_1 f_1(a)) = (z_i f_i)(T_w(a))$$

so $\underline{e}_i$ and $z_i f_i$ coincide on the elements $T_w(a)$, $a \in Z_0^{1,2}$ as desired.

So we are now going to prepare for the proof of 15.4.3.
The root system R_{α_1,α_2} can be of 3 types:

(15.4.3) $$A_1 \times A_1,\ A_2,\ B_2.$$

Let w_0' be the longest element of the Weyl group of R_{α_1,α_2} and let $m = \ell(w_0')$.
In the 3 cases $m = 2, 3, 4$ respectively and a reduced expression J' for w_0' is:

(15.4.4) $$s_1 s_2,\ s_1 s_2 s_1,\ s_1 s_2 s_1 s_2.$$

Lemma. $\ell(ww_0') = \ell(w) + \ell(w_0')$.

Proof. By construction w^{-1} sends the positive roots of R_{α_1,α_2} into positive roots and the reflections s_1, s_2 permute the positive roots not in R_{α_1,α_2} thus $R_{(ww_0')^{-1}} = R_{w^{-1}} \cup R^+_{\alpha_1,\alpha_2}$ and the claim follows from 8.2.

Fix a reduced expression $w = s_{i_1} \dots s_{i_k}$, $\ell(w) = k$. From the computation of the length the expression $ww_0' = s_{i_1} \dots s_{i_k} J'$ is reduced.

Complete this expression to a reduced expression J of $w_0 = ww_0'u$. Let $\{\beta_1, \dots, \beta_N\}$ be the convex ordering of R^+ associated to this reduced expression (see 8.2).

We distinguish 4 sets of roots and have:

(15.4.5)
$$R^1 := \{\beta_1, \dots, \beta_k\} = R_{w^{-1}},\ \beta_{k+1} = w(\alpha_1) = \alpha_i,$$
$$R^2 := \{\beta_{k+2}, \dots, \beta_{k+m}\} = w(R^+_{\alpha_1,\alpha_2} - \{\alpha_1\}) = R^+_{\alpha_i,\alpha} - \{\alpha_i\},$$
$$R^3 := \{\beta_{k+m+1}, \dots, \beta_N\} = ww_0'(R_{u^{-1}}).$$

Define:

(15.4.6) $$\mathfrak{g}^i_{\pm} = \bigoplus_{\gamma \in R^i} \mathbb{C}e_{\pm\gamma},\ i = 1, 2, 3$$

Lemma. *The subspaces $\mathfrak{g}^i_{\pm}$ are subalgebras of the Lie algebra $\mathfrak{g}$ normalized by the 3-dimensional subalgebra $\mathfrak{l} := \mathbb{C}e_i + \mathbb{C}h_i + \mathbb{C}f_i$.*

Proof. Using the Cartan involution, under which $\mathfrak{l}$ is stable, we are reduced to analyze the positive root case.

Given a set S of positive roots, the subspace $\mathfrak{g}_S = \bigoplus_{\gamma \in S} \mathbb{C}e_\gamma$ is a subalgebra if and only if S is closed under addition, i.e. if $\alpha, \beta \in S$, $\alpha + \beta \in R^+$ then $\alpha + \beta \in S$.

$\mathfrak{g}_S$ is normalized by $\mathfrak{l}$ if furthermore $S \pm \alpha_i \cap R \subset S$. Notice that, if for a positive root α we have that $\alpha \pm \alpha_i$ is a root, necessarily this root is positive.

1. $R^1 = R_{w^{-1}} = R^+ \cap w(R^-)$ is clearly closed under addition and $\mathfrak{g}^1_+$ is the Lie algebra of $w(U^-) \cap U^+$. It is normalized by $\mathfrak{l}$ since by the previous remark if $\alpha \in R^1$ and $\alpha \pm \alpha_i$ is a (positive) root, then $w^{-1}(\alpha)$ is negative and $w^{-1}(\alpha \pm \alpha_i) = w^{-1}(\alpha) \pm \alpha_1$ must be negative.

2. $R^2 = w(R^+_{\alpha_1,\alpha_2} - \{\alpha_1\})$ and $\alpha_i = w(\alpha_1)$ thus it suffices to verify that S is closed under addition and $S \pm \alpha_1 \cap R \subset S$ for $S = R^+_{\alpha_1,\alpha_2} - \{\alpha_1\}$ which is clear.

3. As in 2., since $R^3 = ww_0'(R_{a^{-1}})$ it is closed under addition and it suffices to show that $R_{a^{-1}} \pm (ww_0')^{-1}(\alpha_i) \cap R \subset R_{a^{-1}}$. The argument is the same as in 1. since $(ww_0')^{-1}(\alpha_i) = (w_0')^{-1}(\alpha_1)$ is the opposite of a simple root.

Let $U^i_\pm$ be the subgroups of $U_\pm$ corresponding to the algebras $\mathfrak{g}^i_\pm$.

We turn now to the map YZ^2X. We have according to its definition and Proposition 14.3 2), the decompositions of Y and X according to the above decomposition of w_0 and of R^+:

$$YZ^2X = Y_3Y_2(\exp y_i f_i)Y_1Z^2X_1(\exp x_i e_i)X_2X_3 = Y_3Y_2Y_1'((\exp y_i f_i)Z^2(\exp x_i e_i))X_1'X_2X_3,$$

where $Y_1' = (\exp y_1 f_1)Y_1(\exp -y_1 f_1) \subset U^1_-$ and $X_1' = (\exp x_1 e_1)X_1(\exp -x_1 e_1) \in U^1_+$.

In order to prove (15.4.3) note that the action of the vector field $z_i f_i$ on $\Omega = \mathrm{Spec}(B)$ may be calculated as follows.

We work in an open neighborhood A of a point $p \in \Omega$ which is mapped isomorphically under YZ^2X to an open set in G. Consider for $t \in \mathbb{C}$:

$$(\exp tz_i f_i)YZ^2X(\exp -tz_i f_i) = \prod_{s=N}^{1}(\exp y^J_s(t)f^J_s)Z^2(t)\prod_{s=1}^{N}(\exp x^J_s(t)e^J_s).$$

This formula expresses a map from $\mathbb{C} \times A \to G$. Fixing $p \in A$ it defines a complex curve $\mathbb{C} \to G$, its velocity at 0 is the tangent vector $z_i f_i$ computed in p. Then $f_i(x^J_s) = \frac{d}{dt}x^J_s(t)|_{t=0}$, and similarly for the y's.

Observe now that x_α (resp. y_α) occurs only in X_2 (resp. Y_2) and all other factors of YZ^2X lie in subgroups normalized by $\exp tz_i f_i$ and having trivial intersection with U^2_+ (resp. U^2_-). Thus, it suffices to perform the calculation in U^2_+ (resp. U^2_-). We have:

$$\prod_{s=k+2}^{k+m} \exp x^J_\beta(t)e^J_{\beta_s} = (\exp tz_i f_i)\prod_{s=2}^{m}\exp T_w(x^{J'}_s e^{J'}_s)(\exp -tz_i f_i)$$

$$= T_w((\exp tz_1 f_1)(\prod_{s=2}^{m}\exp x^{J'}_s e^{J'}_s)(\exp -tz_1 f_1)),$$

and we can use again the calculation in the rank 2 case. This proves (2). $\square$

15.5 Let

$$\underline{f}_i = T_0\underline{e}_i T_0^{-1}.$$

As in 2.3, we have

$$\underline{f}_i = T_i\underline{e}_i T_i^{-1}.$$

Since $T_i(E_i) = -F_iK_i$ we have that $\underline{f}_i$ is the Hamiltonian vector field given by the function $\frac{-y_i z_i}{d_i}$.

As in 15.4, it suffices to verify the following formula in the rank 1 and 2 cases where we verify it as in 15.3:

Theorem. $\mu_*(\underline{f}_i) = -z_i \dot{e}_i$. □

We still need to compare the Hamiltonian vector fields associated to z_i. From the very definition of the Poisson structure in B we have the formulas for Poisson bracket:

$$\{z_i, z_j\} = 0$$

$$\{z_i, x_\beta\} = \frac{(\alpha_i, \beta_i)}{2} z_i x_\beta$$

$$\{z_i, y_\beta\} = -\frac{(\alpha_i, \beta_i)}{2} z_i y_\beta$$

On the other hand on H we have computed the corresponding Hamiltonian field (14.6.3) to be $\frac{d_i}{2} z_i h_i$. Clearly this field annihilates the functions z_j. let us compute the field $d_i h_i$ on the functions x_β, y_β. By definition the exponential of $d_i h_i$ is the 1-parameter subgroup $\lambda : \mathbf{C}^* \to T$ such that $\alpha(\lambda(t) = t^{(\alpha, \alpha_i)}$ thus its action by conjugation and the formula of YZ^2X imply that $d_i h_i(x_\beta) = (\beta, \alpha_i) x_\beta$, $d_i h_i(y_\beta) = -(\beta, \alpha_i) y_\beta$. This ends the proof of Theorem 14.5.

§16 The geometry of the quantum coadjoint action.

16.1. We have now completely identified the Poisson group Ω with H and we will use these notations in an interchangeable way.

In particular we can consider T, $B^\pm$, $U^\pm$ as subgroups of Ω.

It is proved in [DC–K, §3.5] that the derivations $\underline{e}_i$ (and hence $\underline{f}_i$) of the algebra B integrate to global 1-parameter groups of analytic automorphisms $\exp s\underline{e}_i$ (resp. $\exp s\underline{f}_i$) of the algebraic variety Ω (and in fact for roots of 1 they extend to analytic automorphisms of the corresponding quantum group). Denote by $\tilde{G}$ the (infinite-dimensional) group generated by the groups $\exp se_i$ and $\exp sf_i$, $i = 1, \ldots, n$. The action of $\tilde{G}$ on Ω is called the quantum coadjoint action [DC–K].

Since we integrate vector fields which by Lie brackets generate fields whose value in each point span the tangent space to the corresponding symplectic leaf, Theorems 15.4 and 15.5 and Corollary 14.5 give us immediately

Proposition. *The symplectic leaves coincide with the orbits of the group* $\tilde{G}$.

b) The connected components of the variety $\pi^{-1}(\mathcal{O}^0)$ *are orbits of the group* $\tilde{G}$.

□

Definition. *Denote by F the set of 0 dimensional symplectic leaves, or equivalently fixed points of* $\tilde{G}$.

*A $\tilde{G}$ orbit in Ω is called **unipotent** if its Zariski closure contains a point from F. A $\tilde{G}$–orbit in Ω is called **semisimple** (resp. **regular semisimple**) if it intersects T (resp. T_{reg}).*

REMARK By the previous proposition it follows that F is the subgroup of T consisting of the elements with the square in the center of G.

Theorem. *(a) Every $\tilde{G}$–orbit in Ω contains an element in B^+.*

(b) Every $\tilde{G}$–orbit $\mathcal{O}$ in Ω is Zariski open in its Zariski closure $\overline{\mathcal{O}}$.

(c) A $\tilde{G}$–orbit is closed if and only if it is semisimple.

(d) The union of all regular semisimple orbits $\tilde{G} \cdot T_{\text{reg}}$ is Zariski open in Ω.

(e) There is only a finite number of unipotent orbits.

Proof. follows from Proposition 16.1 and the well–known results on conjugacy classes in simple Lie groups [K]. □

One can deduce from [B–C, Theorem 6.1] a more precise statement than Theorem 16.1a: Every $\tilde{G}$–orbit in Ω contains an element $p \in B^+$ such that:

(i) the set $R_p^+ := \{\beta_k^J \in R^+ | x_k^J(p) \neq 0\}$ is linearly independent (and then it is independent of J),

(iii) if $z_\alpha(p)^2 \neq 1$, then $\alpha \notin R_p^+$.

16.2. The orbits of the action of $\tilde{G}$ on $\Omega = H$ are described by the following

Theorem. *Consider the unramified cover*

$$\sigma : H \to G^0.$$

with Galois group the 2-torsion of T. Let $\mathcal{O}$ be a conjugacy class of a non–central element of G and let $\mathcal{O}_0 = \mathcal{O} \cap G$. Then $(\sigma)^{-1}(\mathcal{O}_0)$ is an orbit of $\tilde{G}$ in Ω.

Proof. We know (Proposition 16.1a) that connected components of $(\sigma)^{-1}(\mathcal{O}_0)$ are orbits. Hence it suffices to show that $(\sigma)^{-1}(\mathcal{O}_0)$ is connected. For that we have to show that the composite homomorphism $\varphi : \pi_1(\mathcal{O}_0) \to Q_2^\vee$ defined by

$$\pi_1(\mathcal{O}_0) \xrightarrow{i^*} \pi_1(G^0) = 2P^* \to Q_2^\vee$$

is surjective. We need the following lemma.

Lemma. *Consider the big cell in the group SL_2:*

$$V = \{\begin{pmatrix} 1 & 0 \\ y & 1 \end{pmatrix}\begin{pmatrix} z & 0 \\ 0 & z^{-1} \end{pmatrix}\begin{pmatrix} 1 & x \\ 0 & 1 \end{pmatrix} | x, y \in \mathbb{C}, z \in \mathbb{C}^\times\},$$

and let $a \in \mathbb{C}$. Let V_a denote the intersection with V of the conjugacy class of non central elements of SL_2 with trace a. Then the inclusion map induces a surjective homomorphism of fundamental groups $\pi_1(V_a) \to \pi_1(V)$.

Proof. The set V_a is a hypersurface in V given by the equation $z + z^{-1} + zxy = a$, with the point $(0, 0, \pm 1)$ deleted if $a = \pm 2$. Consider the m–th cover $\sigma_m :$ Spec $\mathbb{C}[x, y, t, t^{-1}] \to V =$ Spec $\mathbb{C}[x, y, z, z^{-1}]$ given by $(x, y, t) \longmapsto (x, y, t^m)$. Note that $\sigma_m^{-1}(V_a)$ has equation $t^m + t^{-m} + t^m xy = a$, which is irreducible in $\mathbb{C}[x, y, t, t^{-1}]$. Since $\pi_1(V) = \mathbb{Z}$, this proves the lemma. ☐

End of the Proof of Theorem 16.2. It suffices to consider two cases (since the closure of $\mathcal{O}$ contains a semisimple conjugacy class):

a) $\mathcal{O}$ is a conjugacy class of a non–central semisimple element $\exp 2\pi i h, h \in \mathfrak{h}$.

b) $\mathcal{O}$ is a conjugacy class of a non–trivial unipotent element.

In case a) there exists a long root α such that $(\alpha^\vee | h) \notin \mathbb{Z}$, where $\alpha^\vee = 2\alpha/(\alpha|\alpha)$. Let $\gamma_\alpha : SL_2 \to G'$ be the homomorphism corresponding to α. It is well-known that γ_α is injective (see e.g. [K–W, Proposition 2.1]). Then $\mathcal{O} \cap \gamma_2(SL_2)$ is a non-central conjugacy class in SL_2. Due to the lemma, this implies that the image of i^* contains $\alpha^\vee$. Using the Weyl group, we see that the image of i^* contains $\alpha^\vee$ for each long root α, hence contains $Q^\vee$, and hence the map φ is surjective. In case b), it suffices to look at $\mathcal{O} = G' \cdot \exp e_\alpha$, where e_α is a root vector attached to a long root α, since it is well–known that the closure of any non–trivial unipotent conjugacy class contains this one. The same argument as that in case a) now completes the proof. ☐

Let us now consider the subgroup Γ of elements of order 2 in T, under the square map two elements of T go to the same element of G if and only if they are in the same coset modulo Γ. We can thus consider the group $\tilde{W}$ of transformations of T generated by translations by Γ and the action of the Weyl group. Of course $\tilde{W} = W \ltimes \Gamma$ and we have:

Corollary. *(a) Two elements of $T\backslash F$ lie in the same orbit of $\tilde{G}$ if and only if they lie in the same orbit of $\tilde{W}$.*

(b) The closure of a $\tilde{G}$-orbit $\mathcal{O}$ of a non-unipotent point in Ω contains a unique closed $\tilde{G}$-orbit, which is $\mathcal{O}_s$. □

16.3. In this section we study the invariants of the action of $\tilde{G}$ on Ω. Let us recall some elementary facts on the quotient map of G under the adjoint action. In order to avoid confusions let us indicate by T the subgroup of H, its image in G under the map σ is $T/\Gamma := \overline{T}$ a maximal torus in G.

Since we assume that G is simply connected the algebra of functions on G invariant under conjugation is a polynomial algebra isomorphic, by the Chevalley restriction, to the algebra of functions on the maximal torus $\overline{T}$ which are W invariant, in other words to the functions on T which are $\tilde{W} = W \ltimes \Gamma$ invariant and the quotient variety is identified to $T/\tilde{W} = \overline{T}/W$. Consider the composition of σ with the quotient map. It is a map of H to $\overline{T}/W$.

(16.3.1.) $$\pi : H \to \overline{T}/W$$

Let us recall that in G the *regular* elements G_{reg} are the ones whose conjugacy classes have maximal dimension. Let us set:

The quotient map restricted to G_{reg} is a fibration onto $\overline{T}/W$, its fibers are the regular orbits and the complement of G_{reg} is of codimension ≥ 2.

We can similarly define the *regular* symplectic leaves as the leaves of maximal dimension and set:
$$H_{reg} := \text{The union of the regular symplectic leaves.}$$

Theorem. *(a) $H_{reg} = \sigma^{-1}(G_{reg})$.*

(b) The map π restricted to H_{reg} is a fibration onto $\overline{T}/W$, its fibers are the regular symplectic leaves and the complement of H_{reg} is of codimension ≥ 2.

(c) Under the map $\sigma : H \to G$ the ring of conjugation invariant functions $\mathbb{C}[G]^G$ on G is identified to $B^{\tilde{G}}$.

(d) The restriction homomorphism induced by inclusion $i : T \to H$ gives an isomorphism of algebras of invariants:
$$i^* : B^{\tilde{G}} \xrightarrow{\sim} Z_0^{0\tilde{W}}.$$

Proof. (a) and (b) are an immediate consequence of 16.1 and 16.2 and the analogous statements for G_{reg}. Parts (c) and (d) follow from the fact that the symplectic leaves are the $\tilde{G}$ orbits and that the complement of H_{reg} is of codimension ≥ 2.
□

16.4 In this section we want to develop some geometry also for the group B^+ which, as we will see in a moment can be thought as a quotient Poisson group of H.

Consider the quotient:
$$B_+ = H/U^-.$$

Applying the construction given by Lemma 4.2 to $\mathfrak{u}_-\}$ we see that B^+ acquires a structure of Poisson group with Manin triple deduced from the one of H $(\mathfrak{g} \oplus \mathfrak{g}, \mathfrak{h}, \mathfrak{k})$.

The triple we have is $(\mathfrak{g} \oplus \mathfrak{t}, \mathfrak{b}_+, \mathfrak{b}_-)$, where we used identifications
$$\mathfrak{b}_\pm = \{(u_\pm - t, \pm t) | u_\pm \in \mathfrak{u}_\pm, t \in \mathfrak{t}\}.$$

According to Proposition 11.5, the symplectic leaves of the Poisson group B_+ are obtained as follows. We identify the groups $B_\pm$ with the following subgroups of $G \times T$:

$$B_\pm = \{(t^{-1}u_\pm, t^{\pm 1}) | t \in T, u_\pm \in U_\pm\}.$$

The inclusion $B_+ \subset G \times T$ induces an etale morphism

$$\delta : B_+ \to (G \times T)/B_-.$$

Then the symplectic leaves of B_+ are the connected components of the preimages under the map δ of B_-–orbits on $G \times T/B_-$ under the left multiplication.

In order to analyze the B_-–orbits on $G \times T/B_-$, let $\mu_\pm : B_\pm \to T$ denote the canonical homomorphisms with kernels $U_\pm$ and consider the equivariant isomorphism of B_-–varieties $\gamma: G/U^- \to (G \times T)/B_-$ given by $\gamma(gU_-) = (g, 1)B_-$, where B_- acts on G/U^- by

$$b(gU^-) = bg\mu_-(b)U^-. \tag{16.4.1}$$

Then the map δ gets identified with the map $\delta : B_+ \to G/U^-$ given by

$$\delta(b) = b\mu_+(b)U^-.$$

We want to study the orbits of the action of B_- on G/U^-. Consider the action of B_- on G/B_- by left multiplication. Then the canonical map $\pi : G/U^- \to G/B_-$ is B_-–equivariant, hence π maps every B_-–orbit $\mathcal{O}$ in G/U^- to a B_-–orbit in G/B_-, i.e. a Schubert cell $C_w = B_-wB_-/B_-$ for some $w \in W$. We shall say that the orbit $\mathcal{O}$ is associated to w.

REMARK. We have a sequence of maps:

$$B_+ \xrightarrow{\delta} (G \times T)/B_- \xrightarrow[\sim]{\gamma^{-1}} G/U^- \xrightarrow{\pi} G/B_+.$$

Let $\psi = \pi \circ \gamma^{-1} \circ \delta$ and $X_w = B_+ \cap B_-wB_-$. Then:

$$\pi^{-1}(C_w) = B_-wB_-/U^- \text{ and } \psi^{-1}(C_w) = X_w.$$

We can prove now the following

Proposition. *Let $\mathcal{O}$ be a B_-–orbit in G/U^- under the action (16.4.1) associated to $w \in W$. Then the morphism:*

$$\pi|_\mathcal{O} : \mathcal{O} \to C_w$$

is a principal torus bundle with structure group:

$$T^w := \{w^{-1}(t)t^{-1}, \text{ where } t \in T\}.$$

In particular:

$$\dim \mathcal{O} = \dim C_w + \dim T^w = l(w) + \operatorname{rank}(I - w).$$

Proof. For $g \in G$ we shall write $[g]$ for the coset gU^-. The morphism π is clearly a principal T–bundle with T acting on the right by $[g]t := [gt]$. The action (16.4.1) of B_- commutes with the right T–action so that T permutes the B_-–orbits. Each B_-–orbit is a principal bundle whose structure group is the subtorus of T which stabilizes the orbit. This subtorus is independent of the orbit since T is commutative. In order to compute it we proceed as follows. Let $[g_1], [g_2]$ be two elements in $\mathcal{O}$ mapping to $w \in C_w$. We may assume that $g_1 = nh$, $g_2 = nk$ with $h, k \in T$ uniquely determined, where $n \in N_G(T)$ is representative of w. Suppose that $b[nh] = b[nk]$, $b \in B_-$. We can first reduce to the case $b = t \in T$; indeed, writing $b = ut$ we see that u must fix $w \in C_w$ hence $un = nu'$

with $u' \in U^-$ and hence u acts trivially on $t[nh]$. Next we have that, by definition of the T-action,

$$[nk] = [tnht^{-1}] = [n(n^{-1}tnht^{-1})]$$

hence $k = n^{-1}tnht^{-1}$ or $k = h(h^{-1}n^{-1}tnht^{-1}) = h(n^{-1}tnt^{-1})$ as required. □

Lemma. *Let $\mathcal{O} \subset B_+$ be a symplectic leaf associated to w. Then $\mathcal{O}T = X_w$.*

Proof. From our proof we know that the map δ is a principal T-bundle and T permutes transitively the leaves lying over C_w. □

We thus have a canonical stratification of B_+, indexed by the Weyl group, by the subsets X_w. Each such subset is a union of leaves permuted transitively by the right multiplications of the group T.

We say that a point $a \in \operatorname{Spec} Z_0^+ = B_+$ lies over w if $\psi(a) \in C_w$.

16.5 We make a final remark which will be later useful for the representation theory at roots of 1. Let $B^w := B_+ \cap wB_-w^{-1}$ and $U^w := U_+ \cap wU^-w^{-1}$ so that $B^w = U^wT$. Set also $U_w := U_+ \cap wU_+w^{-1}$. One knows that $\dim B^w = n + l(w)$ and that the multiplication map:

$$\sigma : U_w \times B^w \to B_+$$

is an isomorphism of algebraic varieties. We define the map

$$p_w : B_+ \to B^w$$

to be the inverse of σ followed by the projection on the second factor.

Proposition. *The map*

$$p_w|_{X_w} : X_w \to B^w$$

is birational.

Proof. We need to exhibit a Zariski open set $\Omega \subset B^w$ such that for any $b \in \Omega$ there is a unique $u \in U_w$ with $ub \in X_w$.

Let $n \in N_G(T)$ be as above a representative for w so that:

$$X_w = \{b \in B_+ | b = b_1nb_2, \text{ where } b_1, b_2 \in B_-\}.$$

Consider the Bruhat cell $B_-n^{-1}B_- \subset G$. Every element in $B_-n^{-1}B_-$ can be written uniquely in the form:

$$bn^{-1}u, \text{ where } b \in n^{-1}B^wn,\ u \in U^-.$$

The set $B_+U^- = B_+B_-$ is open dense and so it intersects $B_-n^{-1}B_- = n^{-1}B^wU^-$ in a non-empty open set which is clearly B_--stable for the right multiplication, hence $B_+B_- \cap B_-n^{-1}B_- = n^{-1}\Omega U^-$ for some non empty open set $\Omega \subset B^w$. In particular $\Omega \subset nB_+B_- = nU_wB^wU^-$. Take $b \in \Omega$ and write it as $b = nxcv$ with $x \in U_w$, $c \in B^w$, $v \in U^-$. By the remarks made above this decomposition is unique; furthermore, $nxn^{-1} \in U_w$, $ncn^{-1} \in B_-$. For the element $n^{-1}b = xcv$ we have by construction that $xcv \in B_-n^{-1}B_-$ and $nx^{-1}n^{-1}b = (ncn^{-1})nv \in B_-nB_-$ and $nx^{-1}n^{-1} \in U_w$. Thus setting $u := nx^{-1}n^{-1}$ we have found $u \in U_w$ such that $ub \in X_w$. This u is unique since the element x is unique. □

§17 Verma modules.

17.1 We use the decomposition $U = U^- \otimes U^0 \otimes U^+$ to define as for the Lie algebra case the Verma modules. Let $\mathcal{B} = U^0 \otimes U^+$, we can define a 1 dimensional representation Λ for such an algebra, associated to a multiplicative character of M by setting $\Lambda(E_i) = 0, \Lambda(K_\alpha) = \Lambda(\alpha)$. We call v_Λ a generating vector for this representation which we call $k(q)(\Lambda)$ and construct:

Definition. *i)* $V(\Lambda) := U \otimes_{\mathcal{B}} k(q)(\Lambda)$, *the induced module, is called the* **Verma module** *of highest weight* Λ.
ii) $V(\Lambda)$ *is a cyclic module, generated by* $1 \otimes v_\Lambda$ *we let* I_Λ *denote the left ideal annihilating this generator so that* $V(\Lambda) = U/I_\Lambda$.

By abuse of notations we call $v_\Lambda = 1 \otimes v_\Lambda$. Notice that I_Λ is generated, as left ideal, by the elements $E_i, K_\alpha - \Lambda(K_\alpha)$. The basis theorem of 9.3 shows that the vectors

$$F_{\beta_N}^{h_N} \cdots F_{\beta_1}^{h_1} v_\Lambda \tag{17.1.1}$$

are a $k(q)$ basis of $V(\Lambda)$. These vectors are weight vectors for the elements K_α in fact $K_\alpha F_{\beta_N}^{h_N} \cdots F_{\beta_1}^{h_1} v_\Lambda = K_\alpha F_{\beta_N}^{h_N} \cdots F_{\beta_1}^{h_1} K_\alpha^{-1} K_\alpha v_\Lambda = q^{-\sum h_i(\alpha,\beta_i)} < \Lambda|\alpha >> F_{\beta_N}^{h_N} \cdots F_{\beta_1}^{h_1} v_\Lambda$. In particular assume that $\Lambda(\alpha) = q^{<\lambda|\alpha>}$ for an additive character α then the weight of the previous vector is

$$q^{(\lambda - \sum h_i\beta_i|\alpha)}.$$

Dropping the value of α we say that we have a vector of weight $q^{\lambda - \sum h_i\beta_i}$. Thus if μ is in the root lattice the weight $q^{\lambda-\mu}$ appears thus with a multiplicity given by the so called *Kostant partition function* $P(\mu)$ *of* μ i.e.:

$P(\mu)$ equals the number of ways of expressing μ as a positive combination of positive roots.

In this case we will also indicate by $V(\lambda), v_\lambda$ the Verma module and its generating vector.

In particular notice:

Proposition. v_λ *is the unique vector of weight* λ *(the highest weight vector). We denote by* π *the* K_α *equivariant projection on the 1 dimensional space generated by* v_λ.

Notice that, according to the general remark in 9.1 we can tensor any Verma module by a character δ and consider also the Verma modules $V(\Lambda)^\delta$.

The important connection between Verma modules and irreducible representations is obtained by the following:

Theorem. $V(\Lambda)$ *is indecomposable, it contains a unique proper maximal submodule* N_Λ *and so has a unique irreducible quotient* $L_\Lambda := V(\Lambda)/N_\Lambda$.

Proof. We have seen that $V(\Lambda)$ decomposes into finite dimensional weight spaces relative to the K_α. Thus any submodule is also a sum of weight spaces. Clearly then the submodule is proper if and only if it does not contain the vector v_Λ, the unique vector of weight Λ. Thus the sum of all proper submodules is contained in the span of the weight vectors of weight different than Λ (the kernel of the projection π) and thus it is a proper submodule; the rest follows.

17.2 There is a theory of Verma modules which is quite parallel to the classical theory for semisimple Lie algebras. We refer to [L1] for the details and give here a few indications of the main results.

First of all a Verma module is a free module of rank 1 over U_- which is a domain. It follows that a morphism between Verma modules is either 0 or injective.

In order to formulate the next proposition it is convenient to give a name to certain special sets of weights. Given $\beta \in R_+$, $r \in \mathbb{N}$ we set:

$$T_{r\beta} := \{\lambda \in P | 2(\lambda + \varrho|\beta) = r(\beta|\beta)\},$$
$$T^0_{r\beta} := \{\lambda \in T_{r\beta} | 2(\lambda + \varrho|\gamma) \neq m(\gamma|\gamma)$$
$$\text{for all } \gamma \notin R^+ - \beta,\ m \in \mathbb{N}, \text{ and } m\gamma < r\beta\}.$$

Proposition. *If $\lambda \in T^0_{r\beta}$ then $V(\lambda)$ contains a submodule isomorphic to $V(\lambda - r\beta)$.*

The second important result is connected to the irreducible module L_λ quotient of $V(\lambda)$. For every weight $\eta \in P$ let us denote by m_η the dimension of the weight space for the K_α of weight q^η (i.e. the space of vectors v for which $K_\alpha(v) = q^{(\alpha|\eta)}v$ then m_η equals the dimension of the ordinary weight space for the corresponding semisimple Lie algebra and Cartan subalgebra.

17.3 To simplify our notations we will indicate by $F^{\underline{m}}$ the element $F^{h_N}_{\beta_N} \cdots F^{h_1}_{\beta_1}$, where $\underline{m}$ denotes the vector of coordinates h_i. An important tool in the study of Verma modules is the *contravariant form.* Recall that U possesses a canonical anti-automorphism κ defined in §9.2 antilinear relative to the automorphism $^-$ of $\mathbb{C}(q)$ mapping $q \to q^{-1}$.

We decompose U as $U^0 \oplus U'$ where by definition U' is the span of all monomials $F^{\underline{m}_1} K^{\underline{m}_2} E^{\underline{m}_3}$ with $\underline{m}_1$ or $\underline{m}_3$ different from 0 , and consider the projection:

(17.3.1)
$$h : U \to U^0$$

defined by this decomposition. We can use this map to define a U^0 valued κ-Hermitian form on U by the formula:

$$\text{Universal contravariant form} \qquad < a, b > := h(\kappa(a)b).$$

We use now this form on U to define a form on each Verma module.

Lemma. *Given a Verma module V_Λ there exists a unique form (the contravariant form) $< u, v >$ satisfying the following properties:*
i) It is an Hermitian form (rel to $^-$).
ii) $< au, v > = < u, \kappa(a)v >$, $\forall a \in U,\ u, v \in V(\Lambda)$.
iii) $< v_\Lambda, v_\Lambda > = 1$.

Proof. By property ii) it follows that two eigenvectors relative to the K_α and with distinct eigenvalues, must be orthogonal. To define the form it is clearly enough to do it on vectors of type $F^{\underline{m}} v_\Lambda$ and by ii) we must have

$$< F^{\underline{m}} v_\Lambda, F^{\underline{m}'} v_\Lambda > = < v_\Lambda, \kappa(F^{\underline{m}}) F^{\underline{m}} v_\Lambda > = < v_\Lambda, h(\kappa(F^{\underline{m}}) F^{\underline{m}}) v_\Lambda >,$$

by the orthogonality of weight vectors of different weight and the fact that v_Λ is a highest weight vector.

For the existence we start from the universal contravariant form on U and, applying Λ deduce on U a $\mathbb{C}(q)$ valued Hermitian form by the formula $< a, b >_\Lambda := \Lambda(h(\kappa(a)b))$. It is then enough to prove that the left ideal I_Λ is in the kernel of this form. It is clearly enough to show that the generators are in the kernel, by the very definition of the form, for E_i this is clear since $UE_i \subset U'$. For $K_i - \Lambda(K_i)$ we have $U(K_i - \Lambda(K_i) \subset U^0(K_i - \Lambda(K_i)) + U'$, and $\Lambda(h(U^0(K_i - \Lambda(K_i)) + U') = \Lambda(U^0(K_i - \Lambda(K_i))) = 0$ as desired.

An important application of the contravariant form lies in its relation with the module theory developed in 17.3:

Theorem. *The kernel J of the contravariant form coincides with the maximal proper submodule N_Λ.*

Proof. By the adjonction properties of the contravariant form it is clear that the kernel N is a submodule, thus it is enough to show that the quotient $V(\Lambda)/N$ is irreducible. Now on $V(\Lambda)/N$ we have an induced non degenerate contravariant form. If we fix a proper submodule M this is a direct sum of its finite dimensional weight spaces and thus its orthogonal $M^\perp$ with respect to the form is also a complement, i.e. $V(\Lambda)/N = M \oplus M^\perp$. Clearly, by the properties of the form, $M^\perp$ is also a submodule and this contradicts Proposition 17.1.

17.4 A very important tool in the theory of Verma modules is the quantum analogue of Shapovalov's determinant formula [Sh]. Let us restrict now to the case $\Lambda = q^\lambda$, choose μ in the root lattice and consider the subspace $V(\lambda)_\mu \subset V(\lambda)$ relative to the weight $q^{\lambda-\mu}$. We have remarked that $V(\lambda)_\mu$ has dimension the partition function $P(\mu)$ and a basis given by the vectors $F_{\beta_N}^{h_N} \cdots F_{\beta_1}^{h_1} v_\lambda$ with $\lambda - \sum h_i\beta_i = \mu$. One can thus compute the matrix of the contravariant form with respect to this basis, it has as entries the elements:

$$\Lambda(h(\kappa(F^{\underline{m}})F^{\underline{m}'})).$$

Its determinant is

$$\Lambda(\det(h(\kappa(F^{\underline{m}})F^{\underline{m}'}))).$$

Our aim is to compute thus the determinant det_μ of the matrix with entries $h(\kappa(F^{\underline{m}})F^{\underline{m}'})$, where $\underline{m}, \underline{m}'$ run over all the $P(\mu)$ vectors $h_N, \dots, h_1$ giving the weight $\mu = \sum h_i\beta_i$, this is the content of Shapovalov's formula. A priori the entries of this matrix are in U^0 but in fact, due to the special nature of commutation relations between the E_i, F_j we can even say that they lie in the subalgebra generated by the K_i over the ring $k[q, q^{-1}, (q_i - q_i^{-1})^{-1}]$.

In order to formulate the theorem it is useful to introduce a notation:

$$[K_i; n] := (K_i q^n - K_i^{-1} q^{-n})/(q_i - q_i^{-1}).$$

We can now formulate:

Theorem.

$$det_\mu = \prod_{\beta \in R+} \prod_{m \in \mathbb{N}} ([m]_{d_\beta} [K_\beta; (\varrho|\beta) - \frac{m}{2}(\beta|\beta)])^{P(\mu - m\beta)}.$$

The possibility to compute this determinant is a consequence of the representation theory previously developed. In fact, for a given λ, this determinant will be 0 if and only if the given weight subspace has a non zero intersection with the kernel of the contravariant form, which is also the maximal submodule.

We have seen that if $\lambda \in T^0_{r\beta}$ the Verma module $V(\lambda)^\delta$ contains a copy of the Verma module $V(\lambda - r\beta)^\delta$. This implies that det_μ is divisible by the product appearing in the formula.

Next one completes the argument by a degree consideration and an evaluation of a highest term, in order to compute the explicit constants. We refer to [D-K] for further details.

§18 The center.

18.1 We discuss now an analogous of the Harish-Chandra homomorphism, with which we will describe the center of U for q generic. Consider first of all the action by conjugation with respect to elements K_α. An element of U of the form $F^k \varphi E^r$ with $k \in Par(\eta), r \in$

$Par(\gamma)$, $\varphi \in U^0$ is a weight vector under conjugation by K_α of weight $q^{(\alpha,\eta-\gamma)}$ therefore (using the P.B.W. theorem) we have that any element of the center of U is of the form:

$$z = \sum_{\eta\in Q_+} \sum_{k,r\in Par(\eta)} F^k \varphi_{k,r} E^r, \quad \text{where } \varphi_{k,r} \in U^0. \tag{18.1.1}$$

In order to have more information we need:

Lemma. *An element $u \in U$ is in the center if and only if it acts as a scalar on any Verma module.*

Proof. An element z of the type 18.1.1 acts on the highest weight vector of a Verma module by the action of $\varphi_{0,0}$ which is a scalar. If z is in the center it commutes with all the F_i and so it acts by the same scalar on the whole Verma module. Conversely assume that an element u acts as scalar on all Verma modules, then for any element $v \in U$ we have that $[u, v]$ acts as 0 on all Verma modules. So we need to show that an element x which is 0 on all Verma modules is necessarily 0. To see this write x in the P.B.W. basis as

$$x = \sum_{\eta,\mu\in Q^+} \sum_{k\in Par(\eta), r\in Par(\mu)} F^k \varphi_{k,r} E^r, \quad \varphi_{k,r} \in U^0.$$

Set, for $\mu \in Q^+$,

$$x_\mu = \sum_{\eta\in Q^+} \sum_{k\in Par(\eta), r\in Par(\mu)} F^k \varphi_{k,r} E^r.$$

We shall show that $x_\mu = 0$ for each μ. Assume that we have already proven $x_\nu = 0$ for $\nu < \mu$. Then it follows that $x_\mu V(\lambda)_\mu = 0$ for any Verma module $V(\lambda)$.
Write

$$x_\mu = \sum_r F^r x_\mu^{(r)}, \quad x_\mu^{(r)} \in U^{\geq 0}.$$

Since $V(\lambda)$ is a free $U^{<0}$-module we also get that $x_\mu^{(r)} V(\lambda)_\mu = 0$. If we assume that $x_\mu^{(r)} \neq 0$ then we can clearly choose λ in such a way that the element $\kappa(x_\mu^{(r)})v_\lambda$ is a non zero element in $V(\lambda)_\mu$, and that the contravariant form is non degenerate on $V(\lambda)_\mu$. Thus there exists a vector $v \in V(\lambda)_\mu$ such that

$$0 \neq < v, \kappa(x_\mu^{(r)})v_\lambda > = < x_\mu^{(r)} v, v_\lambda > .$$

contradicting the fact that $x_\mu^{(r)}$ kills $V(\lambda)_\mu$. So $x_\mu^{(r)} = 0$ for each r and hence $x_\mu = 0$.

Take now an element z as in 18.1.1, assume it is in the center, we want to show first that the element $\varphi_{0,0}$ determines all the remaining terms, we will work by induction on the height of η. For this let us arrange, for fixed η, the elements $\varphi_{k,r}$, $k, r \in Par(\eta)$ in a matrix which we will call ϕ_η when we evaluate it for a weight q^λ we will call its value $\phi_\eta(\lambda)$. Consider for each γ the element $z_\gamma := \sum_{k,r\in Par(\gamma)} F^k \varphi_{k,r} E^r$ which gives an operator on each space $V(\lambda)_\eta$, and $z = \sum_{\gamma\in Q_+} z_\gamma$, on $V(\lambda)_\eta$ the only operators non zero are the z_γ, $\gamma \leq \eta$. Denote by $H_\eta(\lambda)$ the matrix of the contravariant form on $V(\lambda)_\eta$.

Lemma. *The matrix of z_η on $V(\lambda)_\eta$ is:*

$$G_\eta := \phi_\eta(\lambda) H_\eta(\lambda).$$

Proof. $F^k \varphi_{k,r} E^r (F^h v_\lambda) = F^k \varphi_{k,r} h_{r,h}(\lambda) v_\lambda = \varphi_{k,r}(\lambda) h_{r,h} F^k v_\lambda$.

Since z is assumed to act as the scalar $\varphi_{0,0}(\lambda)$ we have:

$$\phi_\eta(\lambda)H_\eta(\lambda) + \sum_{\gamma<\eta} G_\gamma(\lambda) = \varphi_{0,0}(\lambda)I. \tag{18.1.2}$$

The equality for every λ implies an equality as elements in $U_{0,0}$ and, since by the corollary to Shapolavov's formula one has that the matrix H_η is non singular, one obtains a recursive formula for the matrices ϕ_η:

$$\phi_\eta H_\eta = -\sum_{\gamma<\eta} G_\gamma + \varphi_{0,0}I. \tag{18.1.3}$$

which implies that the element z is determined by $\varphi_{0,0}$, of course in the computation of these matrices one could obtain the determinant of the contravariant form in the denominator and also one might have an infinite series, formally we will get an associated element z which need not lie in U. Thus in order to determine the center one has to understand for which values of $\varphi_{0,0}$ one has no denominators and a finite sum.

18.2 In order to continue, define a homomorphism:

$$\gamma : K_\lambda \to q^{(\varrho|\lambda)}K_\lambda. \tag{18.2.1}$$

Let Γ be the group of order 2^n acting on P by sign exchanges on the basis ω_i and having as fixed subgroup $2P$. Let $\tilde{W}$ be the semidirect product $W \ltimes \Gamma$ whose invariants on $k[P]$ coincide with the invariants $k[2P]^W$. Recall the projection $h : U \to U^0$ taking the term $\varphi_{0,0}$ in its development.

We have:

Lemma. *$\gamma^{-1}(\varphi_{0,0}) \in (U^0)^{\tilde{W}}$ if and only if, for every character δ of $P/2Q_+$, for every $\beta \in R^+$, $r \in \mathbb{N}$ and $\lambda \in T_{r\beta}$ we have $\varphi^\delta_{0,0}(\lambda) = \varphi^\delta_{0,0}(\lambda - r\beta)$.*

Proof. If $\gamma^{-1}(\varphi_{0,0}) \in (U^0)^{\tilde{W}}$, since $s_\beta(\lambda + \varrho) = \lambda + \varrho - r\beta$ if $\lambda \in T_{r\beta}$ we have $\varphi^\delta_{0,0}(\lambda) = \varphi^\delta_{0,0}(\lambda - r\beta)$. On the other and if $\varphi^\delta_{0,0}(\lambda) = \varphi^\delta_{0,0}(\lambda - r\beta)$ for every $\lambda \in T_{r\beta}$ and for every δ, we clearly get that $\gamma^{-1}(\varphi_{0,0})$ is invariant under the action of $P/2Q_+$ and of the s_β's. Since these generate $\tilde{W}$, $\gamma^{-1}(\varphi_{0,0}) \in (U^0)^{\tilde{W}}$.

18.3 We need first a geometric lemma.

Lemma. *Let X be a nonsingular algebraic variety over a field of characteristic zero. Let V and W be two vector bundles of rank n over X.*

a) Given a morphism $\phi : V \to W$ of vector bundles such that $det\phi = \wedge^n\phi \neq 0$. Let S be the divisor in X defined by $det\phi = 0$, p a smooth point in the support of S. Then if $det\phi$ vanishes with multiplicity s on the unique irreducible component Y of $SuppD$ containing p, $n - 1 \geq rank\phi(p) \geq n - s$.

b) Suppose that there exists a Zariski open set $\mathcal{V} \subset X$ with $codX - \mathcal{V} \geq 2$, and such that $X - \mathcal{V}$ contains the singular points in $SuppS$ with the property that if $p \in \mathcal{V}$, denoting by s the order of vanishing of $det\phi$ in p then $rank\phi(p) = n - s$. Then, for all morphisms of vector bundles $\psi : V \to W$ such that $ker(\phi(p)) \subset ker(\psi(p))$, for all $p \in \mathcal{V}$, $\psi\phi^{-1}$ can be uniquely extended to a endomorphism of W defined on X.

Proof. a) Our claim being local we can assume that V and W are trivial and that $X = Spec(k[[x_1, \dots, x_r]])$ with Y defined by $x_1 = 0$ and p by $x_1 = \dots = x_r = 0$. We can also think ϕ represented by an $n \times n$ matrix H with coefficients in $k[[x_1, \dots, x_r]]$ and our hypotheses implies that $\det H = x_1^s P(x_1, \ldots, x_n)$, where $P(x_1, \ldots, x_n)$ is a non zero element in $k[[x_1, \ldots, x_r]]$ with $P(0, \ldots, x_n) \neq 0$.

Let us denote by H_i the i-th column of H. We shall think $\det H$ as $H_1 \wedge \cdots \wedge H_n$. We set $D = \frac{\partial}{\partial x_1}$. Thus applying D^s, we get

$$\text{(18.3.1)} \qquad \sum_{i_1+\cdots+i_n=s} D^{i_1} H_1 \wedge \cdots \wedge D^{i_n} H_n = s!P + x_1 Q$$

Since the right handside of (18.3.1) is non zero in p, it follows that there exists a sequence $I_1, \ldots, i_n$ with $I_1 + \cdots + i_n$ such that the vectors $\{D^{i_s} H_s\}_{s=1}^n$ are linearly independent in p. From this we immediately get that at least $n - s$ among the vectors $H_1 \cdots H_n$ are linearly independent in p. The inequality $\mathrm{rank} H(p) \leq n-1$ being clear everything follows.

b) $\psi\phi^{-1}$ is clearly well defined on $X - Supp(S)$. Since $\mathrm{cod} X - \mathcal{V} \geq 2$ in order to prove our claim it clearly suffices to see that it can be extended to the generic points of each of the irreducible components of $Supp S$. Fix such a component Y. Let $\mathcal{R}_Y$ be the local ring at the generic point of Y. Restricting near that point we can think ϕ and ψ as $n \times n$ matrices with coefficients in $\mathcal{R}_Y$. Call them H and M. Since $\mathcal{R}_Y$ is a discrete valuation ring, by the theory of elementary divisors and our hypotheses there exist two matrices in $Gl(n, \mathcal{R}_Y)$, N_1 and N_2 such that

$$N_1 H N_2 = \begin{pmatrix} I_{n-s} & 0 \\ 0 & tI_s \end{pmatrix},$$

where t is a generator of the maximal ideal in $\mathcal{R}_Y$ and I_h denotes the $h \times h$ identity matrix. Also by our hypotheses the matrix $N_1 H' N_2$ has all the entries in the last s columns lying in the maximal ideal of $\mathcal{R}_Y$. It follows that the matrix $N_1 H' H^{-1} N_1^{-1}$ has coefficients in $\mathcal{R}_Y$. Hence the same holds for $H' H^{-1}$ proving our claim.

Using the map h defined in 17.3.1 we can state:

Theorem. *1) Given $\varphi_{0,0} \in U^0$ the coefficients of the corresponding series z lie in U^0 if and only if:*

$$\gamma^{-1}(\varphi_{0,0}) \in (U^0)^{\tilde{W}}.$$

2) The map:

$$\gamma^{-1} \circ h : Z \to (U^0)^{\tilde{W}}$$

is an isomorphism between the center of U and the ring $k[2P]^W$.

Proof. Suppose given a central element z such that $\varphi_{00} = h(z)$. We have seen that on any Verma module $V(\lambda)$, z acts as the scalar $\varphi_{00}(\lambda)$.

If $\lambda \in T^0_{r\beta}$ then $V(\lambda)^\delta$ contains a submodule isomorphic to $V(\lambda - r\beta)^\delta$ thus $\varphi_{00}(\lambda) = \varphi_{00}(\lambda - r\beta)$ for all such data hence, by the previous Lemma $\gamma^{-1}(\varphi_{0,0}) \in (U^0)^{\tilde{W}}$.

Conversely let us give such an invariant and construct formally the series z as in the beginning of the section. We thus construct the elements $\varphi_{k,r}$ according to the recursive formula 18.1.3 and want to show first that the factors of the determinant of the contravariant form do not appear in the denominator of the resulting expression for ϕ_η. We prove this by induction on η. We consider the matrix H_η as a matrix valued function on $Spec(U^0)$. As before we consider $P/2P \times \Lambda \subset Spec(U^0)$ by considering (δ, λ) as the homomorphism defined by $(\delta, \lambda)(K_\beta) = (-1)^{<\delta,\beta>} q^{<\lambda,\beta>}$. Notice that $P/2P \times \Lambda$ is Zariski dense in $Spec(U^0)$. We now set

$$T^0_{r\beta,\eta} = \{\lambda \in T^0_{r\beta} | <\lambda + \varrho, \gamma> \neq \frac{m}{2} <\gamma, \gamma> \forall \gamma \neq \beta, m \text{ with } m\gamma \leq \eta\}.$$

It is clear that $\det H_\eta$ vanishes exactly of order $Par(\eta - r\beta)$ in (δ, λ). On the other hand we have that $V(\lambda)^\delta$ contains a submodule isomorphic to $V(\lambda - r\beta)^\delta$ so that $V(\lambda)^\delta_\eta \cap J \supset V(\lambda - r\beta)^\delta_{\eta - r\beta}$. Since $dimV(\lambda - r\beta)^\delta_{\eta - r\beta} = Par(\eta - r\beta)$ it follows from Lemma 18.3a) that

$$V(\lambda)^\delta_\eta \cap J = V(\lambda - r\beta)^\delta_{\eta - r\beta}.$$

Set $N_\eta = \varphi_{0,0}(\lambda)I - \sum_{\gamma<\eta} G_\gamma(\lambda)$. By the inductive hypothesis N_η is matrix valued function on $Spec(U^0)$ which has $V(\lambda - r\beta)^\delta_{\eta - r\beta}$ in its kernel. Since $P/2P \times \Lambda$ is Zariski dense in $Spec(U^0)$ we can apply part b) of Lemma 18.3 and conclude that $\phi_\eta = H_\eta N_\eta^{-1}$ has all of its coefficients in U^0 as desired. A simple induction shows that the series z terminates (cf. [D-K]).

18.4 Let us comment on some consequences of the previous theorems. First one has from Lusztig results that for generic q (i.e. q not a root of 1) the irreducible modules $L(\lambda)_\delta$ where λ is a dominant weight exhaust all finite dimensional irreducible representations of U_q, Furthermore one can construct in general the analogue of the Casimir element in the center of U_q and use it as in section 3.1 in order to prove that in this case the finite dimensional representations of U_q are completely reducible (cf. [R1] or [L4]).

CHAPTER 5

Roots of 1

§19 The Frobenius map.

19.1 The quantum group U_ε, ε a primitive l^{th} root of 1 can be defined starting from the $R = k[q, q^{-1}]$ algebra A (introduced in §12.1 and generated by the elements $\overline{E}_i$, $\overline{F}_i$, K_λ and their $\mathcal{B}$ translates) and specializing $q = \varepsilon$ a primitive root of 1. We shall limit ourselves to the case l odd (and for G_2 also relatively prime to 3). The case l even is somewhat different and can be found in the paper of J. Beck [Be].

In the study of U_ε it will play a major role a particular subalgebra of the center. From now on let us assume to be in this algebra. We look at the relations 9.1 and deduce immediately:

The elements $K_\alpha^l = K_{l\alpha}$ are in the center.

From Corollary 5.1 and the relations 9.1.1, since $\ell \geq 1 - a_{ij}$, we deduce also:

The elements $\overline{E}_i^l, \overline{F}_i^l$ are in the center.

We apply now the Braid group $\mathcal{B}$ to these elements and obtain:

Definition. *The smallest subalgebra containing the elements $K_\alpha^l, \overline{E}_i^l, \overline{F}_i^l$ and stable under $\mathcal{B}$ will be denoted by Z_0.*

By construction this algebra is contained in the center of U_ε. We want to describe now its main properties.

It is convenient, for the formulas which will follow to change our notations.

$$z_\beta := K_\beta^\ell \ (\beta \in M), \ z_i = z_{\alpha_i}, \ y_i := \overline{F}_i^\ell, \ x_i := T_0(y_{\overline{i}}) \ (i = 1, \dots, n)$$

From Theorem 9.5, we have $T_i(y_i) = T_0(y_{\overline{i}})$, which implies:

$$x_i = -\overline{E}_i^\ell z_i^{-1}.$$

First of all recall the formulas of the Hopf structure in U which can also be interpreted as formulas in U_ε:

$$\Delta\overline{E}_i = \overline{E}_i \otimes 1 + K_{\alpha_i} \otimes \overline{E}_i, \ \Delta\overline{F}_i = \overline{F}_i \otimes K_{-\alpha_i} + 1 \otimes \overline{F}_i, \ \Delta K_\alpha = K_\alpha \otimes K_\alpha,$$
$$S\overline{E}_i = -K_{-\alpha_i}\overline{E}_i, \ S\overline{F}_i = -\overline{F}_i K_i, \ SK_\alpha = K_{-\alpha},$$
$$\eta\overline{E}_i = 0, \ \eta\overline{F}_i = 0, \ \eta K_\alpha = 1.$$

Notice that the elements $x = \overline{E}_i \otimes 1, y = K_{\alpha_i} \otimes \overline{E}_i$ satisfy $xy = q^{-2}yx$ hence from 5.1 Corollary i) we get:

$$\Delta(\overline{E}_i^\ell) = \overline{E}_i^\ell \otimes 1 + K_{\alpha_i}^\ell \otimes \overline{E}_i^\ell$$

similarly for $\overline{F}_i$, furthermore $(K_{-\alpha_i}\overline{E}_i)^\ell = q^{2(1+2+\ldots+\ell-1)}K_{-\alpha_i}^\ell E_i^\ell$ we deduce in our new notations:

$$\Delta y_i = y_i \otimes 1 + z_i \otimes y_i,\ \Delta x_i = x_i \otimes z_i^{-1} + 1 \otimes x_i,\ \Delta z_\alpha = z_\alpha \otimes z_\alpha,$$
$$Sx_i = -z_i x_i,\ Sy_i = -y_i z_i,\ Sz_\alpha = z_{-\alpha},$$
$$\eta x_i = 0,\ \eta y_i = 0,\ \eta z_\alpha = 1.$$

As an immediate consequence we have:

Corollary. *The algebra $V := k[x_i, y_i, z_\alpha]$ is a sub-Hopfalgebra.*

Recall the definitions of 9.3:

$$E_{\beta_i} = T_{i_1} \cdots T_{i_{t-1}}(E_{\alpha_{i_t}}),\ F_{\beta_i} = T_{i_1} \cdots T_{i_{t-1}}(F_{\alpha_{i_t}})$$

Thus $E_{\beta_i}^l = T_{i_1} \cdots T_{i_{t-1}}(E_{\alpha_{i_t}}^l),\ F_{\beta_i}^l = T_{i_1} \cdots T_{i_{t-1}}(F_{\alpha_{i_t}}^l) \in Z_0$ We wish to apply to U the results of Proposition 11.7 and claim:

Theorem. *i) Z_0 is the Poisson subalgebra generated by V hence it is the coordinate ring of a Poisson group.*

ii) The monomials

$$E_{\beta_1}^{lk_1} \cdots E_{\beta_N}^{lk_N} K_{l\lambda} F_{\beta_N}^{lh_N} \cdots F_{\beta_1}^{lh_1} \tag{19.1.1}$$

are a basis of Z_0.

The monomials

$$E_{\beta_1}^{k_1} \cdots E_{\beta_N}^{k_N} K_\lambda F_{\beta_N}^{h_N} \cdots F_{\beta_1}^{h_1} \tag{19.1.2}$$

with $0 \leq k_i, h_i < \ell$ and λ running in a set of representatives of $P/\ell P$ are a basis of U_ε over Z_0.

iii) Z_0 is canonically isomorphic as a Poisson Hopf algebra with the $\mathcal{B}$ action to the algebra B of the Poisson group H considered in §12. The morphism maps $\overline{E}_i, K_i, \overline{F}_{i\,q=1}$ to $\overline{E}_i^{\,\ell}, K_i^\ell, \overline{F}_{i\,q=\varepsilon}^{\,\ell}$ (a **Frobenius** *morphism.)*

Proof. This theorem will require several steps. For part ii) it is clear that the given monomials are linearly independent, since they are part of a basis of U_ε, and they clearly span a subalgebra, since they commute; since we have characterized the algebra B by a few relations we will verify the minimum number of relations necessary to apply Lemma 14.1.

Step 1 We need to prove that the universal relations described in §13 are satisfied. This will give rise to a $\mathcal{B}$ equivariant map from B to Z_0 which will be an algebra isomorphism by the PBW theorem.

Step 2 Consists in verifying that the elements x_i, z_i give the same Poisson bracket in both algebras, since we have already seen that they have the same Δ all hypotheses of 14.1 will be then verified.

19.2

The following easy lemma will be important in the sequel

Lemma. *Let S be an algebra, q an invertible element in the center of S. $a, b, c \in S$ elements satisfying the following relations,*

$$ca = q^h ac + q^h b$$

$$ba = q^k ab.$$

then for any $1 \leq s \leq l$,

$$c^s a^l = q^{slh} a^l c^s + (l)_{hk}(l-1)_{hk} \cdots (l-s+1)_{hk} a^{l-s} b^s + \sum_{i=1}^{s-1} a^{l-s+i} x_i$$

where $(l)_{hk} = \sum_{t=1}^{l} q^{ih+(l-i)k}$ and the x_i's are non commutative polynomials in b and c.

Proof. We proceed by induction on s and l, the case $s = l = 1$ being trivial. Let us first assume $s = 1$ and make induction on l. We have

$$ca^l = q^{(l-1)h} a^{l-1} ca + (l-1)_{hk} a^{l-2} ba =$$
$$q^{lh} a^l c + q^{lh} a^{l-1} b + (l-1)_{hk} q^k a^{l-1} b = q^{lh} a^l c + (l)_{hk} a^{l-1} b$$

Let us now make induction on s. We have by the above

$$c^s a^l = c^{s-1}(q^{lh} a^l c + (l)_{hk} a^{l-1} b).$$

Now, using the inductive hypothesis we clearly have that

$$c^{s-1} a^l = q^{(s-1)lh} a^l c^{s-1} + \sum_{i=1}^{s-1} a^{l-s+i} y_i,$$

and

$$c^{s-1} a^{l-1} = (l-1)_{hk} \cdots (l-s+1)_{hk} a^{l-s} b^{s-1} + \sum_{i=1}^{s-1} a^{l-s+i} z_i.$$

where the y_i's and z_i's are non commutative polynomials in b and c. Substituting we immediately get our result.

As a consequence of the Lemma we obtain, when l satisfies our restrictions

Proposition. *Suppose that for some $1 \leq i \leq j \leq N$ we have*

$$\overline{E}_{\beta_j} \overline{E}_{\beta_i} = q^h \overline{E}_{\beta_i} \overline{E}_{\beta_j} + q^h \alpha \overline{E}_{\beta_t}^s,$$

where $h = -(\beta_i, \beta_j)$ and $\alpha \in (q - q^{-1})R$ and $t = i+1$ or $j-1$. Set $d = (\beta_i, \beta_i)/2$. Then

$$\overline{E}_{\beta_j}^l \overline{E}_{\beta_i}^l \equiv q^{l^2 h} \overline{E}_{\beta_i}^l \overline{E}_{\beta_j}^l + q^{l(l-1)h - l(l-1)d)} [l]_d! \alpha^l \overline{E}_{\beta_t}^{ls} \quad mod(q-\varepsilon)^2$$

ε being a primitive l-th root of unity.

Proof. Let us recall that

$$\overline{E}_{\beta_t} \overline{E}_{\beta_i} = q^{-((\beta_i, \beta_t))} \overline{E}_{\beta_i} \overline{E}_{\beta_t}$$

and by homogeneity $s\beta_t = \beta_i + \beta_j$. It follows that $k = -(s\beta_t, \beta_i) = -2d + h$. Thus applying the above lemma in this situation we get

$$\overline{E}^l_{\beta_j} \overline{E}^l_{\beta_i} \equiv q^{l^2 h} \overline{E}^l_{\beta_i} \overline{E}^l_{\beta_j} + q^{l(l-1)h - l(l-1)d} . [l]_d! \alpha^l \overline{E}^{ls}_{\beta_t} + \sum_{p=1}^{l-1} \overline{E}^p_{\beta_i} x_p .$$

Notice now that if we write each of the elements x_t as a linear combination of the monomials (11.8.1), each such monomial appearing with non zero coefficient will involve only the elements $\overline{E}_{\beta_m}$ with $m > i$. In particular if we multiply any of those monomials by $\overline{E}^t_{\beta_i}$ we still obtain one of the monomials (11.8.1) and, by Proposition (10.8) none of them will lie in the center when we reduce at ε.

On the other hand we know that $\overline{E}^l_{\beta_j}$, $\overline{E}^l_{\beta_i}$ and $\overline{E}^l_{\beta_t}$ are central at ε, so that the element

$$\frac{[\overline{E}^l_{\beta_j}, \overline{E}^l_{\beta_i}]}{q^l - q^{-l}} \Big|_{q=\varepsilon}$$

is a central element. This clearly implies the claim.

19.3

Corollary. *Assume that $\overline{E}_{\beta_j}, \overline{E}_{\beta_i}$ are as in the previous proposition with $\alpha = \pm q^m (q^d - q_{-d})$. Then for the Poisson bracket of the classes of $\overline{E}^l_{\beta_j} \overline{E}^l_{\beta_i}$ modulo ℓ we have:*

$$\{e_{\beta_j}, e_{\beta_i}\} = \frac{h}{2} e_{\beta_j} e_{\beta_i} \pm d e^s_{\beta_t} .$$

Proof. By definition of the Poisson bracket.

Lemma. *In the rank 2 case consider the elements $e_\alpha := \overline{E}^l_\alpha \in \mathcal{U}_\varepsilon$. The formulas expressing their Poisson brackets are independent of l.*

Proof. In the cases A_2 and B_2 this is an immediate consequence of the above Proposition, due to the special form of the relations $A4-6$. For G_2 we can apply directly this proposition in all 15 cases except for the following two:

$$\{e_2, e_{12}\}, \quad \{e_1, e_{1222}\}.$$

In the first case we use the relation $3e_{12}^3 = \{e_{11222}, e_1\} + \frac{3}{2} e_{11222} e_1$. Applying $\{e_2, -\}$ to both sides we compute:

$$\{e_2, e_{12}\} = 2e_{12}^2 + \frac{1}{2} e_2 e_{12}^3 .$$

In the second case we use the relation $e_{1222} = \frac{1}{2} e_2 e_{122} + \{e_2, e_{122}\}$ we apply $\{e_1, -\}$ and get:

$$\{e_1, e_{1222}\} = -\frac{3}{2} e_1 e_{1222} + 9 e_{12} e_{122} + 3 e_{11222} .$$

In the proof we have given we have used a specific reduced expression of the longest element of the Weyl group to construct the isomorphism of Poisson algebraic groups. We could have used the other expression (i.e. the other term of the braid relation) and the reader will verify that a similar proof holds. Now we know that the Poisson algebra is generated by the elements x_i, y_i, z_i hence the isomorphism being determined on these elements is unique, this implies in particular that the given isomorphism in the rank 2 case is $\mathcal{B}$ equivariant. Since the universal relations 13.1 of the braid group action are in rank 2 we have completed the proof of Step 1.

For step 2 we have to compare Poisson brackets of two elements among the $x_\beta, y_\beta, z_\lambda$. Since these elements will involve 2 roots they can be brought by the Braid group (compare Lemma 15.4) to the rank 2 case which we have verified, this completes the proof of Theorem 19.1 iii).

An implication for the representation theory of U_ε is obtained from Proposition 11.8 and Theorem 11.7 combined with the remarks at the beginning of 16.1 on the existence of the group $\tilde{G}$.

We get thus:

Theorem. *There exists of a group of global analytic automorphisms of U_ε extending the action of $\tilde{G}$ on Z_0.*

As already remarked we use this fact, or equivalently Proposition 11.8 to deduce that, for two points $a, b \in Spec(Z_0)$ in the same symplectic leaf (or $\tilde{G}$ orbit) the corresponding algebras $U_\varepsilon(a), U_\varepsilon(b)$ are isomorphic.

(by $U_\varepsilon(a)$ we mean $U_\varepsilon/m(a)$ where $m(a)$ is the ideal in U_ε generated by the ideal of the point a.

REMARK We do not know whether this group differs from $\tilde{G}$ in any case we will keep the notation $\tilde{G}$ also for this possibly bigger group.

§20 Baby Verma modules and the degree.

20.1 Consider inside the algebra $\mathcal{U}_\varepsilon$ the subalgebra $\mathcal{B}_\varepsilon$ generated by the E_i, K_i and also the subalgebra S obtained by adding to $\mathcal{B}_\varepsilon$ the elements $y_\alpha, \alpha \in R$. By the P.B.W. theorem these elements are algebraically independent over $\mathcal{B}_\varepsilon$ and S is the polynomial algebra $\mathcal{B}_\varepsilon[y_\alpha], \alpha \in R$.

Consider a 1-dimensional representation $k(\sigma)$ of S given by setting $E_i = 0$, $y_\alpha := b_\alpha, K_i = k_i$. We often identify σ with the point of coordinates b_α, k_i. Set:

(20.1.1) $$V_\sigma := \mathcal{U}_\varepsilon \otimes_S k(\sigma)$$

Definition. *V_σ is called the* **Baby Verma module** *of highest weight σ.*

Set $v_\sigma := 1 \otimes 1 \in V_\sigma$.

Theorem. *1) The ℓ^N elements $F_{\beta_N}^{h_N} \cdots F_{\beta_1}^{h_1} v_\sigma$, $0 \le h_i < \ell$ are a basis of V_σ.*

2) $E_i v_\sigma = 0$.

3) The elements of Z_0 act as scalars on V_σ in fact as

$$y_\alpha := b_\alpha, z_i = k_i^\ell, x_\alpha = 0.$$

4) The modules V_σ are generically irreducible.

5) The degree of $\mathcal{U}_\varepsilon$ is ℓ^N.

6) The degree of the center Z of $\mathcal{U}_\varepsilon$ over Z_0 is ℓ^n.

Proof. 2) and 3) follow from the definition and 1) from the P.B.W. basis.

4),5),6) are proved simultaneously.

By 11.7 we have that considering $\mathcal{U}_\varepsilon$ as a sheaf of algebras over Z_0 the isomorphism type of the fibers is constant on symplectic leaves. By Theorem 16.1 each symplectic leaf of Spec(Z_0) meets the subspace A of coordinates $x_\alpha = 0$ therefore the degree of $\mathcal{U}_\varepsilon$ is the dimension of a generic irreducible representation having as central character on Z_0 a point in A. For each such point we have an irreducible module which is a quotient of the Baby Verma module thus we get ℓ^N as upper bound for the degree. From Theorem 10.7 we have the same number as lower bound and this proves 4),5). As for 6) we have that $dim_{Z_0}\mathcal{U}_\varepsilon = \ell^{2N+n}$ and hence the claim.

Notice that generically we can say even more by 16.1d.

Definition. *A baby Verma module is* **diagonal** *if on it the elements y_α act as 0; it is* **diagonalizable** *it it is $\tilde{G}$ equivalent to a diagonal one.*

Remark. *Generically a baby Verma module is diagonalizable.*

Proof. See 16.1d.

20.2 We can apply to the algebra U_ε the ideas and results of §10.1 and consider the filtration there described and the associated graded algebra. We will get an algebra with the same generators and relations 10.1.2-5 except that q is specialized to ε. This follows immediately from the fact proved in 12.1 that the Leverdossky Soibelman relations hold for the algebra A and hence specialize to U_ε.

This fact together with Theorem 6.5 imply:

Theorem. *U_ε is a maximal order.*

Let us denote by Z_ε the center of U_ε. In the next section we will analyze Z_ε in complete detail, for the moment we make the following remark:

Proposition. *i) Z_ε is integrally closed.*
ii) Z_ε is a free Z_0 module of rank ℓ^n.

Proof. i) follows from the theory of maximal orders (§6.5). As for ii) let us remark first of all that the center of U_ε is a direct summand since it is the image of the reduced trace map and we are in characteristic 0. Thus since U_ε is a free Z_0 module we have that also Z_ε is so (by Quillen Suslin proof of the Serre conjecture). To compute its degree we can pass to the quotient fields. If Q is the quotient field of Z_0 we have that $D := U_\varepsilon \otimes_{Z_0} Q$ is a division algebra of dimension ℓ^{2N+n} over Q. $F := Z_\varepsilon \otimes_{Z_0} Q$ is its center and the dimension of D over F is ℓ^{2N} (20.1) thus F has dimension ℓ^n over Q and ii) follows.

20.3. A baby Verma module V_σ is a d–dimensional indecomposable representation of U_ε. Hence all irreducible factors $\rho_1, \ldots, \rho_s$ of $\bar{M}_\varepsilon(\lambda, \nu)$ have the same central character $\chi \in$ Spec Z_ε.

Proposition. *(a) The representation $\bigoplus_{i=1}^{s} \rho_i$ is the semisimple representation of U_ε corresponding to $\chi \in$ Spec Z_ε.*

(b) Given $\chi \in$ Spec Z_ε, choose $g \in \tilde{G}$ such that $\chi_1 = g(\chi)$ has the property that $\chi_1(x_\alpha) = 0$ for all $\alpha \in R^+$ (see Proposition 4.6 below), and let $\nu = \chi_1|_{Z_0^-}$. Then there exists a homomorphism $\lambda : U_\varepsilon^0 \longrightarrow \mathbb{C}$ such that the central character of the module V_σ, is equal to χ_1; let $\rho_1, \ldots, \rho_s$ be all its irreducible factors. Then $\bigoplus_{i=1}^{s} \rho_i^{g^{-1}}$ is the semisimple representation of U_ε corresponding to χ. Clearly the points in SpecZ_ε thus obtained map onto B^-.

(c) Baby Verma modules having the same central character have the same irreducible factors.

Proof. (b) and (c) follows from (a). Due to the discussion relative to Theorem 4.5, it suffices to show that the baby Verma modules V_σ are compatible with the trace map. It is the case if V_σ is irreducible. We know that every $\tilde{G}$ orbit contains a point lying over B^- hence a generic Verma module is irreducible.

Hence, by continuity, all the modules V_σ are compatible with the trace map. □

§21 The center of U_ε.

21.1 We turn now to the center Z of U and the center Z_ε of U_ε. Recall that $U = U^0 \oplus (U^-U + UU^+)$, and denote by h the projection of U on U^0 as in 17.3.1. Similarly, define the projection $h: U_\varepsilon \to U^0_\varepsilon$.

Lemma. *Let $z \in Z_\varepsilon$ be such that*

$$\tilde{G} \cdot z = z \text{ and } h(z) = 0,$$

then $z = 0$.

Proof. From the computation of the degree, the theory developed in Chapter 2 and the discussion on the action of $\tilde{G}$ on U_ε of §19 it is clear that there exists a $\tilde{G}$ stable open set of *Spec*(Z) formed by irreducible representation of maximal degree. It is also clear that the points of *Spec*(Z) corresponding to Verma modules generate under the group $\tilde{G}$ a dense set. Therefore, due to Theorem 16.1d, it suffices to check that the eigenvalue α of z on each diagonal module $\overline{M}_\varepsilon(\lambda)$ (Remark 20.1) is 0. We have

$$zv_\lambda = \alpha v_\lambda = u_- v_\lambda, \text{ where } u_- \in U^-_\varepsilon.$$

Since u_- is a nilpotent endomorphism, we see that $\alpha = 0$. □

Let us go back now to the basic isomorphism of algebras [§18] characterizing the center Z of U for q generic:

$$\gamma^{-1}h: Z \xrightarrow{\sim} U^{0\tilde{W}}. \tag{21.1.1}$$

There it is shown that, for each $\varphi \in U^{0\tilde{W}}$ there exists a unique central element (18.1) of the form:

$$z_\varphi = \gamma(\varphi) + \sum_{\eta > 0} \sum_{k,r \in \mathrm{Par}(\eta)} F^k \varphi_{k,r} E^r, \tag{21.1.2}$$

where the $\varphi_{k,r} \in U^0$ are computed using a recurrent formula, and every central element is obtained in this way. The recurrent formula shows that the only poles of the $\varphi_{k,r}$ are roots of 1 and the poles of φ.

Proposition. *If ε is a primitive ℓ–th root of 1 (with the standard hypotheses) then all the $\varphi_{k,r}$ do not have a pole at $q = \varepsilon$ if φ doesn't.*

Proof. Let $P_\ell(q)$ be the ℓ–th cyclotomic polynomial and let m be the maximum of orders of poles of the $\varphi_{k,r}$. Let $z' = P_\ell(q)^m z$. Then the element $z'|_{q=\varepsilon}$ satisfies all conditions of the lemma and is non–zero, a contradiction. □

Thus, we have a well-defined injective homomorphism of $U^{0\tilde{W}}_\varepsilon$ into Z_ε given by $\varphi \mapsto z_\varphi$. Denote by Z_1 its image.

Corollary. *(a) The subalgebra Z_1 of Z_ε is pointwise fixed under the action of $\tilde{G}$ and the action of $\mathcal{B}$.*
(b) The center Z of U is pointwise fixed under the action of $\mathcal{B}$.

Proof. The first of part of (a) is clear since the inner derivations in U are 0 on Z and this passes to the quotient U_ε and the second part follows from (b). Let now $z \in Z$. If we twist any irreducible finite dimensional representation of U by an element T of $\mathcal{B}$ one obtains by the theory of Lusztig ([L4]) an isomorphic representation which therefore has the same central character. If we write both z and $T(z)$ as in 21.1.2 we see that the elements $\gamma h(z)$ and $\gamma hT(z)$ have the same value for all integral dominant weights. Since these points are Zariski dense in $Spec(U^0)$ b) follows. □

21.2 We return to the study of the quantum coadjoint action. Recall that G is the simply connected.

We have seen that $H = \mathrm{Spec}(Z_0)$ and have constructed in 14.1 a map $\sigma : H \to G^0$ which is an unramified cover of the big cell. We compose σ with the quotient map under conjugation:

$$G^0 \subset G \to G//G,$$

and get an embedding $\mathbb{C}[G]^G \subset \mathbb{C}[H] = Z_0$. We have seen in 16.3 that in this way $\mathbb{C}[G]^G$ is identified to $Z_0^{\tilde{G}}$. We consider the subalgebra U_0 as the coordinate ring of a torus T. The Harish-Chandra isomorphism identifies Z_1 with $U_0^{\dot{W}} = \mathbb{C}[T/(2)]^W$, where we denote by (m) the subgroup of points of m torsion in T.

Notice that on the subalgebra $U_0^{(}\ell) = Z_0^0$ the map γ (used in the Harish-Chandra isomorphism) is the identity.

Lemma. *$Z_0 \cap Z_1 = \mathbb{C}[G]^G$ as subalgebra of Z_0 and $Z_0 \cap Z_1 = \mathbb{C}[T/(2\ell)]^W$ as subalgebra of Z_1 (under the identification given by he Harish-Chandra isomorphism).*

Proof. From Corollary 21.1 we have that $Z_0 \cap Z_1 \subset \mathbb{C}[G]^G$. We have thus to show that $\mathbb{C}[G]^G \subset Z_1$.

Remark that the map i^* introduced in Theorem 16.3 d is the restriction to Z_0 of the projection map h (17.3.1). Given an element $z \in \mathbb{C}[G]^G$ there exists a unique element $z' \in Z_1$ such that $h(z) = \gamma^{-1}h(z') = h(z')$ by the previous remark.

Since z, z' are $\tilde{G}$ invariant by Corollary 21.1 we conclude by Lemma 21.1 that they are equal.

The remaining statement is clear.

This Lemma has an important

Corollary. *The map Spec $Z_0 \to$ Spec $Z_0 \cap Z_1$ induced by the inclusion is smooth in codimension 1.*

Proof. For the simply connected group the quotient map $\varphi : G_c \to$ Spec $\mathbb{C}[G]^G \simeq \mathbb{C}^n$, is given by $g \longmapsto (\chi_1(g), \dots, \chi_n(g))$, where the χ_i are the characters of fundamental representations of G, and it is smooth in codimension 1 [S, §§5 and 8]. Our map is the composition of an unramified covering with this map.

21.3 Our next objective is to prove that Z_ϵ is generated by its subalgebras Z_0 and Z_1. We need first some remarks on weights and Weyl group invariants.

Let $P_+ = \{\lambda \in P | <\lambda, \alpha_i^\vee> \in \mathbb{Z}_+,\ i = 1, \dots, n\}$, $P_+^\ell = \{\lambda \in P_+ | <\lambda, \alpha_i^\vee> < \ell,\ i = 1, \dots, n\}$. For $\Lambda \in P_+$ let $\chi_\Lambda = \sum_{\lambda \in W(\Lambda)} e^\lambda$. Then $\{\chi_\lambda\}_{\lambda \in P_+}$ (resp. $\{\chi_{\ell\lambda}\}_{\lambda \in P_+}$) form a $\mathbb{C}$–basis of A (resp. A_ℓ) and $\{\chi_\lambda\}_{\lambda \in P_+^\ell}$ form a basis of the A_ℓ–module A. For $\lambda = \Sigma n_i \omega_i \in P_+$ let $M_\lambda = \prod_i \chi_{\omega_i}^{n_i}$. Finally, define partial ordering on P_+ by letting $\lambda \geq \mu$ if $\lambda - \mu = \sum_i a_i \alpha_i$ with $a_i \in \mathbb{Q},\ a_i \geq 0$.

Lemma. (a) *If we write $\chi_\lambda = \sum_{\mu \in P_+^\ell} a_{\lambda\mu} \chi_\mu$, where $a_{\lambda\mu} \in A_\ell$, then $a_{\lambda\mu} \neq 0$ only if $\lambda \geq \mu$.*

(b) *If we write $M_\lambda = \sum_{\mu \in P_+^\ell} b_{\lambda\mu} \chi_\mu$, where $b_{\lambda\mu} \in A_\ell$, then $b_{\lambda\mu} \neq 0$ only if $\lambda \geq \mu$. Also, $b_{\lambda\lambda} = 1$ if $\lambda \in P_+^\ell$.*

(c) *If we write $M_\lambda = \sum_{\mu \in P_+^\ell} c_{\lambda\mu} M_\mu$, where $c_{\lambda\mu} \in A_\ell$, then $c_{\lambda\mu} \neq 0$ only if $\lambda \geq \mu$.*

Proof. Write $\lambda = \ell\lambda' + \lambda''$, where $\lambda' \in P_+$, $\lambda'' \in P_+^\ell$. Then $\lambda \geq \lambda''$ and

$$\chi_\lambda = \chi_{\ell\lambda'} \chi_{\lambda''} + \sum_{\mu < \lambda} a'_{\lambda\mu} \chi_\mu$$

Thus (a) follows by induction on the ordering.

Since

$$M_\lambda = \chi_\lambda + \sum_{\mu<\lambda} b'_{\lambda\mu}\chi_\mu,\ b'_{\lambda\mu} \in \mathbb{Z},$$

the first part of (b) follows by induction on the ordering and (a). The second part of (b) is clear.

Thus, the matrix $B := (b_{\lambda\mu})$ in (b) is invertible and $B^{-1} = (c'_{\lambda\mu})$ has the same properties, i.e. we have for $\lambda \in P^\ell_+$:

$$\chi_\lambda = \sum_{\mu\leq\lambda} c''_{\lambda\mu} M_\mu,\ c''_{\lambda\mu} \in A_\ell,\ c''_{\lambda\lambda} = 1.$$

Substituting this in (b) we get (c). □

We can now state:

Proposition. *The ring* $A := \mathbb{C}[P]^W$ *is a complete intersection over its subring* $A_\ell = \mathbb{C}[\ell P]^W$.

Proof. By Lemma 21.3c we have:

$$M^\ell_{\omega_i} = \sum_{\substack{\mu\in P^\ell_+ \\ \mu<\ell\omega_i}} d_{i\mu} M_\mu,\ d_{i\mu} \in A_\ell.$$

Consider the polynomial algebra $A_\ell[x_1, \dots, x_n]$ and let $x_\lambda = x_1^{k_1} \dots x_n^{k_n}$ for $\lambda = \Sigma k_i\omega_i$. Let $P_i = x_i^\ell - \sum_\mu d_{i\mu}x_\mu \in A_\ell[x_1, \dots, x_n]$, let $I = (P_1, \dots, P_n)$ and let $B = A_\ell[x_1, \dots, x_n]/I$. We have a surjective homomorphism $B \to A$ defined by $x_i \longmapsto \chi_{\omega_i}$. We claim that it is injective. This will prove the proposition.

In order to prove the injectivity, it suffices to show that the x_λ, $\lambda \in P^\ell_+$, span B. Take $\lambda \in P_+$ and consider the monomial $x_\lambda = x_1^{k_1} \dots x_n^{k_n}$. If $k_i < \ell$ for all i, we are done; if not, then $k_j > \ell$ for some j and we have:

$$x_\lambda = x_j^\ell x_{\lambda-\ell\omega_j} = \sum_{\substack{\mu<\ell\omega_j \\ \mu\in P^\ell_+}} d_{j\mu} x_{\lambda-\ell\omega_j+\mu}.$$

Since $\lambda - \ell\omega_j + \mu < \lambda$, the proof is completed by induction on the ordering. □

Denote by $\tilde{Z}_\varepsilon$ the algebra:

$$\tilde{Z}_\varepsilon = Z_1 \otimes_{Z_0\cap Z_1} Z_0,$$

where $Z_1 = \mathbb{C}[P]^{\tilde{W}}$ (see 21.1) and $Z_0 \cap Z_1 = \mathbb{C}[\ell P]^{\tilde{W}}$ (by Lemma 21.2). It follows from Proposition 21.3 that Z_1 is a complete intersection over $Z_0 \cap Z_1$. Since $\tilde{Z}_\varepsilon$ is obtained from Z_1 by change of base and Z_0 is a complete intersection ring, we obtain

Theorem. $Z_1 \otimes_{Z_0\cap Z_1} Z_0$ *is a complete intersection ring.* □

21.5 In this section we will complete the analysis of the center of U_ε. In particular we will show that it equals the subring generated by the two subrings Z_0 and Z_1.

Theorem. *The natural map*

$$j : Z_1 \otimes_{Z_0 \cap Z_1} Z_0 \to Z_\varepsilon$$

is an isomorphism.

Proof. By construction $Z_1 \otimes_{Z_0 \cap Z_1} Z_0$ is a free module of rank l^n over Z_0. If we show that it is also an integrally closed domain it will follow (by standard commutative algebra) that j is injective and, by the computation of the degree of l^N of U_ε and the fact that U_ε is free of rank l^{2N+n} over Z_0 we have that j is birational and so an isomorphism as required.

To prove that it is integrally closed we use Serre's theorem [Se, Ch. 4], claiming that a complete intersection variety, which is smooth outside of a subvariety of codimension 2, is normal. Thus, due to Theorem 21.3, it suffices to show that Spec $\tilde{Z}_\varepsilon$ is smooth outside of a subvariety of codimension 2.

As we have seen in the proof of Theorem 21.3, Spec $\tilde{Z}_\varepsilon$ is an open subset in an unramified covering of the fiber product

$$X_G := T/W \times_{T/W} G,$$

where the first map $\varrho_\ell : T/W \to T/W$ is induced by the ℓ'th power map $(t \longmapsto t^\ell)$ and the second map $\sigma : G \to T/W$ is the quotient map.

By [S, §§5 and 8], $G = G_1 \cup G_2$, where G_1 is open, G_2 is a closed subvariety of codimension ≥ 2, and $\sigma : G_1 \to T/W$ is a smooth map (here we use that G is simply connected). Hence $T \times_{T/W} G_1 \to T$ is a smooth map, and since T is smooth, we obtain that $T \times_{T/W} G_1$ is smooth. Since the map $T \to T/W$ is finite, we obtain that $T \times_{T/W} G_2$ has codimension ≥ 2 in $T \times_{T/W} G$, completing the proof. □

Corollary. *(a) The action of $\tilde{G}$ on*

$$\text{Spec } Z_\varepsilon = \text{Spec } Z_1 \times_{\text{Spec } Z_0 \cap Z_1} \text{Spec } Z_0$$

extends from Spec Z_0 by a trivial action on Spec Z_1. Orbits of the action of $\tilde{G}$ on Spec Z_ε are connected components of the preimages of orbits of $\tilde{G}$ on H under the map $\tau :$ Spec $Z_\varepsilon \to H$ induced by inclusion of coordinate rings. □

21.6 We shall give here another construction of the fiber product $G//G \times_{G//G} G$ (cf. Remark 21.3b). Let B be a finitely generated integrally closed algebra without zero divisors and let A be a finitely generated subalgebra of B. Denote by $\overline{A}$ the integral closure of A in B. This gives us a factorization of the map Spec $B \to$ Spec A (induced by the inclusion):

$$\text{Spec } B \to \text{Spec } \overline{A} \to \text{Spec } A,$$

such that Spec $\overline{A}$ is normal, the first map has connected (possibly empty) fibers and the second map is finite, called the *Stein factorization*.

Let now G be a semisimple connected affine algebraic group. Given a positive integer m, denote by $\varrho_m : G \to G$ the map defined by $g \longmapsto g^m$, $g \in G$. Let $G \to X'_G \to G$ be the Stein factorization of this map.

More explicitly this factorization can be constructed as follows (cf. 21.3). Denote by $G//G$ the affine variety with coordinate ring the regular functions on G invariant under conjugation, and let $\sigma : G \to G//G$ be the quotient map. The map ϱ_m induces a map $\varrho_m : G//G \to G//G$, so that we have a commutative diagram:

$$\begin{array}{ccc} G & \xrightarrow{\varrho_m} & G \\ & \searrow \alpha \qquad p_2 \nearrow & \\ \sigma \downarrow & X_G & \downarrow \sigma \\ & p_1 \swarrow & \\ G//G & \xrightarrow[\varrho_m]{} & G//G \end{array}$$

Here X_G denotes the fiber product $G//G \times_{G//G} G$, and p_i is the projection on i-th factor. By the universality property of the fiber product, there exists a morphism $\alpha : G \to X_G$ making the diagram commutative.

Proposition. *$G \xrightarrow{\alpha} X_G \xrightarrow{p_2} G$ is the Stein factorization of $\varrho_m : G \to G$, so that $X'_G \simeq X_G$.*

Proof. Let T be a maximal torus; then we have the canonical isomorphism $G//G \xrightarrow{\sim} T/W$, where W is the Weyl group. Since the map $\varrho_m : T \to T$ is finite, it follows that the map $\varrho_m : G//G \to G//G$ is finite, hence the map $p_2 : X_G \to G$ is finite. Furthermore, it is clear that the map $\alpha : G \to X_G$ is birational. It remains to show that X_G is normal. As shown in 21.3, this is the case if G is simply connected. Hence it is true for arbitrary G. □

§22 The solvable case.

Let us give some simple applications of the Frobenius map for some of the algebras $\mathcal{B}_w$ introduced in §10 using the results and terminology of sections 16.3,4. Recall that T be the usual torus with character group P. For each $t \in T$ we define an automorphism β_t of the algebra B_ε by:

$$\beta_t(K_\alpha) = \alpha(t)K_\alpha, \quad \beta_t(E_\alpha) = \alpha(t)E_\alpha.$$

Note that the automorphisms β_t leave Z_0^+ invariant and permute transitively the leaves of each set $\psi^{-1}(C_w) \subset B_+$.

Given $a \in B_+ = \text{Spec } Z_0^+$, denote by m_a the corresponding maximal ideal of Z_0^+ and let

$$A_a = B_\varepsilon / m_a B_\varepsilon.$$

These are finite-dimensional algebras and we may also consider these algebras as algebras with trace in order to use the theory of §2.

Theorem. *If $a, b \in$ Spec Z_0^+ lie over the same element $w \in W$, then the algebras A_a and A_b are isomorphic (as algebras with trace).*

Proof. We just apply Proposition 12.1 to the vector bundle of algebras A_a over a symplectic leaf and the group T of algebra automorphisms which permutes the leaves in $\psi^{-1}(C_w)$ transitively. □

We are ready now for the concluding theorem (cf. the conjectures in the last section).

Theorem. *Let $p \in X_w$ be a point over $w \in W$ and let A_p be the corresponding algebra. Assume that l is a good integer. Then the dimension of each irreducible representation of A_p is divisible by $= l^{\frac{1}{2}(l(w)+\text{rank}(1-w))}$.*

Proof. Consider the algebra B_ε^w for which we know by Theorem 10.5 that

$$\deg B^w = l^{\frac{1}{2}(l(w)+\text{rank}(1-w))}.$$

The subalgebra $Z_{0,w}$ of Z_0 generated by the elements K_λ^l and E_α^l, where $\lambda \in P$ and $\alpha \in R^+$ is such that $-w^{-1}\alpha \in R^+$, is isomorphic to the coordinate ring of B^w, and $\mathcal{B}_\varepsilon^w$ is a finite free module over $Z_{0,w}$. Thus by 4.5 there is a non empty open set $\mathcal{A}$ of B^w such that for $p \in \mathcal{A}$ any irreducible representation of $\mathcal{B}^w$ lying over p is of maximal dimension, equal to the degree of $\mathcal{B}_\varepsilon^w$. Now the ideal I defining X_w has intersection 0 with $Z_{0,w}$ and so when we restrict a generic representation of $\mathcal{B}_\varepsilon$ lying over points of X_w to the algebra $\mathcal{B}_\varepsilon^w$ we have, as a central character of $Z_{0,w}$, a point in $\mathcal{A}$. Thus the irreducible representation restricted to $\mathcal{B}^w$ has all its composition factors irreducible of dimension equal to $\deg \mathcal{B}^w$. This proves the claim. □

It is possible that the dimension of any irreducible representation of B_ε whose central character restricted to Z_0^+ is a point of X_w is exactly $\ell^{\frac{1}{2}(\ell(w)+\mathrm{rank}(1-w))}$. This fact if true would require a more detailed analysis in the spirit of Section 1.3.

CHAPTER 6

THE REGULAR SHEET

§23 Central characters.

23.1. Let $\lambda \in P_+$ be a dominant weight and let ρ_λ be the finite–dimensional irreducible representation of the group G in a vector space V^λ. Then we have a map: $\mu_\lambda = \rho_\lambda \circ \mu : \Omega = \text{Spec } Z_0 \to GL(V^\lambda)$, where $\mu = YZ^2X$ is as in 14.4. Let

$$\phi_\lambda(u) = tr_{V^\lambda}\mu_\lambda(u), \ u \in Z_0. \tag{23.1.1}$$

From §16.3 these polynomial functions on Spec Z_0 are a basis for the functions constant on symplectic leaves.

One can compute $\phi_\lambda(u)$ from the explicit formula for μ as:

$$\phi_\lambda(u) = tr_{V^\lambda}(exp(y_{\beta_N} f_{\beta_N}) \dots exp(y_{\beta_1} f_{\beta_1}) Z^2 exp(-T_0 y_{\beta_1} t_0 f_{\beta_1}) \dots exp(-T_0 y_{\beta_N} t_0 f_{\beta_N})).$$

Since both $\beta_1, \dots, \beta_N$ and $\beta_N, \dots, \beta_1$ are convex orderings. Which expands (applying the exponential formula) to a linear combination of terms:

$$\sum c_{\underline{h},\underline{k}} y_{\beta_N}^{h_N} \dots y_{\beta_1}^{h_1} T_0 y_{\beta_N}^{k_N} \dots T_0 y_{\beta_1}^{k_1} tr_{V^\lambda}(f_{\beta_N}^{h_N} \dots f_{\beta_1}^{h_1} Z^2 t_0 f_{\beta_1}^{k_1} \dots t_0 f_{\beta_N}^{k_N}). \tag{23.1.2}$$

In each representation only finitely many of these terms contribute since the $t_0 f_\beta$ generate a nilpotent algebra.

Fix a reduced expression of w_0 with associated convex ordering

$$\beta_1, \dots, \beta_N$$

We can also consider the convex ordering:

$$-w_0\beta_1, \dots, -w_0\beta_N.$$

Correspondingly we have the elements:

$$\overline{F}_{\beta_i}, T_0(\overline{F}_{\beta_i})$$

Consider the $\mathbb{Z}_+^{2N+1}$–filtration of the algebra U_ϵ (obtained by the specialization of that of U_A as in §10.1) by comparing the monomials first by height and then placing in increasing order the root vectors as:

$$\overline{F}_{\beta_N}, \dots, \overline{F}_{\beta_1}, T_0\overline{F}_{\beta_N}, \dots, T_0\overline{F}_{\beta_1}.$$

Correspondingly in Z_0 we have the ordered variables:

$$y_{\beta_1}, \dots, y_{\beta_N}, \ T_0 y_{\beta_1}, \dots, T_0 y_{\beta_N}.$$

For an element ϕ of U_ε we denote by $\overline{\phi}$ the monomial of maximal degree in the above filtration that occurs in ϕ with a non-zero coefficient.

We want to compute the leading term of a polynomial $\phi_\lambda(u)$. We will do it for fundamental weights to which one can easily reduce. Fix thus a fundamental weight ω, let α be the corresponding simple root.

Recall that in 10.3 we have introduced a subset I_ω of the indices $1, \dots, N$. We make some further complements to Lemma 10.3. Let ω_i be a fundamental weight V_{ω_i} the corresponding irreducible module with highest weight ω_i and let v_{ω_i} be a highest weight vector. We have defined elements $t_i \in G$ satisfying the braid relations, in particular for $w \in W$ we have an element t_w and $t_w(v_{\omega_i})$ is an extremal weight vector of weight $w(\omega_i)$. Take an $sl(2)$ triple e_j, f_j, h_j, since v_{ω_i} is a highest weight vector we have that $f_j v_{\omega_i} = 0$ if $j \neq i$ and $v_{\omega_i}, f_i v_{\omega_i}$ span the standard 2 dimensional representation otherwise. In particular $f_i v_{\omega_i} = -t_i v_{\omega_i}$.

Given a reduced expression $w = s_{i_1} s_{i_2} \dots s_{i_l} = u s_{i_l}$ of an element $w \in W$, define recursively an element $f_i(w)$ in the universal enveloping algebra of the Lie algebra $\mathfrak{g}$:

$$\begin{aligned} \text{if } w = 1, \quad & f_i(w) := 1 \\ \text{if } i_l \neq i, \quad & f_i(w) := f_i(u) \\ \text{if } i_l = i, \quad & f_i(w) := -t_u(f_i) f_i(u) \\ & e_i(w) := t_0(f_i(w)) \end{aligned}$$

Lemma.

(a)
$$t_w(f_j) t_w v_{\omega_i} = \begin{cases} 0 & \text{if } i \neq j \\ t_w t_i v_{\omega_i} & \text{if } i = j \end{cases}$$

(b)
$$f_i(w)(v_{\omega_i}) = t_w(v_{\omega_i}), \quad e_i(w)(t_0 v_{\omega_i}) = t_0 t_w(v_{\omega_i}).$$

Proof. (a) By compatibility $t_w(f_j) t_w v_{\omega_i} = t_w(f_j v_{\omega_i}) = 0$ if $i \neq j$ or $t_w(f_i v_{\omega_i}) = t_w t_i v_{\omega_i}$.

(b) By induction using the definition and the fact that $t_j v_{\omega_i} = v_{\omega_i}$ if $j \neq i$. If $w = u s_i$ we have $f_i(w)(v_{\omega_i}) = t_u(f_i) f_i(u)(v_{\omega_i})$ by induction equals $t_u(f_i) t_u(v_{\omega_i}) = t_u t_i(v_{\omega_i}) = t_w(v_{\omega_i})$ by part (a). For the $e_i(w)$ one applies t_0.

Proposition. *Let ϕ_ω be the element of Z_0 defined by (23.1.1) for $\lambda = \omega$. Then*

$$\overline{\phi}_\omega = \pm z_\omega^2 \prod_{t \in I_\omega} T_0 y_{\beta_t} y_{\beta_t}.$$

Proof. Fix a basis $\{v_j\}$, $j = 1, \dots, t = \dim V^\omega$, of V^ω such that v_j is a weight vector of weight μ_j and $i < j$ if $\mu_i > \mu_j$, so that v_t has the highest weight ω and v_1 has the lowest weight $w_0(\omega)$, we can also assume that $v_1 = t_0(v_t)$.

The operator $t_0 f_{\beta_N}^{m_N} \dots t_0 f_{\beta_1}^{m_1}$ applied to a vector of weight ν gives a vector of weight $\sum_{j=1}^N m_j \beta_j$ or 0, similarly for the operator $f_{\beta_N}^{h_N} \dots f_{\beta_1}^{h_1}$ while Z^2 is diagonal, hence the only non zero contributions to 23.1.2 come from choices of the exponents for which $\sum_{j=1}^N m_j \beta_j = \sum_{j=1}^N h_j \beta_j \leq \omega - w_0(\omega)$. Let us call $\sum_{j=1}^N m_j \beta_j$ the weight of the contributing monomial.

We claim thus that the term in which the exponents h_i, m_i for the roots β_i are 0 except for the indices k_i in I_ω for which they are 1 gives the highest term in the sum.

We use the previous Lemma plus a few remarks. From Lemma 23.1 the element $\prod_{t \in I_\omega} f_{\beta_t} = f_i(w_0)$ (where the product is taken in increasing order) and for the convex ordering $\beta_1, \dots, \beta_N$ while $\prod_{t \in I_\omega} t_0 f_{\beta_t} = e_i(w_0)$ (where the product is taken in increasing order) for the convex ordering: $-w_0 \beta_1, \dots, -w_0 \beta_N$ thus $\prod_{t \in I_\omega} t_0 f_{\beta_t} v_1 = v_t$ and

$\prod_{t\in I_\omega} f_{\beta_t} Z^2 \prod_{t\in I_\omega} t_0 f_{\beta_t} v_1 = z_\omega^2 v_1$ hence the given term contributes to the required monomial. It remains to show that all other terms contribute with lower monomials. It suffices to look at the contributions of weight $\omega - w_0(\omega)$, which as operators give a possibly non zero contribution only when applied to the lowest weight vector; for these we use lexicographic order. We need to look at the possible contributing monomials and choose inductively the highest exponent for β_1 then β_2 and so on, recalling that if we have already chosen $t_0 f_{\beta_s}^{m_s} \dots t_0 f_{\beta_1}^{m_1}$ we must have $t_0 f_{\beta_s}^{m_s} \dots t_0 f_{\beta_1}^{m_1} v_1 \neq 0$ and we must choose the highest exponent m_{s+1} for which $t_0 f_{\beta_{s+1}}^{m_{s+1}} t_0 f_{\beta_s}^{m_s} \dots t_0 f_{\beta_1}^{m_1} v_1 \neq 0$. By induction we assume that we have constructed the monomial $t_0 f_{\beta_{k_s}} \dots t_0 f_{\beta_{k_1}}$. We are thus exactly in the situation described by our previous lemma, at the next step the one parameter group $exp(\tau t_0 f_{\beta_j})$ fixes $t_0 f_{\beta_{k_s}} \dots t_0 f_{\beta_{k_1}} v_1$ unless $j = k_{s+1} \in I_\omega$, in this case the maximal exponent m for which $t_0 f_{\beta_{k_{s+1}}}^m t_0 f_{\beta_{k_s}} \dots t_0 f_{\beta_{k_1}} v_1 \neq 0$ is 1. $\square$

§24. The regular elements.

24.1. We return now to the algebra U_ε of degree $d = \ell^N$ (§20.1). Let us denote by $\chi : Spec Z_\varepsilon \to \Omega = Spec Z_0$ the natural map. Recall that $\Omega = H$ and in 16.3 we have identified its regular elements. Since the map χ is finite and compatible with the Poisson structure it is clear that $Spec Z_{\varepsilon,\, reg} = \chi^{-1}(\Omega_{reg})$.

Theorem. *The points in $Spec Z_{\varepsilon,\, reg}$ parametrize irreducible U_ε-modules of maximal dimension ($= d$).*

Proof. Let $\Omega^0 = \{u \in \text{Spec } Z_0 |$ all representations from $\chi_0^{-1}(u)$ have dimension $d\}$, as introduced in Remark 4.5. We are claiming that, given a $\tilde{G}$-orbit $\mathcal{O}$ of maximal dimension, $\mathcal{O} \subset \Omega^0$. Using the action of $\tilde{G}$ (cf. §16) it suffices to see that the Zariski closure $\overline{\mathcal{O}}$ of $\mathcal{O}$ has non empty intersection with Ω^0. We wish to apply to our situation Lemma 4.10 with I the ideal of $\overline{\mathcal{O}}$. From Theorem 10.7 each of the algebras $U_\varepsilon^{(j)}$ introduced in Remark 10.1 has degree ℓ^N. We are then reduced to show that if $\overline{I}$ is the associated graded ideal of I in $\overline{U}_\varepsilon = U_\varepsilon^{(2N)}$, and $\mathcal{O}_1$ is its set of zeroes, then

$$\mathcal{O}_1 \cap \Omega^0_{\overline{U}_\varepsilon} \neq \emptyset. \tag{24.1.1}$$

We will describe $\overline{I}$ using the method of Proposition 4.9. It is well known that the Zariski closure of $\pi(\mathcal{O})$, being the Zariski closure of the conjugacy class of maximal dimension in G, is given by the equations (for some $c_i \in \mathbb{C}$):

$$tr_{V_{\omega_i}} g = c_i,\ i = 1, \dots, u,\ g \in \overline{\pi(\mathcal{O})}.$$

It follows from Proposition 19.2b and d that the Zariski closure of $\mathcal{O}$ in Spec Z_0 is given by the equations

$$\phi_{\omega_i} = c_i,\ i = 1, \dots, n. \tag{24.1.2}$$

Consider the elements $\overline{\phi}_{\omega_i}$, images of ϕ_{ω_i} in the graded algebra $\overline{U}_\varepsilon$. We want to show that these elements form a regular sequence of $\overline{Z}_0 \subset \overline{U}_\varepsilon$ so that they generate $\overline{I}$ (cf. Proposition 4.9). The elements $\overline{\phi}_{\omega_i}$ have been computed in Lemma 23.1 where we have seen that they are monomials in disjoint sets of indeterminates. Hence they form a regular sequence by trivial reasons. In order to complete the proof we have to show the set $\mathcal{O}_1$ of solutions of the system of equations

$$\overline{\phi}_{\omega_i} = 0,\ i = 1, \dots, n, \tag{24.1.3}$$

intersects nontrivially the set $\Omega^0_{\overline{U}_\varepsilon}$.

The variety given by the equations (24.1.3) is a union of subvarieties given as follows: we choose from each monomial $\overline{\phi}_{\omega_i}$ a factor $T_0 y_{\pm\beta}$ and letting them 0 we define a component of the variety under study. It is enough to prove that each one of these subvarieties intersects the open set $\Omega^0_{\overline{U}_\varepsilon}$ non trivially. But since $\overline{U}_\varepsilon$ is essentially a twisted polynomial algebra this is a consequence of the structure of the associated skew matrix performed in §10. □

§25 Some open problems.

25.1 In conclusion, we present some conjectures.

Let π be a finite-dimensional irreducible representation of a quantum group U_ε in a vector space V. Let $\chi_\pi : Z_0 \to \mathbb{C}$ be the corresponding point of Ω and let $\mathcal{O}$ be the $\tilde{G}$-orbit of χ_π.

CONJECTURE 1. $\dim V$ is divisible by $\ell^{\dim \mathcal{O}/2}$.

We have seen in fact that $\dim V = \ell^N$ if $\mathcal{O}$ is an orbit of maximal dimension.

We can reduce, in order to study these problems, to analyze a point p in the spectrum of Z_ε lying over B^+. Taking its semisimple part s consider $R^\pi = \{\alpha \in R | \alpha(s^2) = 1\}$ and let U^π denote the subalgebra of U_ε generated by U^0_ε and the $T_w E_i$ such that $w(\alpha_i) \in R^\pi$. Let $v \in V$ be an eigenvector of $U^0_\varepsilon U^+_\varepsilon$, and let $V^0 = U^\pi v$. CONJECTURE 2. V^0 is irreducible with respect to U^π and $\dim V = \ell^{|R_+ \setminus R^\pi|} \dim V_0$.

Note that Conjecture 1 is similar to a conjecture in [W-K] and Conjecture 2 is similar to Theorem 2 from [W-K] on representations of Lie algebras in characteristic p.

It is possible that the dimension of any irreducible representation of B_ε whose central character restricted to Z_0^+ is a point of X_w is exactly $\ell^{\frac{1}{2}(\ell(w)+\mathrm{rank}(1-w))}$. This fact if true would require a more detailed analysis in the spirit of Section 1.3.

We would like, in conclusion, to propose a more general conjecture, similar to one of the results of [WK] on solvable Lie algebras of characteristic p.

Let A be an algebra over $\mathbb{C}[q, q^{-1}]$ on generators $x_1, \ldots, x_n$ satisfying the following relations:

$$x_i x_j = q^{h_{ij}} x_j x_i + P_{ij} \text{ if } i > j,$$

where (h_{ij}) is a skew-symmetric matrix over $\mathbb{Z}$ and $P_{ij} \in \mathbb{C}[q, q^{-1}][x_1, \ldots, x_{i-1}]$. Let $\ell > 1$ be an integer relatively prime to all elementary divisors of the matrix (h_{ij}) and let ε be a primitive ℓ-th root of 1. Let $A_\varepsilon = A/(q - \varepsilon)$ and assume that all elements x_i^ℓ are central. Let $Z_0 = \mathbb{C}[x_1^\ell, \ldots, x_n^\ell]$; this algebra has a canonical Poisson structure.

CONJECTURE 3. Let π be an irreducible representation of the algebra A_ε and let $\mathcal{O}_\pi \subset \operatorname{Spec} Z_0$ be the symplectic leaf containing the restriction of the central character of π to Z_0. Then the dimension of this representation is equal to $\ell^{\frac{1}{2}\dim \mathcal{O}_\pi}$.

This conjecture of course holds if all P_{ij} are 0, and it is in complete agreement with our results.

Appendix

In this appendix we shall treat all the results for the rank one and rank 2 cases that we left unproven before. In the rank 2 case whenever the are two root lengths α_1 will be the long simple root. Also we set $E_i = E_{\alpha_i}$, $i = 1, 2$.

One has the following formulas (see §9) which one can check by direct computation.

Case A_2**.** $w_0 = s_1 s_2 s_1 = s_2 s_1 s_2$.

$$\begin{aligned} T_1(E_2) &= -E_1 E_2 + q^{-1} E_2 E_1 \\ T_1 T_2(E_1) &= E_2 \\ T_2(E_1) &= -E_2 E_1 + q^{-1} E_2 E_1 \\ T_2 T_1(E_2) &= E_1 \end{aligned} \tag{A1}$$

Case B_2**.** $w_0 = s_1 s_2 s_1 s_2 = s_2 s_1 s_2 s_1$.

$$\begin{aligned} T_1(E_2) &= -E_1 E_2 + q^{-2} E_2 E_1 \\ T_1 T_2(E_1) &= q^{-2} E_2^{(2)} E_1 - q^{-1} E_2 E_1 E_2 + E_1 E_2^{(2)} \\ T_1 T_2 T_1(E_2) &= E_2 \\ T_2(E_1) &= E_2^{(2)} E_1 - q^{-1} E_2 E_1 E_2 + q^{-2} E_1 E_2^{(2)} \\ T_2 T_1(E_2) &= -E_2 E_1 + q^{-1} E_2 E_1 \\ T_2 T_1 T_2(E_1) &= E_1 \end{aligned} \tag{A2}$$

Case G_2**.** $w_0 = s_1 s_2 s_1 s_2 s_1 s_2 = s_2 s_1 s_2 s_1 s_2 s_1$.

$$\begin{aligned} T_1(E_2) &= -E_1 E_2 + q^{-3} E_2 E_1 \\ T_1 T_2(E_1) &= [3]^{-1}(q T_1 T_2 T_1(E_2) T_1(E_2) - T_1(E_2) T_1 T_2 T_1(E_2)) \\ T_1 T_2 T_1(E_2) &= q^{-4} E_2^{(2)} E_1 - q^{-2} E_2 E_1 E_2 + E_1 E_2^{(2)} \\ T_1 T_2 T_1 T_2(E_1) &= q^{-3} E_2^{(3)} E_1 - q^{-2} E_2^{(2)} E_1 E_2 + q^{-1} E_2 E_1 E_2^{(2)} - E_1 E_2^{(3)} \\ T_1 T_2 T_1 T_2 T_1(E_2) &= E_2 \\ T_2(E_1) &= -E_2^{(3)} E_1 + q^{-1} E_2^{(2)} E_1 E_2 - q^{-2} E_2 E_1 E_2^{(2)} + q^{-3} E_1 E_2^{(3)} \\ T_2 T_1(E_2) &= E_2^{(2)} E_1 - q^{-2} E_2 E_1 E_2 + q^{-4} E_1 E_2^{(2)} \\ T_2 T_1 T_2(E_1) &= [3]^{-1}(q T_2 T_1 T_2 T_1(E_2) T_2 T_1(E_2) - T_2 T_1(E_2 T_2 T_1 T_2 T_1(E_2)) \\ T_2 T_1 T_2 T_1(E_2) &= -E_2 E_1 + q^{-3} E_1 E_2 \\ T_2 T_1 T_2 T_1 T_2(E_1) &= E_1 \end{aligned} \tag{A3}$$

From these formulas one than gets by a boring but straightforward computation the following identities:

Case A_2**.** We set $T_1(\overline{E}_2) = \overline{E}_{12}$, $T_2(\overline{E}_1) = \overline{E}_{21}$. Then we have,

$$\begin{aligned} \overline{E}_2 \overline{E}_1 &= q \overline{E}_1 \overline{E}_2 + q(q - q^{-1}) \overline{E}_{12}, \\ \overline{E}_{12} \overline{E}_1 &= q^{-1} \overline{E}_1 \overline{E}_{12} \end{aligned} \tag{A4}$$

Also

(A5) $$\overline{E}_{21} = -q\overline{E}_{12} - \overline{E}_1\overline{E}_2.$$

Case B_2**.** We set $T_1(\overline{E}_2) =, T_1T_2(\overline{E}_1) =, T_2(\overline{E}_1) =, T_2T_1(\overline{E}_2) =$. Then we have,

(A6)
$$\begin{aligned}
\overline{E}_2\overline{E}_1 &= q^2\overline{E}_1\overline{E}_2 + q^2(q^2 - q^{-2})\overline{E}_{12},\\
\overline{E}_2\overline{E}_{12} &= \overline{E}_{12}\overline{E}_2 + (q - q^{-1})\overline{E}_{122},\\
\overline{E}_2\overline{E}_{122} &= q^{-2}\overline{E}_{122}\overline{E}_2\\
\overline{E}_{122}\overline{E}_1 &= \overline{E}_1\overline{E}_{122} + q(q^2 - q^{-2})\overline{E}_{12}^2,\\
\overline{E}_{122}\overline{E}_{12} &= q^{-2}\overline{E}_{12}\overline{E}_{122}\\
\overline{E}_{12}\overline{E}_1 &= q^{-2}\overline{E}_1\overline{E}_{12}.
\end{aligned}$$

Also

(A7)
$$\begin{aligned}
\overline{E}_{221} &= q^2\overline{E}_{122} + q^2(q + q^{-1})\overline{E}_{12}\overline{E}_2 + q\overline{E}_1\overline{E}_2^2\\
\overline{E}_{21} &= -q^2\overline{E}_{12} - \overline{E}_1\overline{E}_2,
\end{aligned}$$

Case G_2**.** We set:

$$T_1(\overline{E}_2) = \overline{E}_{12}, \quad T_1T_2(\overline{E}_1) = \overline{E}_{11222},$$
$$T_1T_2T_1(\overline{E}_2) = \overline{E}_{122}, \quad T_1T_2T_1T_2(\overline{E}_1) = \overline{E}_{1222}, \; T_2(\overline{E}_1) = \overline{E}_{2221},$$
$$T_2T_1(\overline{E}_2) = \overline{E}_{221}, \; T_2T_1T_2(\overline{E}_1) = \overline{E}_{22211}, \; T_2T_1T_2T_1(\overline{E}_2) = \overline{E}_{21}.$$

Then we have,

(A8)
$$\begin{aligned}
\overline{E}_2\overline{E}_1 &= q^3\overline{E}_1\overline{E}_2 + q^3(q^3 - q^{-3})\overline{E}_{12},\\
\overline{E}_2\overline{E}_{12} &= q\overline{E}_{12}\overline{E}_2 + q(q^2 - q^{-2})\overline{E}_{122},\\
\overline{E}_2\overline{E}_{11222} &= \overline{E}_{11222}\overline{E}_2 - q(q^3 - q^{-3})\overline{E}_{122}^2,\\
\overline{E}_2\overline{E}_{122} &= q^{-1}\overline{E}_{122}\overline{E}_2 + q^{-1}(q - q^{-1})\overline{E}_{1222},\\
\overline{E}_2\overline{E}_{1222} &= q^{-3}\overline{E}_{1222}\overline{E}_2,\\
\overline{E}_{1222}\overline{E}_1 &= q^3\overline{E}_1\overline{E}_{1222} - q^3[3](q^3 - q^{-3})\overline{E}_{12}\overline{E}_{122} +\\
&\quad (q^3 - q^{-3})(1 - q^4 - q^2)\overline{E}_{11222},\\
\overline{E}_{1222}\overline{E}_{12} &= \overline{E}_{12}\overline{E}_{1222} + q(q^3 - q^{-3})\overline{E}_{122}^2,\\
\overline{E}_{1222}\overline{E}_{11222} &= q^{-3}\overline{E}_{11222}\overline{E}_{1222} + (q^3 - q^{-3})\overline{E}_{122}^3,\\
\overline{E}_{1222}\overline{E}_{122} &= q^{-3}\overline{E}_{122}\overline{E}_{1222},\\
\overline{E}_{122}\overline{E}_1 &= \overline{E}_1\overline{E}_{122} - q(q^3 - q^{-3})\overline{E}_{12}^2,\\
\overline{E}_{122}\overline{E}_{12} &= q^{-1}\overline{E}_{12}\overline{E}_{122} + q^{-1}(q - q^{-1})\overline{E}_{11222},\\
\overline{E}_{122}\overline{E}_{11222} &= q^{-3}\overline{E}_{11222}\overline{E}_{122},\\
\overline{E}_{11222}\overline{E}_1 &= q^{-3}\overline{E}_1\overline{E}_{11222} + (q^3 - q^{-3})\overline{E}_{12}^3,\\
\overline{E}_{11222}\overline{E}_{12} &= q^{-3}\overline{E}_{12}\overline{E}_{11222},\\
\overline{E}_{12}\overline{E}_1 &= q^{-3}\overline{E}_1\overline{E}_{12}.
\end{aligned}$$

Also

$$\overline{E}_{2221} = -q^3\overline{E}_{1222} - q^4[3]\overline{E}_{122}\overline{E}_2 - q^4[3]\overline{E}_{12}\overline{E}_2^2 - q^3\overline{E}_1\overline{E}_2^3,$$

$$\overline{E}_{221} = q^4\overline{E}_{122} + q^3[2]\overline{E}_{12}\overline{E}_2 + \overline{E}_1\overline{E}_2^2$$

(A9)
$$\overline{E}_{22211} = q^6\overline{E}_{11222} + q^4(q^2 - q^{-2} - 1)\overline{E}_1\overline{E}_{1222} + (q - q^{-1})q^5[3]!\overline{E}_1\overline{E}_{122}\overline{E}_2 + q^7[3]\overline{E}_1\overline{E}_{12}\overline{E}_2^2 + q^6\overline{E}_1^2\overline{E}_2^3 + q^6(2q + q^{-1})[3]\overline{E}_{12}\overline{E}_2 + q^7[3]\overline{E}_{12}\overline{E}_{122},$$

$$\overline{E}_{21} = -q^3\overline{E}_{12} - \overline{E}_1\overline{E}_2,$$

Finally we remark that using the antiautomorphism κ we obtain completely analogous formulas for the $\overline{F}_\beta$'s.

Let us now prove Theorem 12.1 in the rank one and rank 2 cases.

Theorem A1. *The algebra A is a quantum deformation of $k[H]$, H being the Poisson algebraic group defined in example 2 of section 11.4.*

Following the line of Section 12 we shall divide the proof in steps and in each step prove only the statements which have are needed to complete that proof.

Step 1. If we fix a reduced expression for w_0, and we consider the corresponding elements $\overline{E}_{beta}$'s and F_β's, then the monomials

$$\overline{E}^k K_\lambda \overline{F}^t$$

are a basis for A.

We set A' equal to the R span of the monomials $\overline{E}^k K_\lambda \overline{F}^t$, A^+ equal to the span of those among these monomials of the form $\overline{E}_{\beta_1}^{k_1} \cdots \overline{E}_{\beta_N}^{k_N}$, A^- the span of those of the form $\overline{F}_{\beta_N}^{t_N} \cdots \overline{F}_{\beta_1}^{t_1}$, A^0 equal to the span of the K_λ's. Looking at the proof given in section 11 we see that we only nee to verify that A^+ and A^- are closed under product and do not depend on the reduced expression. This is clear from the above relations.

We then have

Lemma. $adE_i(A^+) \subset A^+$; $\tilde{ad}F_i(A^-) \subset A^-$.

Proof. This statement is obvious by inspection on the above relations.

From this lemma as in section 12 we then get

Step 2. $A/(q-1)$ *is a commutative algebra.*

The next step requires more proof.

Step 3. $\Delta(A) \subset A \otimes A$, $S(A) \subset A$, $\varepsilon(A) \subset R$.

Proof. The proof given in Section 11 works in all cases except G_2. Even in this case let us remark that

$$\overline{E}_{12} = -adE_1(\overline{E}_2)$$

$$\overline{E}_{2221} = -qadE_2(\overline{E}_{221})$$

Thus, using the arguments in section 11 we will be done if we can show that $\Delta(\overline{E}_{122})$ and $\Delta(\overline{E}_{11222})$ both lie in $A \otimes A$. A little computation shows

$$\Delta(\overline{E}_{122}) = \overline{E}_{122} \otimes 1 + K_{\alpha_1+2\alpha_2} \otimes \overline{E}_{122} + q\overline{E}_1 K_{2\alpha_2} \otimes \overline{E}_2^2 - (q+q^{-1})K_{\alpha_2}\overline{E}_{12} \otimes \overline{E}_2.$$

As for $\overline{E}_{11222}$, notice that

$$\overline{E}_{11222} = q^{-1} adE_1(\overline{E}_{1222}) - (q + q^{-1})\overline{E}_{12}\overline{E}_{122} - q^{-2}\overline{E}_{122}\overline{E}_{12},$$

so again everything follows as in Section 12. The cases of the antipode and augmentation are much easier and left to the reader.

Step 4. *There is a isomorphism of Poisson groups between SpecB and the Poisson group H.*

For this one has to proceed as in §12 and verify explicitly (as sketched there by explicit matrix computations) the necessary identities for the group H, we leave the details to thc reader.

REFERENCES

[A] M.Artin, On Azumaya algebras and finite-dimensional representations of rings, J. Algebra 11(1969), 532-563.

[B-C] P. Bala, R.W. Carter, Classes of unipotent elements in simple algebraic groups II Math. Proc. Cambridge Phil. Soc. 80 1976 pp. 1-18

[B-D] A. Belavin, V. Drinfeld, On the solutions of the classical Yang-Baxter equations for simple Lie algebras, Funk. Anal. Pri. 16 (1982), n0. 3, 1-29.

[Be] J. Beck, Thesis

[B] N. Bourbaki, Algebre, Ch X, Algèbre homologique, Paris, 1980.

[C] R.W. Carter, Conjugacy classes in the Weyl group, Compositio Math. 25 (1972), 1-59.

[DC-K] C. De Concini, V.G. Kac, Representations of quantum groups at roots of 1, in Progress in Math., 92, Birhäuser, 1950, pp. 471-506.

[DK2] C. De Concini, V.G. Kac, Representations of quantum groups at roots of 1: reduction to the exceptional case, RIMS 792 preprint, 1991.

[DC-K-P] C. De Concini, V.G. Kac, C. Procesi, Quantum coadjoint action, Journal of AMS, 5 (1992), 151-190.

[DKP2] C. De Concini, V.G. Kac, C. Procesi, Some remarkable degenerations of quantum groups, preprint, 1991.

[D1] V.G. Drinfeld, Hopf algebras and the quantum Yang-Baxter equation, Soviet. Math. Dokl. 32 (1985), n. 1,254-258.

[D2] V.G. Drinfeld, Quantum groups, Proc. ICM Berkeley 1 (1986), 789-820

[GL] K.R. Goodearl, E.S. Letzter, Prime ideals in skew and q-skew polynomial rings, preprint, 1991.

[J1] M. Jimbo, A q-difference analogue of $U(q)$ and the Yang-Baxter equation, Lett. Math. Phys. 10 (1985),63-69.

[J2] M. Jimbo, A q-analogue of $U(gl(N+1))$, Hecke algebras and the Yang-Baxter equation, Lett. Math. Phys. 11 (1986),247-252.

[KP] V.G. Kac, D.H. Peterson, Generalized invariants of groups generated by reflections, in Progress in Math. 60, Birkhäuser, 1985, 231-250.

[KP2] V.G. Kac, D.H. Peterson, Defining relations of certain infinite-dimensional groups Asterisque, hors série 1985 pp.165-208

[KW] V.G. Kac, B. Yu. Weisfeiler, Coadjoint action of a semi-simple algebraic group and the center of the enveloping algebra in characteristic p

[K] B. Kostant, Lie group representations of polynomial rings Amer. J. Math. 86 1963 pp. 327-402

[K-R] P. Kulish, N. Reshetikhin, J. Soviet math 23, n.4 (1983).

[LS] S.Z. Levendorskii, Ya. S. Soibelman, Algebras of functions on compact quantum groups, Schubert cells and quantum tori, Comm. Math. Physics 139 (1991), 141-170.

[Lu-R] J.-H. Lu, T. Ratiu On the nonlinear convexity theorem of Kostant, Journal of AMS, 1991

[LW] J.-H. Lu, A Weinstein, Poisson Lie groups, dressing transformations and Bruhat decompositions, J. Diff. Geom. 31 (1990), 501-526.

[L1] G. Lusztig, Quantum deformations of certain simple modules over enveloping algebras, Adv. in Math. 70 (1988),237-249.

[L2] G. Lusztig, Quantum groups at roots of 1, Geom. Ded. 35 (1990), 89–114.

[L3] G. Lusztig, Finite dimensional Hopf algebras arising from quantum groups, J. Amer. Math. Soc. 3 (1990),257-296.

[L4] G. Lusztig, Canonical bases arising from quantized enveloping algebras, J.Amer. Math.Soc. 3 (1990) 447-498.

[M] Matsumoto,

[Pa] P. Papi, On certain convex orderings associated to root sysyems Proc. A.M.S.

[MC-R] J.C. McConnell, J.C. Robson, Noncommutative Noetherian rings, John Wiley & sons (1987).

[P1] C.Procesi, Rings with polynomial identities. Pure and Applied Mathematics 17, 1973, M. Dekker.

[P2] C.Procesi, Finite dimensional representations of algebras. Israel Journal of Mathematics 19 (1974).

[P3] C.Procesi, Trace identities and standard diagrams. Ring theory. Lecture notes in Pure and Appl. Mathematics 51 M. Dekker, (1979), pp. 191-218.

[P4] C.Procesi, A formal inverse to the Cayley Hamilton theorem. Journal of Algebra 107, (1987), pp. 63-74.

[Re] N. Reshetikhin, dquantizzazione

[Ri] C. Ringel, Hall algebras and quantum groups, 1988

[R1] M. Rosso, Finite dimensional representations of the quantum analogue of the enveloping algebra of a semisimple Lie algebra, Comm. Math. Phys. 117 (1988),581-593.

[R2] M. Rosso, Analogues de la forme de Killing et du théorème d'Harish–Chandra pour les group es quantiques, Ann. Sci. Ec. Norm. Sup. 23 (1990), 445–467.

[STS] M.A. Semenov–Tian–Shansky, Dressing transformations and Poisson group actions, Publ. RIMS 21 (1985), 1237-1260.

[Se] J.-P. Serre, Algebre locale Multiplicites Lecture Notes in Math 11, 1975, Springer Verlag

[Sh] N. Shapovalov, On a bilinear form on the universal enveloping algebra of a complex semisimple Lie algebra, Func. Anal. Appl. 6 (1972), 307-312.

[Sk] E. Sklyanin, Uspekhi Mat. Nauk 40, n.2 (242) (1985), 214.

[S] R. Steinberg, Regular elements of semi-simple algebraic groups Publ. Math. IHES vol 25 1965 pp. 49–80

[Ta] T. Takesaki, Killing forms, Harish Chandra isomorphisms, and universal R–matrices for quantum algebras, RIMS–816 preprint (1991).

[T] J. Tits, Sur les constants de structure et le théorème d'existence des algebres de Lie semi-simple, Publ. Math. IHES 31 (1966), 21–58.

[WK] B. Yu, Weisfeiler, V.G. Kac, On irreducible representations of Lie p–algebras, Funct. Anal. Appl. 5:2 (1971), 28–36.

INDEX THEOREMS FOR R-CONSTRUCTIBLE SHEAVES AND FOR $\mathcal{D}$-MODULES

PIERRE SCHAPIRA   JEAN-PIERRE SCHNEIDERS

0. INTRODUCTION

In these lectures we shall construct characteristic classes in order to calculate Euler-Poincaré indices. In a first step, as an introduction, we consider **R**-constructible sheaves on a real analytic manifold. Next we treat the more general case of an "elliptic pair" on a complex manifold, which includes the case of $\mathcal{D}$-modules.

These classes are constructed using classical procedures: diagonal embedding and duality, as Grothendieck-Illusié-Verdier did for the Lefschetz fixed point formula, but here we construct these classes "microlocally", that is, in the cotangent bundle. This is possible thanks to the "microlocal theory of sheaves" of Kashiwara-Schapira (see [K-S]) whose main features we recall in §1. This microlocal point of view will appear as necessary when performing inverse images and products.

Let us explain with more details one of our main results (see [S-Sc 1,2]). Let X be an n-dimensional complex manifold, $\mathcal{M}$ a coherent $\mathcal{D}$-module, F an **R**-constructible sheaf (more generally: objects of derived categories). One says $(\mathcal{M}, F)$ is elliptic if $\mathrm{char}(\mathcal{M})$, the characteristic variety of $\mathcal{M}$, and $SS(F)$, the micro-support of F, do not intersect outside of the zero-section of T^*X. To such a pair, one can associate what we call its "microlocal Euler class", $\mu\mathrm{eu}(\mathcal{M}, F)$, a cohomology class of $H^{2n}_{\mathrm{char}(\mathcal{M})+SS(F)}(T^*X; \mathbf{C}_{T^*X})$. If $\mathrm{supp}(\mathcal{M}) \cap \mathrm{supp}(F)$ is compact, and if one denotes by $\mathrm{eu}(\mathcal{M}, F)$ the restriction of $\mu\mathrm{eu}(\mathcal{M}, F)$ to the zero-section, then one shows that:

$$\chi(X; F \odot \mathcal{M}) = \int_X \mathrm{eu}(\mathcal{M}, F)$$

where the left-hand side denotes the Euler-Poincaré index of the complex $R\Gamma(X; F \otimes \mathcal{M} \overset{L}{\underset{\mathcal{D}_X}{\otimes}} \mathcal{O}_X)$, which is proven to have finite dimensional cohomology.

This formula is in fact a particular case of a more general result which asserts that, under suitable hypotheses the microlocal Euler class "commutes" to direct images and to inverse images. In particular one gets that $\mu\mathrm{eu}(\mathcal{M}, F)$ is the "product" (we give a precise meaning to it here) of $\mu\mathrm{eu}(\mathcal{M}, \mathbf{C}_X)$ and $\mu\mathrm{eu}(\mathcal{O}_X, F)$, this last class being nothing but the "characteristic cycle" of F, as defined by Kashiwara [K2].

In the last section we define the microlocal Chern character $\mu ch_\Lambda(\mathcal{M})$ of a coherent $\mathcal{D}_X$-module $\mathcal{M}$ endowed with a good filtration and show that this construction also commutes to the usual operations. The conjecture that $\mu ch^{2n}_\Lambda(\mathcal{M}) = \mu\mathrm{eu}(\mathcal{M})$ would provide a direct proof and a wide generalization of the Atiyah-Singer theorem.

I. Microlocal Study of Sheaves

In this section we fix some notations and recall a few results of [K-S].

Let X be a real analytic manifold. One denotes by $\tau : TX \to X$ and by $\pi : T^*X \to X$ its tangent and cotangent bundles, respectively. If M is a submanifold, one denotes by T_MX and T^*_MX the normal and conormal bundles to M in X, respectively. In particular T^*_XX is the zero-section of T^*X. If Λ is a subset of T^*X, Λ^a will denote its image by the antipodal map on T^*X.

One denotes by $\delta : \Delta \hookrightarrow X \times X$ the diagonal embedding and we identify X to Δ and T^*X to $T^*_\Delta(X \times X)$ by the first projection defined on $X \times X$ and on $T^*(X \times X) \simeq T^*X \times T^*X$, respectively.

If X and Y are two manifolds, one denotes by q_1 and q_2 the first and second projection, defined on $X \times Y$.

Let A be a commutative unitary ring with finite global homological dimension (e.g. $A = \mathbf{Z}$). One denotes by $D(X)$ the derived category of the category of sheaves of A-modules on X, and by $D^b(X)$ the full subcategory consisting of objects with bounded cohomology. If Z is a locally closed subset of X, one denotes by A_Z the sheaf on X which is constant with stalk A on Z and zero on $X \setminus Z$.

One denotes by or_X the orientation sheaf on X, and by ω_X the dualizing complex on X. Hence :

$$\omega_X \simeq or_X[\dim X],$$

where $\dim X$ is the dimension of X. More generally, if f is a morphism from Y to X, one sets:

$$\omega_{Y/X} = \omega_Y \otimes f^{-1}\omega_X^{\otimes -1}.$$

The "six operations" (as says Grothendieck), that is, the operations $\otimes^L$, $R\mathcal{H}om$, Rf_*, $Rf_!$, f^{-1}, $f^!$, are now classical tools that we shall not recall. We simply introduce some notations. For $F \in Ob(D^b(X))$ and $G \in Ob(D^b(Y))$, one sets:

$$F \overset{L}{\boxtimes} G = q_1^{-1}F \overset{L}{\otimes} q_2^{-1}G,$$

$$D'F = R\mathcal{H}om(F, A_X),$$
$$DF = R\mathcal{H}om(F, \omega_X).$$

There are other operations of interest on sheaves. If M is a closed submanifold of X and $F \in Ob(D^b(X))$, the specialization of F along M, $\nu_M(F)$, is an object of $D^b(T_MX)$ and the microlocalization of F along M, $\mu_M(F)$, an object of $D^b(T^*_MX)$. Sato's functor μ_M has been generalized as follows (see [K-S]). For F and G in $D^b(X)$ one sets:

$$\mu\mathrm{hom}(G, F) = \mu_\Delta R\mathcal{H}om(q_2^{-1}G, q_1^!F).$$

Then:

$$R\pi_*\mu\mathrm{hom}(G, F) \simeq R\mathcal{H}om(G, F),$$

$$\mu\mathrm{hom}(A_M, F) \simeq \mu_M(F).$$

After the introduction of the functor μ_M it became natural to work in T^*X, and M. Kashiwara and one of the authors introduced in 1982 (see [K-S]) the micro-support $SS(F)$ of an object F of $D^b(X)$. This is a closed conic subset of T^*X which roughly speaking describes the set of codirections of non-propagation of F. More precisely:

Definition 1.1. We say that an open subset U of T^*X does not meet $SS(F)$ if for any real C^1-function φ on X and any $x_0 \in X$ such that $(x_0; d\varphi(x_0)) \in U$, one has:

$$\left(R\Gamma_{\{x;\varphi(x)\geq\varphi(x_0)\}}(F)\right)_{x_0} = 0.$$

Examples 1.2.

(a) One has:

$$SS(A_X) = T^*_X X.$$

More generally, if F belongs to $D^b(X)$, then $SS(F) \subset T^*_X X$ if and only if for all j, $H^j(F)$ is locally constant.

(b) If X is vector space and if γ is a closed cone with vertex at 0, then:

$$SS(A_\gamma) \cap \pi^{-1}(0) = \gamma^\circ.$$

In particular, if M is a closed submanifold of X, then $SS(A_M) = T^*_M X$.

(c) Note that if Z is any locally closed subset of X, then it is now natural to define its conormal bundle $T^*_Z X$ as $SS(A_Z)$.

Example 1.3. Assume X is a complex manifold, let $\mathcal{D}_X$ denote the sheaf of rings of differential operators on X (cf §3 below) and let $\mathcal{M}$ be a coherent $\mathcal{D}_X$-module. Then:

$$SS(R\mathcal{H}om_{\mathcal{D}_X}(\mathcal{M}, \mathcal{O}_X)) = \mathrm{char}(\mathcal{M}) \tag{1.1}$$

where char($\mathcal{M}$) denotes the characteristic variety of $\mathcal{M}$.

Note that the inclusion $* \subset *$ in (1.1) is easily deduced from the precise Cauchy-Kowalevski theorem of Leray [Le]. In fact, using purely algebraic arguments, one is reduced to prove the inclusion in case $\mathcal{M} = \mathcal{D}_X/\mathcal{D}_X.P$, for a single differential operator P. Then one has to show that if ϕ is a real function on X such that the principal symbol of P does not vanishes at $d\phi(x_0)$, then P induces an isomorphism on the spaces $(H^j_{\{x;\varphi(x)\geq\varphi(x_0)\}}(\mathcal{O}_X))_{x_0}$ for all j.

In order to state the involutivity theorem below we need to recall some notions on normal cones and symplectic geometry.

If A and B are two subsets of X, the normal cone of B along A, $C(A,B)$, is the closed cone of TX defined in a local coordinate system by:

$$\begin{cases} (x_0, v_0) \in C(A,B) \Longleftrightarrow \text{ there exists a} \\ \text{sequence } \{(x_n, y_n, c_n)\} \subset A \times B \times \mathbf{R}^+ \text{ such} \\ \text{that } x_n \xrightarrow[n]{} x_0, y_n \xrightarrow[n]{} x_0, c_n(x_n - y_n) \xrightarrow[n]{} v_0. \end{cases} \tag{1.2}$$

If M is a closed submanifold, one denotes by $C_M(B)$ the image of $C(M,B)$ in the normal bundle $T_M X$.

The symplectic isomorphism $-H$ identifies T^*T^*X and TT^*X. In a local symplectic coordinate system $(x;\xi)$ on T^*X:

$$-H(\lambda dx_i + \mu d\xi_j) = \lambda\frac{\partial}{\partial \xi_i} - \mu\frac{\partial}{\partial x_j}.$$

By this isomorphism, if Λ is a Lagrangian submanifold of T^*X one identifies $T^*\Lambda$ and $T_\Lambda T^*X$.

Definition 1.4. A subset S of T^*X is involutive at $p \in S$ if for each $\theta \in T_p^*T^*X$ such that $C_p(S, S) \subset \{v \in T_pT^*X : \langle v, \theta\rangle = 0\}$, one has $-H(\theta) \in C_p(S)$.

One says S is involutive if it is so at each $p \in S$.

Involutivity Theorem 1.5. *Let $F \in Ob(D^b(X))$. Then $SS(F)$ is involutive.*

Note that in view of (1.1) one recovers the theorem of [S-K-K] which asserts that if X is a complex manifold and if $\mathcal{M}$ is a coherent $\mathcal{D}_X$-module, then char($\mathcal{M}$) is involutive.

In order to study the functorial properties of the micro-support, we use the following notations.

Let $f : Y \longrightarrow X$ be a morphism of manifolds. To f one associates the maps:

(1.3)
$$TY \xrightarrow[f']{} Y \underset{X}{\times} TX \xrightarrow[f_\tau]{} TX$$

and

(1.4)
$$T^*Y \xleftarrow[{}^tf']{} Y \underset{X}{\times} T^*X \xrightarrow[f_\pi]{} T^*X.$$

One also sets:

(1.5)
$$T_Y^*X = {}^tf'^{-1}(T_Y^*Y).$$

Definition 1.6. Let Λ be a closed conic subset of T^*X. One says f is non characteristic with respect to Λ if:

$$f_\pi^{-1}(\Lambda) \cap T_Y^*X \subset Y \underset{X}{\times} T_X^*X.$$

If $F \in Ob(D^b(X))$, one says f is non characteristic with respect to F if f is non characteristic with respect to $SS(F)$.

Theorem 1.7. *Let $F \in Ob(D^b(X))$ and $G \in Ob(D^b(Y))$.*

(a) $SS(G \overset{L}{\boxtimes} F) \subset SS(G) \times SS(F)$, $SS(R\mathcal{H}om(q_2^{-1}G, q_1^{-1}F)) \subset SS(G)^a \times SS(F)$.

(b) *Assume f is non characteristic for F. Then:*

$$SS(f^{-1}F) \subset {}^tf' f_\pi^{-1}(SS(F))$$

and moreover the natural morphism

$$f^{-1}F \otimes \omega_{Y/X} \longrightarrow f^!F$$

is an isomorphism.

(c) *Assume f is proper on supp(G). Then:*

$$SS(Rf_*G) \subset f_\pi {}^tf'^{-1}(SS(G)).$$

(d) *Assume $Y = X$. Then:*

$$SS(\mu hom(G, F)) \subset C(SS(F), SS(G)).$$

In particular:

$$\operatorname{supp} \mu hom(G, F) \subset \operatorname{supp}(G) \cap \operatorname{supp}(F).$$

Notes. The "microlocal point of view" appeared in the 70's with the introduction by M. Sato of the functor μ_M of microlocalization. Next Kashiwara and the author introduced the micro-support of sheaves and developed the microlocal study of sheaves (including theorems 1.5 and 1.7, the functor μhom and the equality (1.1)). We refer to [K-S] for detailed historical Notes.

II. Kashiwara's index theorem for R-constructible sheaves

The results of this section are explained with all details in [K-S, Chapters VIII and IX].

We do not review here the theory of subanalytic sets. Let us only recall that the family of subanalytic subsets of X contains that of semi-analytic sets, is stable by finite union or finite intersection, difference, inverse image and proper direct image.

If Λ is a conic subanalytic subset of T^*X, one says Λ is isotropic if the canonical 1-form of T^*X vanishes on Λ_{reg}, the smooth part of Λ. If Λ is both isotropic and involutive, one says Λ is Lagrangian.

In this section we assume the base ring A is a field that we denote by k. An object F of $D^b(X)$ is called weakly $\mathbf{R}$-constructible ($w-\mathbf{R}$-constructible for short) if there exists a subanalytic stratification $X = \bigsqcup_{\alpha} X_\alpha$ such that for all α, all $j \in \mathbf{Z}$, the sheaves $H^j(F)_{|X_\alpha}$ are locally constant. If moreover of each $x \in X$, each $j \in \mathbf{Z}$, the stalk $H^j(F)_x$ is finite dimensional, one says F is $\mathbf{R}$-constructible. One denotes by $D^b_{w-\mathbf{R}-c}(X)$ (resp. $D^b_{\mathbf{R}-c}(X)$) the full subcategory of $D^b(X)$ consisting of $w-\mathbf{R}$-constructible (resp. $\mathbf{R}$-constructible) objects.

The involutivity theorem allows us to characterize microlocally $w-\mathbf{R}$-constructible objects.

Theorem 2.1. *Let $F \in Ob(D^b(X))$. The following conditions are equivalent.*

(a) *F is $w-\mathbf{R}$-constructible.*
(b) *$SS(F)$ is contained in a closed conic subanalytic isotropic subset of T^*X.*
(c) *$SS(F)$ is a closed conic subanalytic Lagrangian subset of T^*X.*

Remark 2.2. Note that if X is a complex manifold one defines similarly the notions of (weakly) $\mathbf{C}$-constructible sheaves by assuming that the strata X_α are complex manifolds. Then one shows that an object F of $D^b_{w-\mathbf{R}-c}(X)$ is weakly $\mathbf{C}$-constructible if and only if $SS(F)$ is conic for the action of $\mathbf{C}^\times$ on T^*X.

Now let $f: Y \to X$ be a morphism of manifolds and let Λ_X (resp. Λ_Y) be a closed conic subanalytic isotropic subset of T^*X (resp. T^*Y). One shows easily that:

(i) ${}^tf' f_\pi^{-1}(\Lambda_X)$ is a closed conic subanalytic isotropic subset of T^*Y,
(ii) if f_π is proper on ${}^tf'^{-1}(\Lambda_Y)$ (i.e. if f is proper on $\Lambda_Y \cap T^*_YY$), then $f_\pi {}^tf'^{-1}(\Lambda_Y)$ is a closed conic subanalytic isotropic subset of T^*X.

Moreover, if Λ is a (locally closed) smooth Lagrangian submanifold of T^*X and if S is a conic subanalytic isotropic subset of T^*X, one proves that the normal cone $C_\Lambda(S)$ is subanalytic and isotropic in $T^*\Lambda$.

Using these results, and Theorem 2.1 one easily deduces:

Theorem 2.2. *Let $F \in Ob(D^b_{w-\mathbf{R}-c}(X))$ and $G \in Ob(D^b_{w-\mathbf{R}-c}(Y))$. Then:*

(a) *$G \boxtimes F \in Ob(D^b_{w-\mathbf{R}-c}(Y \times X))$.*
(b) *$f^{-1}F$ and $f^!F$ belong to $D^b_{w-\mathbf{R}-c}(Y)$.*
(c) *If f is proper on $\mathrm{supp}(G)$, $Rf_*(G) \in Ob(D^b_{w-\mathbf{R}-c}(X))$.*
(d) *Assume $Y = X$. Then:*

$$\mu hom(G, F) \in Ob(D^b_{w-\mathbf{R}-c}(T^*X))$$

and in particular

$$R\mathcal{H}om(G, F) \in Ob(D^b_{w-\mathbf{R}-c}(X)).$$

Moreover all there results remain true replacing everywhere "$w-\mathbf{R}-c$" by "$\mathbf{R}-c$".

Let $F \in Ob(D^b_{\mathbf{R}-c}(X))$ and assume F has compact support. The complex $R\Gamma(X;F)$ has finite dimensional cohomology, and a natural question then arises : to calculate the Euler-Poincaré index:

$$\chi(X;F) = \sum_j (-1)^j \dim_k H^j(X;F). \tag{2.1}$$

We shall describe Kashiwara's answer to this question (see [K 2], [K-S, Chapter IX]).

Consider the chain of morphisms:

$$R\mathcal{H}om(F,F) \xrightarrow{\sim} \delta^!(DF \boxtimes F) \longrightarrow DF \otimes F \underset{\mathrm{tr}}{\longrightarrow} \omega_X \tag{2.2}$$

where the second morphism is induced by $\delta^! \to \delta^{-1}$ and the third one is the trace morphism:

$$R\mathcal{H}om(F,\omega_X) \otimes F \to \omega_X.$$

Applying the functor $H^0(R\Gamma(X;\cdot))$ to (2.2) we get the morphism:

$$\mathrm{Hom}(F,F) \longrightarrow H^o_{\mathrm{supp}(F)}(X;\omega_X). \tag{2.3}$$

Definition 2.3. The Euler class of F, denoted $\mathrm{eu}(F)$, is the image of id_F in $H^o_{\mathrm{supp}(F)}(X;\omega_X)$ by the morphism (2.3).

Consider the diagram below where $\{\mathrm{pt}\}$ is the set with one element, and a_X is the canonical map $X \longrightarrow \{\mathrm{pt}\}$:

$$\begin{array}{ccccccc}
Ra_{X*}R\mathcal{H}om(F,F) & \simeq & Ra_{X*}\delta^!_X(DF \boxtimes F) & \to & Ra_{X*}(DF \otimes F) & \to & Ra_{X*}\omega_X \\
\downarrow & & & & \uparrow & & \downarrow \\
R\mathcal{H}om(Ra_{X!}F, Ra_{X!}F) & & \xleftarrow{\quad\sim\quad} & & D(Ra_{X!}F) \otimes Ra_{X!}F & \longrightarrow & k
\end{array}$$

One can prove that this diagram commutes, which implies:

Theorem 2.4. *Let $F \in Ob(D^b_{\mathbf{R}-c}(X))$ with compact support. Then:*

$$\chi(X;F) = \int_X \mathrm{eu}(F).$$

Recall that $\int_X : H^o_c(X;\omega_X) \to k$ is the map deduced from $Ra_{X!}\omega_X = Ra_{X!}a^!_X k_{\{pt\}} \to k_{\{pt\}}$.

When $k = \mathbf{R}$, if one calculate $H^o_c(X;\omega_X) \simeq H^n_c(X;\mathrm{or}_X)$ using the De Rham complex, the morphism $\int_X$ coincides with classical integration (cf. [K-S, Exercise VI 20]).

To calculate $\mathrm{eu}(F)$, it is worth to make it "microlocal". For that purpose one uses the relation:

$$\begin{aligned} R\mathcal{H}om(F,F) &\simeq R\pi_*\,\mu\mathrm{hom}(F,F) \\ &\xleftarrow{\sim} R\pi_*\,R\Gamma_{SS(F)}\mu\mathrm{hom}(F,F). \end{aligned}$$

Set $\Lambda = SS(F)$. We get the chain of morphisms:

$$\begin{aligned} R\mathcal{H}om(F,F) &\simeq R\pi_* R\Gamma_\Lambda \mu\mathrm{hom}(F,F) \\ &\simeq R\pi_* R\Gamma_\Lambda \mu_\Delta(DF \boxtimes F) \\ &\to R\pi_* R\Gamma_\Lambda \mu_\Delta(\delta_* DF \otimes F) \\ &\to R\pi_* R\Gamma_\Lambda \mu_\Delta(\delta_* \omega_X) \\ &\simeq R\pi_* R\Gamma_\Lambda \pi^{-1}\omega_X. \end{aligned}$$

The third arrow is induced by $\mathrm{id} \to \delta_*\delta^{-1}$ and the fourth is induced by the trace map. Applying $H^0 R\Gamma(X;\cdot)$ we get the map:

$$\mathrm{Hom}(F,F) \to H^o_{SS(F)}(T^*X;\pi^{-1}\omega_X). \tag{2.4}$$

Definition 2.5. The microlocal Euler class, denoted $\mu\mathrm{eu}(F)$, is the image of id_F in $H^o_{SS(F)}(T^*X;\pi^{-1}\omega_X)$ by the morphism (2.4).

Of course $\mathrm{eu}(F)$ is the restriction of $\mu\mathrm{eu}(F)$ to X, the zero-section of T^*X.

Since $SS(F)$ is Lagrangian, it is purely n-dimensional in T^*X and a section of $H^o_{SS(F)}(\pi^{-1}\omega_X) = H^n_{SS(F)}(\pi^{-1}\mathrm{or}_X)$ is a "Lagrangian cycle", with values in $\pi^{-1}\mathrm{or}_X$.

Then $\int_X \mathrm{eu}(F)$ may be interpreted as the intersection number of $\mu\mathrm{eu}(F)$ and of the cycle $[T^*_XX]$ associated to the zero-section of T^*X. Since this intersection number is invariant by homotopy, one can replace $[T^*_XX]$ by a deformation of it which will intersect $SS(F)$ at smooth points and transversally. Then the calculation of the intersection number is a local problem, and is quite easy. We refer to [K-S, Chapter IX] for a detailed study.

Now we shall study the functorial properties of the microlocal Euler class. (The same formalism will appear in the next section.)

Let $f : Y \to X$ be a morphism of manifolds and let Λ_X (resp. Λ_Y) be a closed conic subset of T^*X (resp. T^*Y). (We do not ask Λ_X and Λ_Y to be subanalytic or isotropic.)

First assume f_π is proper on ${}^t f'^{-1}(\Lambda_Y)$. Then the morphism:

$$Rf_{\pi!}\pi^{-1}\omega_Y \xrightarrow{\sim} \pi^{-1}Rf_!\omega_Y \longrightarrow \pi^{-1}\omega_X$$

defines the morphism:

$$f_\mu : H^o_{\Lambda_Y}(T^*Y;\pi^{-1}\omega_Y) \longrightarrow H^o_{f_\pi {}^t f'^{-1}(\Lambda_Y)}(T^*X;\pi^{-1}\omega_X). \tag{2.5}$$

Similarly, one can canonically construct a morphism for all $j \in \mathbf{Z}$:

$$R{}^t f'_!\, \pi^{-1} f^{-1} \omega_X \longrightarrow \pi^{-1}\omega_Y \tag{2.6}$$

from which one deduces that if ${}^t f'$ is proper on $f_\pi^{-1}(\Lambda_X)$, i.e. if f is non characteristic for Λ_X, then there is a canonical morphism:

$$f^\mu : H^o_{\Lambda_X}(T^*X;\pi^{-1}\omega_X) \longrightarrow H^o_{{}^t f' f_\pi^{-1}(\Lambda_X)}(T^*Y;\pi^{-1}\omega_Y). \tag{2.7}$$

To understand (2.6), one may consider the isomorphisms:

$$\pi^{-1} f^{-1} \omega_X \xrightarrow{\sim} \mathbf{C}_{Y\times_X T^*X} \otimes f_\pi^{-1} \pi^{-1} \mathrm{or}_X[n_X]$$
$$\xrightarrow{\sim} \omega_{Y\times_X T^*X} \otimes \pi^{-1} \mathrm{or}_Y[-n_Y]$$

from which we get:

$$R^t f_!' f^{-1} \omega_X \longrightarrow R^t f_!' \omega_{Y\times_X T^*X} \otimes \pi^{-1} \mathrm{or}_Y[-n_Y]$$
$$\longrightarrow \omega_{T^*Y} \otimes \pi^{-1} \mathrm{or}_Y[-n_Y]$$
$$\xrightarrow{\sim} \pi^{-1}\omega_Y$$

Here n_X (resp. n_Y) denotes the dimension of X (resp. Y).

Finally, there is a natural morphism:

(2.8)
$$\boxtimes : H^o_{\Lambda_X}(T^*X; \pi^{-1}\omega_X) \times H^o_{\Lambda_Y}(T^*Y; \pi^{-1}\omega_Y)$$
$$\longrightarrow H^o_{\Lambda_X \times \Lambda_Y}(T^*(X\times Y); \pi^{-1}\omega_{X\times Y}).$$

Combining the inverse image and the external product we can define a product. Let Λ_0 and Λ_1 be two closed conic subsets of T^*X satisfying the condition:

$$\Lambda_0 \cap \Lambda_1^a \subset T^*_X X,$$

where Λ_1^a is the image of Λ_1 by the anti-antipodal map. Then there is a natural morphism:

(2.9)
$$\underset{\mu}{*} : H^o_{\Lambda_0}(T^*X; \pi^{-1}\omega_X) \times H^o_{\Lambda_1}(T^*X; \pi^{-1}\omega_X)$$
$$\longrightarrow H^o_{\Lambda_0 + \Lambda_1}(T^*X; \pi^{-1}\omega_X).$$

It is defined as $\delta^* \circ \boxtimes$, the inverse image by the diagonal embedding composed with the external product.

Theorem 2.6. *Let $F \in Ob(D^b_{\mathbf{R}-c}(X))$ and $G \in Ob(D^b_{\mathbf{R}-c}(Y))$.*

(i) *Assume f is proper on* supp(G). *Then:*

$$\mu\mathrm{eu}(Rf_* G) = f_\mu \mu\mathrm{eu}(G).$$

(ii) *Assume f is non characteristic with respect to F. Then:*

$$\mu\mathrm{eu}(f^{-1}F) = f^\mu \mu\mathrm{eu}(F).$$

(iii) $\mu\mathrm{eu}(G \boxtimes F) = \mu\mathrm{eu}(G) \boxtimes \mu\mathrm{eu}(F)$.

(iv) *In particular, when $Y = X$ and $SS(F) \cap SS(G)^a \subset T^*_X X$, one has:*

$$\mu\mathrm{eu}(G \otimes F) = \mu\mathrm{eu}(G) \underset{\mu}{*} \mu\mathrm{eu}(F).$$

Note that Theorem 2.4 is a particular case of Theorem 2.6 (i) when applied to $a_X : X \to \{\mathrm{pt}\}$.

Notes. The microlocal study of stratifications was initiated by Kashiwara [K 1], and the microlocal approach to constructible sheaves, with the tool of the micro-support, is due to Kashiwara-Schapira (see [K-S]).

The characteristic cycle of **R**-constructible sheaves (what we call here the microlocal Euler class) and the index theorem are due to Kashiwara [K 2] and is developed in [K-S].

In case of **C**-constructible sheaves, let us quote related works due to MacPherson, Dubson, Ginsburg, Sabbah, Dubson-Brylinsky-Kashiwara. See the historical Notes of [K-S, Chapter IX].

The notations f^μ, f_μ, $\underset{\mu}{*}$ are different from that of [K-S].

III. ELLIPTIC PAIRS

In this section all manifolds and morphisms of manifolds are complex analytic and the base field k is the field of complex numbers. We shall sometimes identify a complex manifold X to the real underlying manifold that we still denote by X.

We refer to [K 1] and [S] for an introduction to $\mathcal{D}$-modules theory.

Let $(X, \mathcal{O}_X)$ be a complex manifold of complex dimension n. One denotes by Ω_X the sheaf of holomorphic n-forms on X and by $\mathcal{D}_X$ (resp. $\mathcal{D}_X^\infty$) the sheaf of finite order (resp. infinite order) holomorphic differential operators on X. Recall Sato's formula:

$$\mathcal{D}_X^\infty \simeq \delta^! \mathcal{O}_{X\times X}^{(0,n)}[n] \tag{3.1}$$

where $\mathcal{O}_{X\times X}^{(0,n)} = \mathcal{O}_{X\times X} \underset{q_2^{-1}\mathcal{O}_X}{\otimes} q_2^{-1}\Omega_X$.

One denotes by $D(\mathcal{D}_X)$) (resp. $D(\mathcal{D}_X^{op})$) the derived category of the abelian category of left (resp. right) $\mathcal{D}_X$-modules, and by $D^b(\mathcal{D}_X)$ (resp. $D^b_{\mathrm{coh}}(\mathcal{D}_X)$) the full triangulated subcategory of $D(\mathcal{D}_X)$ consisting of objects with bounded (resp. bounded and coherent) cohomology. One defines similarly $D^b(\mathcal{D}_X^{op})$ and $D^b_{\mathrm{coh}}(\mathcal{D}_X^{op})$.

If $\mathcal{M}$ is an object of $D^b_{\mathrm{coh}}(\mathcal{D}_X)$, its characteristic variety, denoted char$(\mathcal{M})$, is a closed conic analytic subset of T^*X, which is involutive ([S-K-K]). Recall (formula (1.1)) that char$(\mathcal{M})$ equals the micro-support of the complex $R\mathcal{H}om_{\mathcal{D}_X}(\mathcal{M}, \mathcal{O}_X)$ of holomorphic solutions of $\mathcal{M}$.

Let $f : Y \to X$ be a morphism of complex manifolds. One denotes by $\mathcal{D}_{Y\to X}$ the sheaf $\mathcal{O}_Y \underset{f^{-1}\mathcal{O}_X}{\otimes} f^{-1}\mathcal{D}_X$ endowed with its natural structure of a $(\mathcal{D}_Y, f^{-1}\mathcal{D}_X)$-bimodule.

Let $\mathcal{M} \in Ob(D^b(\mathcal{D}_X))$. One sets:

$$\underline{f}^{-1}\mathcal{M} = \mathcal{D}_{Y\to X} \overset{L}{\underset{f^{-1}\mathcal{D}_X}{\otimes}} f^{-1}\mathcal{M}.$$

Let $\mathcal{N} \in Ob(D^b(\mathcal{D}_Y^{op}))$. One sets:

$$\underline{f}_*\mathcal{N} = Rf_*(\mathcal{N} \overset{L}{\underset{\mathcal{D}_Y}{\otimes}} \mathcal{D}_{Y\to X}).$$

If $\mathcal{M} \in Ob(D^b(\mathcal{D}_X))$ and $\mathcal{N} \in Ob(D^b(\mathcal{D}_Y))$ one sets:

$$\mathcal{M} \underline{\boxtimes} \mathcal{N} = \mathcal{D}_{X\times Y} \underset{\mathcal{D}_X \boxtimes \mathcal{D}_Y}{\otimes} (\mathcal{M} \boxtimes \mathcal{N}),$$

and there is a similar formula for right modules.

Finally one sets:

$$\underline{D}'\mathcal{M} = R\mathcal{H}om_{\mathcal{D}_X}(\mathcal{M}, \mathcal{D}_X),$$

$$\underline{D}\mathcal{M} = R\mathcal{H}om_{\mathcal{D}_X}(\mathcal{M}, \Omega_X \underset{\mathcal{O}_X}{\otimes} \mathcal{D}_X[n]).$$

(In this last formula, $\mathcal{M}$ is a right module.)

Definition 3.1. An elliptic pair $(\mathcal{M}, F)$ on X is the data of $\mathcal{M} \in Ob(D^b_{\mathrm{coh}}(\mathcal{D}_X))$ and $F \in Ob(D^b_{\mathbb{R}-c}(X))$ satisfying: char$(\mathcal{M}) \cap SS(F) \subset T^*_X X$.

We use the same terminology for objects of $D^b_{\mathrm{coh}}(\mathcal{D}_X^{op})$.

Examples 3.2.

(i) Let M be a real analytic manifold, X a complexification of M. A coherent $\mathcal{D}_X$-module $\mathcal{M}$ is elliptic on M if and only if $(\mathcal{M}, \mathbf{C}_M)$ is an elliptic pair.

(ii) $(\mathcal{M}, \mathbf{C}_M)$ and $(\mathcal{O}_X, F)$ are elliptic pairs on X for all $\mathcal{M}$ and all F.

(iii) More generally, let $X = \bigsqcup_\alpha X_\alpha$ be a μ-stratification (cf. [K-S, Chapter VIII]). Let $F \in Ob(D^b_{\mathbf{R}-c}(X))$ such that $H^j(F)_{|X_\alpha}$ is locally constant for each j, each α and let $\mathcal{M} \in Ob(D^b_{\mathrm{coh}}(\mathcal{D}_X))$ such that X_α is non characteristic for $\mathcal{M}$ for all α. Then $(\mathcal{M}, F)$ is an elliptic pair.

Theorem 3.3. *Let $(\mathcal{M}, F)$ be an elliptic pair.*

(i) *The natural morphism:*

$$R\mathcal{H}om_{\mathcal{D}_X}(\mathcal{M}, D'F \otimes \mathcal{O}_X) \to R\mathcal{H}om_{\mathcal{D}_X}(\mathcal{M} \otimes F, \mathcal{O}_X) \tag{3.2}$$

is an isomorphism.

(ii) *Assume* $\mathrm{supp}(\mathcal{M}) \cap \mathrm{supp}(F)$ *is compact. Then for all $j \in \mathbf{Z}$, the $\mathbf{C}$-vector spaces $H^j R\Gamma(X; R\mathcal{H}om_{\mathcal{D}_X}(\mathcal{M} \otimes F, \mathcal{O}_X))$ are finite dimensional.*

Sketch of proof. (i) Follows from a general result of [K-S] which asserts that for F and G in $D^b(X)$ the natural morphism

$$R\mathcal{H}om(F, A_X) \overset{L}{\otimes} G \to R\mathcal{H}om(F, G)$$

is an isomorphism as soon as $SS(F) \cap SS(G)$ is contained in $T^*_X X$.

(ii) Using techniques of [Sc] one can reduce the problem to the case where $\mathcal{M}$ admits a free presentation. Then one represents F by a bounded complex whose components are direct sums of sheaves $\mathbf{C}_U$, U open, relatively compact, subanalytic in X and such that $D'\mathbf{C}_U = \mathbf{C}_{\overline{U}}$.

Next one calculates $R\Gamma(X; \mathbf{C}_{\overline{U}} \otimes \mathcal{O}_X)$ using a Dolbeault resolution of $\mathcal{O}_X$ by real analytic forms and $R\Gamma(X; R\mathcal{H}om(\mathbf{C}_U, \mathcal{O}_X))$ using a Dolbeault resolution of $\mathcal{O}_X$ by C^∞-forms. Then the left-hand side of (3.2) is represented by a complex of topological vector spaces of type DFS and the right-hand side by FS-spaces. Since these are quasi-isomorphic by (i), the cohomology is finite dimensional by classical functional analysis. □

Example 3.4. In the situation of Example 3.2 (1), isomorphism (3.2) reads as:

$$R\mathcal{H}om_{\mathcal{D}_X}(\mathcal{M}, \mathcal{A}_M) \xrightarrow{\sim} R\mathcal{H}om_{\mathcal{D}_X}(\mathcal{M}, \mathcal{B}_M)$$

where $\mathcal{A}_M$ (resp. $\mathcal{B}_M$) is the sheaf of real analytic functions (resp. hyperfunctions) on M. If M is compact, one recovers the finiteness of the spaces $\mathrm{Ext}^j_{\mathcal{D}_X}(M; \mathcal{M}, \mathcal{A}_M)$.

Theorem 3.3 (ii) may be generalized to the relative case.

Let $f : X \to S$ be a smooth morphism of complex manifolds. The ring $\mathcal{D}_{X/S}$ of relative differential operators is the subsheaf of $\mathcal{D}_X$ generated by $\mathcal{O}_X$ and the vector fields tangent to the fibers of f.

Let $T^*(X/S)$ denote the relative cotangent bundle, defined by the exact sequence:

$$0 \to X \underset{X}{\times} T^*S \to T^*X \underset{p}{\to} T^*(X/S) \to 0. \tag{3.3}$$

If $\mathcal{M}$ is a coherent $\mathcal{D}_{X/S}$-module one defines, similarly as in the case of $\mathcal{D}_X$-modules, its characteristic variety in $T^*(X/S)$ and denote it by $\mathrm{char}_{X/S}(\mathcal{M})$ (cf. [S, Chapter III, §1.3]).

4. Estimates for the Solutions in terms of the Main Parameters

The arguments described in the last two paragraphs give a proof of Theorem 2.1 for forms of degree $n \geq 5$ and non-exceptional forms of degree $n = 4$. We also obtained estimates for the values c_{14} and c_{15} in terms of the regulator of the field $\mathbb{K} = \mathbb{Q}(\theta)$ (cf.(2.27)). In this paragraph we apply Lemmas 2.1, 2.2 of Ch.II and Lemma 8.2, Ch.III to obtain a bound for the solutions of the equations (1.1) which is a little worse than (2.6) in terms of its dependence on $|A|$ and H_f, but which is better than (2.27) in respect of its dependence on R. Moreover, our arguments deal with all irreducible forms of degree $n \geq 3$ (cf. [227]). The assertion of Theorem 2.1 for forms of degree 3 and exceptional forms of degree 4 is bettered by the general Theorem 5.1 of the next section.

Theorem 4.1 *Under the conditions of Theorem 2.1 we have*

$$X = \max(|x|, |y|) < \exp(c_{31} R \ln R^* \cdot T \ln T), \tag{4.1}$$

where $T = \ln|A| + \ln H_f + R^*$, $R^* = \max(R, e)$, *and* c_{31} *depends only on* n.

Proof. We return to the arguments employed in §1 but bring to them some additional estimates.

Let U be a group of units of the field $\mathbb{K}$ constructed by Lemma 2.1 Ch.II, and let η be a unit in U determined by Lemma 2.2, Ch.II and satisfying (2.11). Then because of (1.5), Ch.II we have

$$h(\nu) \leq 2^n |a| \max(1, \overline{|\nu|})^n \leq 2^n |A| e^{c_{19} R}.$$

Since $\overline{|\mu|} \leq |x| + \overline{|\theta|}\,|y| < 2nH_f X$, we find

$$\overline{|\eta|} = \overline{|\mu|}/\overline{|\nu|} < 2nH_f X |a|^{1/n} e^{c_{19} R} \leq 2nH_f^2 X e^{c_{19} R}.$$

Hence

$$\sum_{j=1}^{r} h_j \ln|\eta_j^{(i)}| \leq \ln X + c_{19} R + 2 \ln H_f + \ln 2n \qquad (i = 1, 2, \ldots, n),$$

from which, relying on Lemma 2.1 of Ch.II, we obtain

$$H = \max_{(j)} |hj| < c_{32}(\ln X + R + \ln H_f) \qquad (j = 1, 2, \ldots, r). \tag{4.2}$$

The inequalities (1.7) imply

$$\left| \left(\frac{\eta_1^{(2)}}{\eta_1^{(3)}}\right)^{h_1} \cdots \left(\frac{\eta_r^{(2)}}{\eta_1^{(3)}}\right)^{h_r} \frac{\nu^{(2)}}{\nu^{(3)}} \frac{(\theta^{(3)} - \theta^{(1)})}{(\theta^{(2)} - \theta^{(1)})} - 1 \right| < (2nH_f)^{n^2+1} c_4 |y|^{-n}. \tag{4.3}$$

Observe now that

$$h\left(\frac{\eta_j^{(2)}}{\eta_j^{(3)}}\right) \le 2^n \max(1, \overline{|\eta_j|}^n \overline{|\eta_j^{-1}|}^n) \le 2^n \overline{|\eta_j|}^{2n^2}.$$

Using Lemmas 1.1 and 2.1 of Ch.II we find that

$$\prod_{j=1}^{r} \ln h(\eta_j^{(2)}/\eta_j^{(3)}) < c_{33} \prod_{j=1}^{r} \ln \overline{|\eta_j|} < c_{34} R.$$

We note this inequality and the estimate (2.13), and now apply Lemma 8.2, Ch.III to the left-hand side of (4.3). That gives

$$H^{-c_{33} R \ln R^* (\ln(|A| H_f)+R)} < (2nH_f)^{n^2+1} c_4 |y|^{-n}.$$

From (4.2) we obtain

$$|y| < \exp\{c_{36} R \ln R^* \cdot (\ln|A| + \ln H_f + R) \ln(\ln X + R + \ln H_f),$$

from which (4.1) follows, since one may suppose $|y| > c_3^{1/n}$ and $X \le 2nH_f|y|$, whence

$$|y| < \exp\{c_{37} R \ln R^* .(\ln|A| + \ln H_f + R) \times \\ \times (\ln R^* + \ln(\ln|A| + \ln H_f + R))\} < \exp\{c_{31} R \ln R^* . T \ln T\}.$$

In Theorem 4.1 the regulator R may be replaced by a more accessible parameter, that is, by the discriminant D of the field $\mathbb{K}$. By Lemma 2.4, Ch.II

$$R < c_{38} |D|^{1/2} (\ln|D|)^{n-1}, \qquad c_{38} = c_{38}(n),$$

so it follows from (4.1) that

$$X < \exp\{c_{39} |D|^{1/2} (\ln|D|)^n T' \ln T'\}, \tag{4.4}$$

where $T' = \ln|A| + \ln H_f + |D|^{1/2}(\ln|D|)^{n-1}$. Further, given the discriminant D of the field, we can pass to the discriminant D_f of the form $f = f(x,y)$ by noting the magnitude of the leading coefficient $a > 0$ of the polynomial $f(x,1)$. Indeed, D divides the discriminant of the number $a\theta$, and it is

$$a^{n(n-1)} \prod_{1 \le i < j \le n} (\theta^{(i)} - \theta^{(j)})^2 = a^{(n-1)(n-2)} D_f.$$

Since prime numbers exceeding n occur in D with exponent not greater than $n-1$, and all the rest occur with exponent not greater than $n((\ln n/\ln 2)+1)$ (this follows from the results of Ore [150]), we have

$$|D| \le n^{n((\ln n/\ln 2)+1)} a^{n-1} |D_f|. \tag{4.5}$$

Finally, by viewing the discriminant D_f as a polynomial in the coefficients of the form f, we can estimate $|D_f|$ in terms of H_f. Namely,

$$|D_f| < c_{40} H_f^{2(n-1)}, \quad c_{40} = c_{40}(n). \tag{4.6}$$

Definition 3.5. Let $\mathcal{M}$ be a coherent $\mathcal{D}_{X/S}$-module and let $F \in Ob(D^b_{\mathbf{R}-c}(X))$. One says $(\mathcal{M}, F)$ is a relative elliptic pair if $\rho^{-1}\mathrm{char}(\mathcal{M}) \cap SS(F) \subset T^*_X X$.

This is equivalent to saying that $(\mathcal{D}_X \underset{\mathcal{D}_{X/S}}{\otimes} \mathcal{M}, F)$ is an elliptic pair.

Theorem 3.6. *Let $(\mathcal{M}, F)$ be a relative elliptic pair and assume f is proper on* $\mathrm{supp}(\mathcal{M}) \cap \mathrm{supp}(F)$. *Then for all $j \in \mathbf{Z}$, $H^j(Rf_* R\mathcal{H}om_{\mathcal{D}_{X/S}}(\mathcal{M} \otimes F, \mathcal{O}_X))$ is $\mathcal{O}_S$-coherent.*

Using Theorem 3.6, one could state a direct image theorem replacing $\mathcal{M}$ by a $\mathcal{D}_X$-module $\mathcal{N}$ (assuming $\mathcal{N}$ is generated by a $\mathcal{D}_{X/S}$-module) generalizing the direct image theorem of [H-S].

Next we shall adapt to the case of elliptic pairs the construction of the microlocal Euler class.

Let $(\mathcal{M}, F)$ be an elliptic pair, $\mathcal{M}$ being a right $\mathcal{D}_X$-module. Set for short:

$$\mathcal{P} = F \otimes \mathcal{M}, \quad \check{D}\mathcal{P} = D'F \otimes \underline{D}\mathcal{M}. \tag{3.4}$$

Isomorphism (3.1) allows us to construct the natural morphism:

$$\delta_* R\mathcal{H}om_{\mathcal{D}_X}(\mathcal{P}, \mathcal{P}) \to R\mathcal{H}om_{q_2^{-1}\mathcal{D}_X}(q_2^{-1}\mathcal{P}, q_1^{-1}\mathcal{P} \underset{q_1^{-1}\mathcal{D}_X}{\otimes} \mathcal{O}^{(0,n)}_{X\times X}[n]).$$

Using the hypothesis that $(\mathcal{M}, F)$ is an elliptic pair we find that the right-hand side is isomorphic to

$$H = \check{D}\mathcal{P} \boxtimes \mathcal{P} \underset{\mathcal{D}_{X\times X}}{\overset{L}{\otimes}} \mathcal{O}_{X\times X}. \tag{3.5}$$

Next we consider the $\mathcal{D}_X \otimes \mathcal{D}_X$-linear contraction

$$\check{D}\mathcal{P} \otimes \mathcal{P} \to \mathcal{D}_X \underset{\mathcal{O}_X}{\otimes} \Omega_X[n] \tag{3.6}$$

and the natural morphism:

$$(\mathcal{D}_X \underset{\mathcal{O}_X}{\otimes} \Omega_X[n]) \underset{\mathcal{D}_X \otimes \mathcal{D}_X}{\overset{L}{\otimes}} \delta^{-1}\mathcal{O}_{X\times X} \to \omega_X. \tag{3.7}$$

Let

$$\Lambda = \mathrm{char}(\mathcal{M}) + SS(F). \tag{3.8}$$

It follows from (1.1) and the hypothesis that $(\mathcal{M}, F)$ is elliptic that $SS(H) \subset \Lambda \times \Lambda^a$. Hence, by combining all the above morphisms, we get the chain of morphisms:

$$\begin{aligned} R\mathcal{H}om_{\mathcal{D}_X}(\mathcal{P}, \mathcal{P}) &\to \delta^! H \\ &\simeq R\pi_* R\Gamma_\Lambda \mu_\Delta(H) \\ &\to R\pi_* R\Gamma_\Lambda \mu_\Delta(\delta_*\delta^{-1}H) \\ &\to R\pi_* R\Gamma_\Lambda \mu_\Delta(\delta_*\omega_X) \\ &\simeq R\pi_* R\Gamma_\Lambda \pi^{-1}\omega_X. \end{aligned}$$

Applying the functor $H^0 R\Gamma(X; \cdot)$ we get the morphism:

$$\mathrm{Hom}_{\mathcal{D}_X}(F \otimes \mathcal{M}, F \otimes \mathcal{M}) \to H^0_\Lambda(T^*X; \pi^{-1}\omega_X). \tag{3.9}$$

Definition 3.7. The microlocal Euler class of $(\mathcal{M}, F)$, denoted $\mu\mathrm{eu}(\mathcal{M}, F)$, is the image of $\mathrm{id}_{F \odot \mathcal{M}}$ by the morphism (3.9).
The restriction of $\mu\mathrm{eu}(\mathcal{M}, F)$ to X, the zero-section of T^*X, is called the Euler class of $(\mathcal{M}, F)$, and denoted $\mathrm{eu}(\mathcal{M}, F)$.

If $\mathcal{M}$ is a left $\mathcal{D}_X$-module, there is a similar construction of $\mu\mathrm{eu}(\mathcal{M}, F)$, replacing H in (3.5) by $\Omega_{X\times X} \underset{\mathcal{D}_{X\times X}}{\overset{L}{\odot}} \mathcal{P} \boxtimes \tilde{D}\mathcal{P}$.

One shows easily that $\mu\mathrm{eu}(F)$ defined in §2 equals $\mu\mathrm{eu}(\mathcal{O}_X, F)$. Similarly, we set:

$$\mu\mathrm{eu}(\mathcal{M}) = \mu\mathrm{eu}(\mathcal{M}, \mathbf{C}_X), \tag{3.10}$$

and call $\mu\mathrm{eu}(\mathcal{M})$ the microlocal Euler class of $\mathcal{M}$.

Then one proves a similar result to Theorem 2.6, in the case of elliptic pairs. Since the hypothesis for direct images and inverse images are rather technical (see [S-Sc2]) we restrict ourselves to a particular case.

Theorem 3.8.

(i) *Assume* $\mathrm{supp}(\mathcal{M}) \cap \mathrm{supp}(F)$ *is compact. Then:*

$$\chi(X; F \odot \mathcal{M}) = \int_X \mathrm{eu}(\mathcal{M}, F)$$

where the left-hand side denotes the Euler-Poincaré index of the complex $R\Gamma(X; F \odot \mathcal{M} \underset{\mathcal{D}_X}{\odot} \Omega_X)$.

(ii) *One has:*

$$\mu\mathrm{eu}(\mathcal{M}, F) = \mu\mathrm{eu}(\mathcal{M}) \underset{\mu}{*} \mu\mathrm{eu}(F)$$

where the product is defined by (2.9).

Remark 3.9. Let Λ_0 be a closed conic $\mathbf{C}$-analytic subset of T^*X. Denote by $K_{\Lambda_0}(\mathcal{D}_X^{op})$ the Grothendieck group of the full triangulated subcategory of $D^b_{\mathrm{coh}}(\mathcal{D}_X^{op})$ consisting of objects with characteristic variety contained in Λ_0. Similarly for a closed conic subanalytic subset Λ_1 of T^*X, denote by $K_{\Lambda_1,\mathbf{R}-c}(X)$ the Grothendieck group of the full triangulated subcategory of $D^b_{\mathbf{R}-c}(X)$ consisting of objects with micro-support contained in Λ_1. Then $\mu\mathrm{eu}(\cdot)$ should be (this point is not proved yet) a well-defined group homomorphism from $K_{\Lambda_0}(\mathcal{D}_X^{op})$ to $H^0_{\Lambda_0}(T^*X; \pi^{-1}\omega_X)$ and from $K_{\Lambda_1,\mathbf{R}-c}(X)$ to $H^0_{\Lambda_1}(T^*X; \pi^{-1}\omega_X)$.

Denote by χ_{an} the map which associates to an elliptic pair $(\mathcal{M}, F)$ with compact support the Euler-Poincaré index of the complex $R\Gamma(X; F \odot \mathcal{M} \underset{\mathcal{D}_X}{\odot} \mathcal{O}_X)$. Then χ_{an} is additive with respect to $\mathcal{M}$ and to F. We could then sum up the results of Theorem 3.8 by the commutative diagram below, assuming:

$$\Lambda_0 \cap \Lambda_1 \quad \text{is contained in} \quad T^*_X X \quad \text{and is compact.} \tag{3.11}$$

$$\begin{array}{ccc}
K_{\Lambda_0}(\mathcal{D}_X^{op}) \times K_{\Lambda_1,\mathbf{R}-c}(X) & \xrightarrow{\mu\mathrm{eu}} & H^0_{\Lambda_0}(T^*X; \pi^{-1}\omega_X) \times H^0_{\Lambda_1}(T^*X; \pi^{-1}\omega_X) \\
 & & \downarrow \times \\
\chi_{an} \downarrow & & H^0_{\Lambda_0+\Lambda_1}(T^*X; \pi^{-1}\omega_X) \\
 & & \downarrow \\
\mathbf{Z} & \xleftarrow[\int_X]{} & H^0_c(X; \omega_X)
\end{array}$$

Remark 3.10. Let $\mathcal{M}, F)$ be an elliptic pair with compact support. One can also formulate Theorem 3.8 by the formula :

$$\chi(X; F \otimes \mathcal{M}) = \int_{T^*X} \mu\mathrm{eu}(\mathcal{M}) \smile \mu\mathrm{eu}(F) \tag{(3.12)}$$

where $\smile$ denotes the cup product.

IV. Microlocal Chern classes of $\mathcal{D}$-modules (after [S-Sc 2])

Let X be an algebraic manifold over $\mathbf{C}$, Z a complex analytic subset and denote by $K_Z^{an}(X)$ the Grothendieck group of the full subcategory of $D^b_{\mathrm{coh}}(\mathcal{O}_X)$ consisting of objects supported by Z. We do not recall here the definition of the local Chern character (see [B-F-M]):

$$ch_Z : K_Z^{an}(X) \longrightarrow H_Z^{ev}(X; \mathbf{C}_X).$$

Let us define the local Euler character:

$$\mathrm{eu}_Z : K_Z^{an}(X) \longrightarrow H_Z^{ev}(X; \mathbf{C}_X)$$

both related by the formula:

$$\mathrm{eu}_Z(\mathcal{F}) = ch_Z(\mathcal{F}) \smile td_X(TX) \tag{4.1}$$

where $\smile$ is the cup product and $td_X(TX)$ is the Todd class of the tangent bundle.

As it is well-known, the local Chern character is compatible to external product and inverse image and the local Euler character is compatible to external product and proper direct image, this last point being the Grothendieck-Riemann-Roch theorem.

Here, for sake of convenience, we shall make as if such a theory also exists in the complex analytic case.

Now consider a left coherent $\mathcal{D}_X$-module $\mathcal{M}$ endowed with a good filtration and whose characteristic variety is contained in a closed conic analytic subset Λ_X of T^*X. Let $\mathrm{gr}(\mathcal{M})$ denote the associated graded module and set :

$$\widetilde{\mathrm{gr}}(\mathcal{M}) = \mathcal{O}_{T^*X} \otimes_{\pi^{-1}\mathrm{gr}(\mathcal{D}_X)} \pi^{-1}\mathrm{gr}(\mathcal{M}). \tag{4.2}$$

The element $\sigma_\Lambda(\mathcal{M})$ of $K_\Lambda^{an}(T^*X)$ defined by $\widetilde{\mathrm{gr}}(\mathcal{M})$ locally depends only on $\mathcal{M}$, not on the choice of the good filtration (see [K1]).

Let $f : Y \to X$ be a morphism of complex manifolds, and assume f is non-characteristic with respect to Λ_X (i.e. ${}^tf'$ is proper over $f_\pi^{-1}(\Lambda_X)$). Then the morphisms

$$f_\pi^* : K_{\Lambda_X}^{an}(T^*X) \to K_{f_\pi^{-1}\Lambda_X}(Y \underset{X}{\times} T^*X)$$

and

$${}^tf'_* : K_{f_\pi^{-1}\Lambda_X}(Y \underset{X}{\times} T^*X) \to K_{{}^tf' f_\pi^{-1}(\Lambda_X)}(T^*Y)$$

are well-defined and one can show (see [K1]) that :

$${}^tf'_* f_\pi^* \sigma_{\Lambda_X}(\mathcal{M}) = \sigma_{{}^tf' f_\pi^{-1}(\Lambda_X)}(\underline{f}^{-1}\mathcal{M}). \tag{4.3}$$

Similarly if $\mathcal{N}$ is a right coherent $\mathcal{D}_Y$-module whose characteristic variety is contained in Λ_Y and if f is proper on $\Lambda_Y \cap T^*_Y Y$ (i.e. f_π is proper on ${}^tf'^{-1}(\Lambda_Y)$), then ${}^tf'^* : K^{an}_{\Lambda_Y}(T^*Y) \to K^{an}_{{}^tf'^{-1}(\Lambda_Y)}(Y \underset{X}{\times} T^*X)$ and $f_{\pi *} : K^{an}_{{}^tf'^{-1}(\Lambda_Y)}(Y \underset{X}{\times} T^*X) \to K^{an}_{f_\pi {}^tf'^{-1}(\Lambda_Y)}(T^*X)$ are well-defined, and it is shown in [La] that :

$$f_{\pi *}{}^tf'^*\sigma_{\Lambda_Y}(\mathcal{N}) = \sigma_{f_\pi {}^tf'^{-1}(\Lambda_Y)}(\underline{f}_*\mathcal{N}). \tag{4.4}$$

Finally one shows easily that:

$$\sigma_{\Lambda_X}(\mathcal{M}) \boxtimes \sigma_{\Lambda_Y}(\mathcal{N}) = \sigma_{\Lambda_X \times \Lambda_Y}(\mathcal{M} \underline{\boxtimes} \mathcal{N}). \tag{4.5}$$

Definition 4.1. Let $\mathcal{M}$ (resp. $\mathcal{N}$) be a left (resp. right) coherent $\mathcal{D}_X$-module endowed with a good filtration and whose characteristic variety is contained in a closed conic analytic subset Λ of T^*X. The microlocal Chern character of $\mathcal{M}$ (resp. $\mathcal{N}$) along Λ is :

$$\mu ch_\Lambda(\mathcal{M}) = ch(\sigma_\Lambda(\mathcal{M})) \smile \pi^* td_X(T^*X)$$

(resp.

$$\mu ch_\Lambda(\mathcal{N}) = ch(\sigma_\Lambda(\mathcal{N})) \smile \pi^* td_X(TX)).$$

We denote by $\mu ch^j_\Lambda(\mathcal{M})$ the j-th component of $\mu ch_\Lambda(\mathcal{M})$ in $H^j_\Lambda(T^*X; \mathbf{C}_{T^*X})$, and similarly for $\mathcal{N}$.

Note that the microlocal Chern character of the right $\mathcal{D}_X$-module $\mathcal{M} \otimes_{\mathcal{O}_X} \Omega_X$ is the microlocal Chern character of the left $\mathcal{D}_X$-module $\mathcal{M}$.

Denote by n_X the complex dimension of the manifold X. Using the constructions (2.5), (2.7), (2.8), (2.9) we get (under the same hypotheses on Λ_X, Λ_Y, Λ_0, Λ_1) the morphisms :

$$f^\mu : H^{j+2n_X}_{\Lambda_X}(T^*X; \mathbf{C}_{T^*X}) \longrightarrow H^{j+2n_Y}_{{}^tf' f_\pi^{-1}(\Lambda_X)}(T^*Y; \mathbf{C}_{T^*Y})$$

$$f_\mu : H^{j+2n_Y}_{\Lambda_Y}(T^*Y; \mathbf{C}_{T^*Y}) \longrightarrow H^{j+2n_X}_{f_\pi {}^tf'^{-1}(\Lambda_Y)}(T^*X; \mathbf{C}_{T^*X})$$

$$\boxtimes : H^j_{\Lambda_X}(T^*X; \mathbf{C}_{T^*X}) \times H^k_{\Lambda_Y}(T^*Y; \mathbf{C}_{T^*Y}) \longrightarrow H^{j+k}_{\Lambda_X \times \Lambda_Y}(T^*X \times Y; \mathbf{C}_{T^*X \times Y})$$

$$\underset{\mu}{*} : H^j_{\Lambda_0}(T^*X; \mathbf{C}_{T^*X}) \times H^k_{\Lambda_1}(T^*X; \mathbf{C}_{T^*X}) \longrightarrow H^{j+k-2n_X}_{\Lambda_0 + \Lambda_1}(T^*X; \mathbf{C}_{T^*X}).$$

Using (4.3), (4.4), (4.5) and the Grothendieck-Riemann-Roch theorem one proves easily :

Theorem 4.2.

(i) *In the situation of (4.3) we have:*

$$f^\mu \mu ch_{\Lambda_X}(\mathcal{M}) = \mu ch_{{}^tf' f_\pi^{-1}(\Lambda_X)}(\underline{f}^{-1}\mathcal{M}).$$

(ii) *In the situation of (4.4) we have:*

$$f_\mu \mu ch_{\Lambda_Y}(\mathcal{N}) = \mu ch_{f_\pi {}^tf'^{-1}(\Lambda_Y)}(\underline{f}_*\mathcal{N}).$$

(iii) *In the situation of (4.5) we have:*

$$\mu ch_{\Lambda_X}(\mathcal{M}) \boxtimes \mu ch_{\Lambda_Y}(\mathcal{N}) = \mu ch_{\Lambda_X \times \Lambda_Y}(\mathcal{M} \underline{\boxtimes} \mathcal{N}).$$

As a corollary we get that if $\mathcal{M}$ and $\mathcal{N}$ are two left $\mathcal{D}_X$-modules with characteristic variety contained in Λ_0 and Λ_1 respectively, and if $\Lambda_0 \cap \Lambda_1 \subset T^*_X X$, then :

$$\mu ch_{(\Lambda_0+\Lambda_1)}(\mathcal{M} \overset{L}{\underset{\mathcal{O}_X}{\otimes}} \mathcal{N}) = \mu ch_{\Lambda_0}(\mathcal{M}) \underset{\mu}{*} \mu ch_{\Lambda_1}(\mathcal{N}).$$

In view of Theorem 4.2, the microlocal Chern character has the same functorial properties as the microlocal Euler class.

Let us come back to the situation of Definition 4.1 and set :

$$d = \operatorname{codim}_{\mathbf{C}}\Lambda, \quad n = \dim_{\mathbf{C}} X.$$

It is clear that:

$$\mu ch^j_\Lambda(\mathcal{M}) = 0 \quad \text{for} \quad j < 2d.$$

Moreover one can prove:

Proposition 4.3. *$\mu ch^{2d}_\Lambda(\mathcal{M})$ is the characteristic cycle of $\mathcal{M}$ along Λ (see* [K1] *or* [S]*). In particular, if $\mathcal{M}$ is holonomic :*

$$\mu ch^{2n}_\Lambda(\mathcal{M}) = \mu\mathrm{eu}(\mathcal{M}).$$

Now we make the:

Conjecture 4.4.
(i) $\mu ch^j_\Lambda(\mathcal{M}) = 0$ *for* $j \notin [2d, 2n]$.
(ii) $\mu ch^{2n}_\Lambda(\mathcal{M}) = \mu\mathrm{eu}(\mathcal{M})$.

Note that in the case of coherent $\mathcal{O}_X$-modules, the link between a diagonal class similar to our Euler class and the Chern class has been made by O'Brian-Toledo-Tong [0-T-T].

Conjecture 4.4(ii) would provide an entirely new proof and a generalization of the Atiyah-Singer theorem [A-S] as we shall see in Example 4.5 below. Moreover, since one proves [S-Sc 2] that under natural conditions, $\mu\mathrm{eu}(\cdot)$ commutes to direct images, one would also obtained a direct proof and a generalization of the relative version of the Atiyah-Singer theorem due to Boutet de Monvel-Malgrange [B-M].

Example 4.5. Let M be a compact n-dimensional real analytic manifold, X a complexification of M, $\mathcal{M}$ a right coherent $\mathcal{D}_X$-module, elliptic on M. Applying Theorem 3.8 and Remark 3.10, we get :

$$\chi(R\Gamma(M; \mathcal{M} \overset{L}{\underset{\mathcal{D}_X}{\otimes}} \mathcal{O}_X)) = \int_{T^*X} \mu\mathrm{eu}(\mathcal{M}) \smile \mu\mathrm{eu}(\mathbf{C}_M). \tag{4.7}$$

Denote by σ_M the zero-section embedding $M \hookrightarrow T^*_M X$ and by j the embedding $T^*_M X \hookrightarrow T^*X$. Using the notations of Remark 3.10 and noticing that $T^*_M X \cap \mathrm{char}(\mathcal{M})$ is contained in M, we get :

$$\int_{T^*X} \mu\mathrm{eu}(\mathcal{M}) \smile \mu\mathrm{eu}(\mathbf{C}_M) = \int_{T^*_M X} j^*\mu\mathrm{eu}(\mathcal{M}) = \int_M \sigma^*_M j^* \mu\mathrm{eu}(\mathcal{M}).$$

Now assume Conjecture 4.4 (ii) is true. We get, with $\Lambda = \mathrm{char}(\mathcal{M})$:

$$\begin{aligned}\chi(R\Gamma(M; \mathcal{M} \overset{L}{\underset{\mathcal{D}_X}{\otimes}} \mathcal{O}_X)) &= \int_M \sigma^*_M j^*[ch(\sigma_\Lambda(\mathcal{M})) \smile \pi^* td_X TX] \\ &= \int_M \sigma^*_M [j^* ch(\sigma_\Lambda(\mathcal{M}))] \smile td_M(TM^{\mathbf{C}}).\end{aligned}$$

This is the Atiyah-Singer theorem.

References

[A-S] M. Atiyah, I.M. Singer, *The index of elliptic operators on compact manifolds*, Bull. Am. Math. Soc. **69** (1963), 422–433.

[B-F-M] P. Baum, W. Fulton, R. MacPherson, *Riemann-Roch and topological K-theory for singular varieties*, Acta Math. **143** (1979), 155–211.

[B-M] M. Boutet de Monvel, B. Malgrange, *Le théorème de l'indice relatif*, Ann. Sc. Ec. Norm. Sup. **23** (1990), 151-192.

[H-S] C. Houzel, P. Schapira, *Images directes des modules différentiels*, C.R. Acad. Sci. Paris **298** (1984), 461–464.

[K 1] M. Kashiwara, *Systems of microdifferential equations*, Progress in Math., Birkhauser, 1983.

[K 2] M. Kashiwara, Index theorem for constructible sheaves, Systèmes différentiels et singularités, Astérisque 130, 1985, pp. 193–209; *Character, character cycle, fixed point theorem and group representation*, vol. 14, 1988, pp. 369-378.

[K-S] M. Kashiwara, P. Schapira, *Sheaves on manifolds*, Grundlehren der Math. Wissenschaften 292, Springer-Verlag, 1990.

[La] G. Laumon, Sur la catégorie dérivée des $\mathcal{D}$-modules filtrés, LNM 1016, Springer-Verlag, 1983, pp. 151–237.

[Le] J. Leray, *Problème de Cauchy I*, Bull. Soc. Math. France **85** (1957), 389–430.

[O-T-T] N. O'Brian, D. Toledo, Y.L. Tong, *Hirzebruch-Riemann-Roch for coherent sheaves*, Amer. J. Math. **103** (1981), 253-271.

[S-K-K] M. Sato, T. Kawai, M. Kashiwara, Hyperfunctions and pseudo-differential equations, Lect. Notes Math. Springer 287, 1973, pp. 265–529.

[S] P. Schapira, *Microdifferential systems in the complex domain*, Grundlehren der Mathematischen Wissenschaften 269, Springer-Verlag, 1985.

[S-Sc 1] P. Schapira, J-P. Schneiders, *Paires elliptiques I - Finitude et dualité*, C.R. Acad. Sci. **311** (1990), 83–86; *II Classes d'Euler et indice* **312** (1991), 81–84.

[S-Sc 2] P. Schapira, J-P. Schneiders, *Elliptic Pairs I. Relative finiteness and duality. II. Euler class and relative index theorem*, Preprint RIMS, Kyoto Univ. (1993) (to appear).

[Sc] J-P. Schneiders, *Un théorème de dualité relative pour les modules différentiels*, Thèse Univ. Liège (1987).

The equivariant Chern character and index of G-invariant operators. Lectures at CIME, Venise 1992

Nicole Berline and Michèle Vergne

Introduction

Let $\sigma : \mathcal{E}^+ \to \mathcal{E}^-$ be a homomorphism of complex vector bundles over a smooth manifold U. If σ is invertible outside a compact set of U, then σ determines an element $[\sigma]$ of the K-group $K(U)$. The *excision* construction associates to σ a formal difference of bundles $[\mathcal{F}^+]-[\mathcal{F}^-]$ on any compact manifold M which contains U as an open subset. Using this construction, one can define the Chern character $\mathrm{ch}([\sigma]) \in H^\bullet_{cpt}(U)$ as a de Rham cohomology class with compact support on U: a differential form, with compact support contained in U, representing $\mathrm{ch}([\sigma])$ is given by $\mathrm{ch}(\mathcal{F}^+, \nabla^+) - \mathrm{ch}(\mathcal{F}^-, \nabla^-)$, where the bundles with connection $(\mathcal{F}^+, \nabla^+)$ and $(\mathcal{F}^-, \nabla^-)$ coincide on $M - U$.

Following Quillen [21], consider on the superbundle $\mathcal{E} = \mathcal{E}^+ \oplus \mathcal{E}^-$ the odd endomorphism

$$v(\sigma) = \begin{pmatrix} 0 & \sigma^* \\ \sigma & 0 \end{pmatrix}$$

where σ^* is the adjoint of σ for a choice of Hermitian metrics on $\mathcal{E}^\pm$. Let ∇ be a connection on $\mathcal{E}$ which preserves $\mathcal{E}^+$ and $\mathcal{E}^-$. Then $\mathbb{A}(\sigma) = iv(\sigma) + \nabla$ is a superconnection on $\mathcal{E}$ with [0]-exterior degree term $iv(\sigma)$. Its Chern character form $\mathrm{ch}(\mathbb{A}(\sigma))$, defined as the supertrace of the matrix valued differential form $\exp(\mathbb{A}(\sigma)^2)$, is a de Rham closed differential form on U. When U is the total space of a real vector bundle on a compact base, one can define the de Rham complex $\mathcal{A}^\bullet_{rapid}(U)$ of rapidly decreasing differential forms. Its cohomology $H^\bullet_{rapid}(U)$ coincides with $H^\bullet_{cpt}(U)$. When σ satisfies some growth conditions (we will say that σ is good, see Definition 16), the Chern character form $\mathrm{ch}(\mathbb{A}(\sigma))$ is a rapidly decreasing differential form. In these notes, we give a proof of the following result of Quillen [21]: the class of $\mathrm{ch}(\mathbb{A}(\sigma))$ in $H^\bullet_{rapid}(U)$ coincides with $\mathrm{ch}([\sigma])$, (Theorem 19):

$$\mathrm{ch}([\sigma]) \equiv \mathrm{ch}(\mathbb{A}(\sigma)).$$

This equality means that one can use superconnections instead of excision in order to define the Chern character. Our motivation is to apply to the Atiyah-Singer index formula this way of computing the Chern character. The index formula thus obtained will generalize nicely to the equivariant case, including the case of *transversally* elliptic pseudodifferential operators.

Let B be a compact manifold and let $U = T^*B$ be the cotangent bundle to B. Denote by $p : T^*B \to B$ the projection. Let $\mathcal{E}^\pm$ be complex vector bundles over

B and let $\Gamma(B, \mathcal{E}^{\pm})$ be the space of smooth sections of the vector bundles $\mathcal{E}^{\pm}$. Let $P : \Gamma(B, \mathcal{E}^{+}) \to \Gamma(B, \mathcal{E}^{-})$ be an elliptic pseudo-differential operator with principal symbol $\sigma(P)$. Without changing the class $[\sigma(P)]$ in $K(T^*B)$, we may assume that P has order ≥ 1. Then we obtain a good representative of $[\sigma(P)]$ by taking any smooth bundle map $\sigma : p^*\mathcal{E}^+ \to p^*\mathcal{E}^-$ which coincides with $\sigma(P)$ outside a compact set. Let ∇ be a direct sum connection on $\mathcal{E} = \mathcal{E}^+ \oplus \mathcal{E}^-$. Then $\mathbb{A}(\sigma) = iv(\sigma) + p^*\nabla$ is a superconnection on $p^*\mathcal{E}$ and its Chern character form $\mathrm{ch}(\mathbb{A}(\sigma))$ is a de Rham closed differential form which is rapidly decreasing on the fibers of T^*B. Let $\mathrm{Ker}\, P$ be the space of solutions of P and let $\mathrm{Coker}\, P$ be the cokernel of the map P. Atiyah-Singer formula for $\mathrm{index}(P) = \dim \mathrm{Ker}\, P - \dim \mathrm{Coker}\, P$ can be written

$$\mathrm{index}(P) = (2i\pi)^{-\dim B} \int_{T^*B} \mathrm{ch}(\mathbb{A}(\sigma)) J(B)^{-1},$$

where $J(B)$ is an invertible de Rham cohomology class on B (independent of P).

Section 1 is intended to be elementary. The main technical tool is the result of ([9]; Chapter 9 , Section 9.1) on the Chern character of the kernel of a bundle map of constant rank. We give a proof in full details which is even slightly simpler than that of [9] and which extends easily to the equivariant case.

Section 2 concerns the case of G-equivariant vector bundles. Let G be a compact Lie group (with Lie algebra $\mathfrak{g}$) acting on a manifold U. A G-equivariant differential form is a G-invariant smooth map from $\mathfrak{g}$ to the space $\mathcal{A}(U)$ of differential forms on U. The space $\mathcal{A}_G^\infty(\mathfrak{g}, U)$ of G-equivariant forms has an equivariant exterior differential $d_\mathfrak{g}$. The G-equivariant cohomology space $\mathcal{H}_G^\infty(\mathfrak{g}, U)$ is defined to be the space $\mathrm{Ker}\, d_\mathfrak{g} / \mathrm{Im}\, d_\mathfrak{g}$.

Let $\mathcal{E}^{\pm}$ be G-equivariant vector bundles over M. If $\sigma : \mathcal{E}^+ \to \mathcal{E}^-$ is a G-equivariant bundle map which is invertible outside a compact set, it determines an element $[\sigma]$ of the equivariant K-group $K_G(U)$. Using excision $U \subset M$, the equivariant Chern character $\mathrm{ch}([\sigma]) \in \mathcal{H}^\infty_{G,cpt}(\mathfrak{g}, U)$ is defined as a G-equivariant cohomology class with compact support on U.

Let ∇ be a G-invariant direct sum connection on $\mathcal{E} = \mathcal{E}^+ \oplus \mathcal{E}^-$. Then $\mathbb{A}(\sigma) = iv(\sigma) + \nabla$ is a G-invariant superconnection. Its equivariant Chern character form $\mathrm{ch}(\mathbb{A}(\sigma))$ is defined similarly to the non equivariant case, as the supertrace of the exponential of the equivariant curvature of $\mathbb{A}(\sigma)$. It is an equivariantly closed differential form on U. When U is the total space of a real vector bundle on a compact base, we introduce the space of rapidly decreasing equivariant differential forms and the corresponding cohomology space $\mathcal{H}^\infty_{G,rapid}(\mathfrak{g}, U)$. We prove (Theorem 29) that the class of $\mathrm{ch}(\mathbb{A}(\sigma))$ in $\mathcal{H}^\infty_{G,rapid}(\mathfrak{g}, U)$ is equal to $\mathrm{ch}([\sigma])$, when σ is a good bundle map.

This part of section 2 reproduces section 1, using $\TeX$ with the addition of G-equivariance. Proofs are similar to the ones of section 1, therefore we give only short indications. In the remaining of the article we only sketch results.

For $s \in G$, let $G(s)$ be the centralizer of $s \in G$, let $\mathfrak{g}(s)$ be the Lie algebra of $G(s)$ and let $U(s)$ be the fixed point submanifold for the action of s on U. A bouquet of equivariant differential forms is a family $\alpha = (\alpha_s)_{s \in G}$ where, for each $s \in G$, α_s is a closed $G(s)$-equivariant differential form on $U(s)$. The family (α_s) must satisfy some compatibility conditions (Definition 25). A global analogue of $\mathcal{H}_G^\infty(\mathfrak{g}, U)$, denoted by $\mathcal{K}_G(U)$, has been defined in [19] [20] (for any real almost-algebraic group) and [15].

When U is a point, $\mathcal{K}_G(\bullet) = C^\infty(G)^G$. For any G-manifold, $\mathcal{K}_G(U)$ is a module over $C^\infty(G)^G$. An element $[\alpha] = (\alpha_s)_{s\in G}$ of $\mathcal{K}_G(U)$ is a family of equivariant cohomology classes where, for each $s \in G$, α_s is a $G(s)$-equivariant cohomology class on $U(s)$. The family (α_s) must satisfy some compatibility conditions (see [19] [20]). A bouquet α of equivariant differential forms determines an element $[\alpha]$ of $\mathcal{K}_G(M)$. The space $\mathcal{K}_{G,cpt}(M)$ is similarly defined (see [19] [20]). To a G-equivariant superbundle with a G-invariant superconnection $\mathbb{A}$ is associated a bouquet of equivariant differential forms which we call the *bouquet* of Chern characters of the superconnection (Formula 39) and denote by $\mathrm{bch}(\mathbb{A}) = (\mathrm{ch}_s(\mathbb{A}))_{s\in G}$. Using excision $U \subset M$, this construction induces a map

$$[\sigma] \mapsto \mathrm{bch}([\sigma]) : \ \mathcal{K}_G(U) \to \mathcal{K}_{G,cpt}(U).$$

When U is the total space of a real vector bundle on a compact base, we can introduce $\mathcal{K}_{G,rapid}(U)$. When σ is a good bundle map, the equality of classes in $\mathcal{H}^\infty_{G,rapid}(\mathfrak{g}, U)$,

$$\mathrm{ch}([\sigma]) \equiv \mathrm{ch}(\mathbb{A}(\sigma)),$$

can be extended to a global equality in $\mathcal{K}_{G,rapid}(U)$

$$\mathrm{bch}([\sigma]) \equiv \mathrm{bch}(\mathbb{A}(\sigma)).$$

When U is the cotangent bundle T^*B of a compact manifold B, one can define the integral $\int_b \alpha$ of a bouquet $\alpha = (\alpha_s)_{s\in G}$ of rapidly decreasing equivariant differential forms on U. It is a G-invariant C^∞-function on the group G. The function $\int_b \alpha$ is described in [19] [20]. Its definition is strongly motivated by the notion of direct images in K-theory and by the corresponding index formulas. For each $s \in G$, the function $(\int_b \alpha)$ is given in a *neighborhood* of s as the integral over $T^*B(s)$ of a $G(s)$-equivariant differential form on $T^*B(s)$ related to α_s.

Consider a G-invariant elliptic pseudodifferential operator P with principal symbol $\sigma(P)$. Then $\mathrm{Ker}\, P$ and $\mathrm{Coker}\, P$ are finite dimensional representation spaces of G. The difference of the traces of the action of G in $\mathrm{Ker}\, P$ and $\mathrm{Coker}\, P$ is a C^∞-function on G called the equivariant index of P:

$$\mathrm{index}(P)(g) = \mathrm{Tr}(g, \mathrm{Ker}\, P) - \mathrm{Tr}(g, \mathrm{Coker}\, P).$$

Using the method of [12], it is easy to deduce from the fixed point formula of Atiyah-Segal-Singer the following formula

$$\mathrm{index}(P) = \int_b \mathrm{bch}(\mathbb{A}(\sigma)).$$

Indeed, for each $s \in G$, the "s"-part of this integral generalizes the fixed point formula: we have, for each $s \in G$ and $Y \in \mathfrak{g}(s)$,

$$\mathrm{index}(P)(s \exp Y) = \int_{T^*B(s)} (2i\pi)^{-\dim B(s)} \frac{\mathrm{ch}_s(\mathbb{A}(\sigma))(Y)}{D_s(\mathcal{N}(B/B(s))(Y) J(B(s))(Y)}$$

where $J(B(s))$ and $D_s(\mathcal{N}(B/B(s))$ are certain equivariant forms on $B(s)$ independent of P. For $Y = 0$, one recovers the fixed point formula of Atiyah-Segal-Singer.

In section 3, we explain how to generalize the formula above to the case of transversally elliptic operators. Let $P : \Gamma(B, \mathcal{E}^+) \to \Gamma(B, \mathcal{E}^-)$ be a *transversally* elliptic pseudodifferential operator with symbol $\sigma(P)$. Let ω be the canonical 1-form on T^*B. We modify the superconnection which we used in the case of an elliptic operator and introduce the following G-invariant superconnection on $p^*\mathcal{E}$:

$$\mathbb{A}^\omega(\sigma) = iv(\sigma) + p^*\nabla + i\omega.$$

Then the Chern character forms $(\mathrm{ch}_s(\mathbb{A}^\omega(\sigma)))_{s\in G}$ are mean rapidly decreasing (definition 37), and the integral of the bouquet $\mathrm{bch}(\mathbb{A}^\omega(\sigma)$ is well defined as aa G-invariant *generalized* function on the group G. From the assumption that P is transversally elliptic it follows [2] that $\mathrm{Ker}\, P$ and $\mathrm{Coker}\, P$ are trace class representations of G. Thus the equivariant index of P is now a generalized function on G denoted by $\mathrm{index}(P)$. We have

$$\mathrm{index}(P) = \int_b \mathrm{bch}(\mathbb{A}^\omega(\sigma)). \tag{1}$$

For each point $s \in G$, the "s"-part of this formula expresses the value of the generalized function $Y \mapsto \mathrm{index}(P)(s \exp Y)$ as an integral over $T^*B(s)$, by a formula similar to the formula given above for the elliptic case. However now both sides have to be integrated against a test function on $\mathfrak{g}(s)$ for the formula to be meaningful. The case where s is the identity of G was announced in [23]. The proof of (1) will appear elsewhere.

1 Chern characters

1.1 The difference bundle

Let M be a topological space. Let $\mathcal{E} \to M$ be a (real or complex) vector bundle. We denote by $\mathcal{E}_x$ the fiber of $\mathcal{E}$ at x and we will sometimes note an element of $\mathcal{E}$ as (x, v) with $x \in M, v \in \mathcal{E}_x$. Let $\mathcal{E}^\pm$ be two vector bundles over M. Consider a bundle map $\mathcal{E}^+ \xrightarrow{\sigma} \mathcal{E}^-$. The *support* of $\mathcal{E}^+ \xrightarrow{\sigma} \mathcal{E}^-$ is the set of points $x \in M$ where $\sigma(x)$ is *not* invertible.

An important example of construction of vector bundles is the "gluing" construction. The space $M = U \cup V$ is covered by two open sets, $\mathcal{E}^+$ is a vector bundle over U , $\mathcal{E}^-$ a vector bundle over V and $\mathcal{E}^+|_{U\cap V} \xrightarrow{\sigma} \mathcal{E}^-|_{U\cap V}$ is a bundle isomorphism. The glued bundle over M

$$\mathcal{G}(\mathcal{E}^+, \sigma, \mathcal{E}^-)$$

is defined as the quotient of $\mathcal{E}^+ \bigsqcup \mathcal{E}^-$ by the equivalence relation where the point $(x, v), x \in U \cap V, v \in \mathcal{E}^+_x$ is identified with the point $(x, \sigma(x)v)$ of $\mathcal{E}^-_x$.

Let $\mathcal{E}^\pm$ be two vector bundles over the space M. Let U be an open subset of M. Let $\mathcal{E}^+|_U \xrightarrow{\sigma} \mathcal{E}^-|_U$ be a bundle map (defined only over U) and assume that σ has compact support $F \subset U$. The manifold M is covered by U and $V = M - F$. Over $U \cap V = U - F$, the bundle map σ is an isomorphism. Thus we can form the glued bundle

$$\mathcal{G}(\sigma) = \mathcal{G}(\mathcal{E}^+|_U, \sigma|_{U-F}, \mathcal{E}^-|_V). \tag{2}$$

If σ extends to M and remains an isomorphism on $M - F$, then the bundle $\mathcal{G}(\sigma)$ is isomorphic to $\mathcal{E}^+$.

It will be useful to realize $\mathcal{G}(\sigma)$ as a subbundle of $\mathcal{E}^+ \oplus \mathcal{E}^-$. We choose a smooth function ϕ with compact support contained in U such that $\phi(x) = 1$ for $x \in F$. We will say that ϕ is a cut-off function adapted to σ. Consider the map

$$u(\sigma,\phi) : (\mathcal{E}^+ \oplus \mathcal{E}^-)|_U \to \mathcal{E}^-|_U \tag{3}$$

given for $(v_0, v_1) \in \mathcal{E}^+_x \oplus \mathcal{E}^-_x$ by

$$u(\sigma,\phi)(v_0,v_1) = \sigma(x)v_0 - \phi(x)v_1.$$

It is easy to see that $u(\sigma,\phi)$ is surjective. Thus the kernel of the bundle map $u(\sigma,\phi)$ is a subbundle of $(\mathcal{E}^+ \oplus \mathcal{E}^-)|_U$. At points $x \in U$ where $\phi(x) = 0$, the map $\sigma(x)$ is invertible, therefore $\operatorname{Ker}(u(\sigma,\phi)_x) = \mathcal{E}^-_x$. Thus $\operatorname{Ker}(u(\sigma,\phi))$ extends to a bundle over the whole of M.

Definition 1 *The bundle $\mathcal{F}(\sigma,\phi) \subset \mathcal{E}^+ \oplus \mathcal{E}^-$ is the bundle over M defined by $\mathcal{F}(\sigma,\phi)_x = \operatorname{Ker}(u(\sigma,\phi)_x)$ for $x \in U$ and $\mathcal{F}(\sigma,\phi)_x = \mathcal{E}^-_x$ outside the compact support of ϕ.*

Lemma 2 *The vector bundles $\mathcal{G}(\sigma)$ and $\mathcal{F}(\sigma,\phi)$ are isomorphic.*

Proof. Let $\mathcal{F} = \mathcal{F}(\sigma,\phi)$. It is easy to see that we define an isomorphism $\mathcal{E}^+|_U \to \mathcal{F}|_U$ by:

$$v_0 \mapsto (\phi(x)v_0, \sigma(x)v_0) \quad \text{for } x \in U, v_0 \in \mathcal{E}^+_x.$$

We define an isomorphism $\mathcal{E}^-|_{M-F} \to \mathcal{F}|_{M-F}$ by

$$\begin{aligned} v_1 &\mapsto (\phi(x)\sigma(x)^{-1}v_1, v_1) && \text{for } x \in U - F, v_1 \in \mathcal{E}^-_x \\ v_1 &\mapsto (0, v_1) && \text{for } x \in M - \operatorname{supp}\phi, v_1 \in \mathcal{E}^-_x. \end{aligned}$$

Through these two isomorphisms the bundle map σ becomes the identity. Thus $\mathcal{G}(\sigma)$ is isomorphic to $\mathcal{F}(\sigma,\phi)$.

Let M be a compact topological space. The isomorphism classes of complex vector bundles over M form an abelian semigroup under direct sums. The associated abelian group is denoted by $K(M)$. If $\mathcal{E} \to M$ is a vector bundle over M, we denote by $[\mathcal{E}]$ the corresponding element of $K(M)$.

In the case of a locally compact space, which we will denote by U, there is a description of the "K-theory with compact supports" $K(U)$ which is best fitted to the index theory of pseudo-differential operators. Let us briefly recall it. We consider bundle maps $\mathcal{E}^+ \xrightarrow{\sigma} \mathcal{E}^-$ where $\mathcal{E}^+$ and $\mathcal{E}^-$ are two complex vector bundles over U and σ a bundle map with compact support. Two such bundle maps $\mathcal{E}^+ \xrightarrow{\sigma} \mathcal{E}^-$ and $\mathcal{F}^+ \xrightarrow{\nu} \mathcal{F}^-$ are said to be homotopic if there is a bundle map over $U \times [0,1]$ with restriction to $U \times \{0\}$ isomorphic to $\mathcal{E}^+ \xrightarrow{\sigma} \mathcal{E}^-$ and restriction to $U \times \{1\}$ isomorphic to $\mathcal{F}^+ \xrightarrow{\nu} \mathcal{F}^-$ (here we consider only bundle maps with compact support). The homotopy classes of bundle maps with compact support over U form an abelian semi-group $C(U)$ under direct sums. It contains a sub-semigroup $C_0(U)$ represented by bundle maps with empty support. The quotient $K(U) = C(U)/C_0(U)$ is a group. Thus two bundle maps $\mathcal{E}^+ \xrightarrow{\sigma} \mathcal{E}^-$ and $\mathcal{F}^+ \xrightarrow{\nu} \mathcal{F}^-$ define the same class in $K(U)$ if and only if they are stably homotopic: there exist vector bundles $\mathcal{M}, \mathcal{N}$ such that the bundle maps $\mathcal{E}^+ \oplus \mathcal{M} \stackrel{\sigma\oplus I_{\mathcal{M}}}{\longrightarrow} \mathcal{E}^- \oplus \mathcal{M}$ and $\mathcal{F}^+ \oplus \mathcal{N} \stackrel{\nu\oplus I_{\mathcal{N}}}{\longrightarrow} \mathcal{F}^- \oplus \mathcal{N}$ are homotopic. The class of a bundle map $\mathcal{E}^+ \stackrel{\sigma}{\longrightarrow} \mathcal{E}^-$ is denoted by $[\mathcal{E}^+, \sigma, \mathcal{E}^-]$ or simply by $[\sigma]$.

Two bundle maps $\mathcal{E}^+ \stackrel{\sigma}{\longrightarrow} \mathcal{E}^-$ and $\mathcal{E}^+ \stackrel{\nu}{\longrightarrow} \mathcal{E}^-$ which differ only on a compact subset of U are homotopic. Furthermore we can assume that σ is defined only outside a compact subset F of U. Indeed if $\phi \in C^\infty(U)$ is a continuous function with compact support and equal to 1 on F, then $(1-\phi)\sigma$ extends continuously to U. Clearly the class $[(1-\phi)\sigma]$ is independent of the choice of the cut-off function ϕ. We still denote it by $[\sigma]$. If U is compact, $[\mathcal{E}^+, \sigma, \mathcal{E}^-] = [\mathcal{E}^+, 0, \mathcal{E}^-] = [\mathcal{E}^+] - [\mathcal{E}^-]$.

Consider the case where U is an open subset of a compact *manifold* M. A neighborhood of $M - U$ will be called a neighborhood of infinity. If a bundle map $\mathcal{F}^+ \stackrel{\nu}{\longrightarrow} \mathcal{F}^-$ over M has its support contained in U, it defines by restriction an element of $K(U)$. Let us now show that every element of $K(U)$ has a representative of this form.

Proposition 3 *(Excision) Let $U \subset M$ be an open subset of a compact manifold M. Every element of $K(U)$ has a representative of the form $\mathcal{F}^+|_U \stackrel{\nu|_U}{\longrightarrow} \mathcal{F}^-|_U$ where $\mathcal{F}^\pm$ are bundles over M and $\mathcal{F}^+ \stackrel{\nu}{\longrightarrow} \mathcal{F}^-$ is a bundle map over M with support contained in U.*

Proof. Let $\mathcal{E}^+ \stackrel{\sigma}{\longrightarrow} \mathcal{E}^-$ be a bundle map over U with compact support. The bundle $\mathcal{E}^-$ is a direct summand of a trivial bundle on U (this property may be false if no assumption is made about the topology of U), therefore, by adding an element of $C_0(U)$, we may assume that $\mathcal{E}^-$ itself is trivial, in particular extends to a bundle defined over M. The bundle $\mathcal{E}^+$ being isomorphic to $\mathcal{E}^-$ in a neighborhood of infinity extends also to M. Thus we have shown that every element of $K(U)$ has a representative of the form $\mathcal{E}^+|_U \stackrel{\sigma}{\longrightarrow} \mathcal{E}^-|_U$ where $\mathcal{E}^\pm$ are bundles defined over M. Let ϕ be a cut-off function adapted to σ. Consider the bundle $\mathcal{G}(\sigma)$, realized as a sub-bundle (Lemma 2):

$$\mathcal{F} = \mathcal{F}(\sigma, \phi) \subset \mathcal{E}^+ \oplus \mathcal{E}^-.$$

Let π be the second projection of $\mathcal{F}$ on $\mathcal{E}^-$. It is easy to see that π has the same support as σ. Thus $\mathcal{F} \stackrel{\pi}{\longrightarrow} \mathcal{E}^-$ defines by restriction to U an element of $K(U)$. Proposition 3 follows from the next lemma.

Lemma 4 *$\mathcal{F}|_U \stackrel{\pi|_U}{\longrightarrow} \mathcal{E}^-|_U$ and $\mathcal{E}^+|_U \stackrel{\sigma}{\longrightarrow} \mathcal{E}^-|_U$ are stably homotopic.*

Proof. (see [22]). Recall (Definition 1) that the subbundle $\mathcal{F}$ of $\mathcal{E}^+ \oplus \mathcal{E}^-$ is isomorphic over U to the kernel of the surjective map $u = u(\sigma, \phi)$. From the exact sequence

$$0 \to \mathcal{F}|_U \to \mathcal{E}^+ \oplus \mathcal{E}^- \stackrel{u}{\longrightarrow} \mathcal{E}^- \to 0,$$

we construct an isomorphism $T : \mathcal{E}^+ \oplus \mathcal{E}^- \to \mathcal{F} \oplus \mathcal{E}^-$ in the following way. We fix metrics on $\mathcal{E}^+$ and $\mathcal{E}^-$. This defines an isomorphism between $\mathcal{E}^-$ and the orthogonal complement of $\mathcal{F}$ in $\mathcal{E}^+ \oplus \mathcal{E}^-$. In this way we obtain the isomorphism $T^{-1} : \mathcal{F} \oplus \mathcal{E}^- \to \mathcal{E}^+ \oplus \mathcal{E}^-$. Let us write explicitly the formula for T. Denote by $\sigma(x)^*$ the adjoint of $\sigma(x)$. Then $G(x) = \sigma(x)\sigma(x)^* + \phi(x)^2 I_{\mathcal{E}_x^-}$ is invertible for every $x \in U$. For $(v_0, v_1) \in \mathcal{E}_x^+ \oplus \mathcal{E}_x^-$, then $T(v_0, v_1) = (z, w) \in \mathcal{F}_x \oplus \mathcal{E}_x^-$ is equal to :

$$\begin{aligned} w &= u(v_0, v_1) = \sigma(x)v_0 - \phi(x)v_1 \in \mathcal{E}_x^- \\ z &= (v_0 - \sigma(x)^* G(x)^{-1} w, v_1 + \phi(x) G(x)^{-1} w) \in \mathcal{F}_x. \end{aligned}$$

Next, we show that the bundle maps $\mathcal{E}^+ \oplus \mathcal{E}^- \xrightarrow{(\pi\oplus I_{\mathcal{E}^-})\circ T} \mathcal{E}^- \oplus \mathcal{E}^-$ and $\mathcal{E}^+ \oplus \mathcal{E}^- \xrightarrow{\sigma\oplus I_{\mathcal{E}^-}} \mathcal{E}^- \oplus \mathcal{E}^-$ are homotopic with a homotopy invertible over $U - F$. For $(v_0, v_1) \in \mathcal{E}_x^+ \oplus \mathcal{E}_x^-$,

$$(\pi \oplus I_{\mathcal{E}^-})T(v_0, v_1) = (v_1 + \phi(x)G(x)^{-1}w, w).$$

Now consider

$$\tau_t(v_0, v_1) = (v_1 + t\phi(x)G(x)^{-1}w_t, w_t)$$

with $w_t = \sigma(x)v_0 - t\phi(x)v_1 \in \mathcal{E}_x^-$. It is easy to see that τ_t remains invertible outside F. For $t = 0$, $\tau_0(v_0, v_1) = (v_1, \sigma(x)v_0)$. Consider the isomorphism S of $\mathcal{E}^- \oplus \mathcal{E}^-$ obtained by exchanging its two identical factors. Then $S \circ \tau_0$ is equal to $\sigma \oplus I_{\mathcal{E}}$ and we obtain the lemma.

Remark 5 Consider the canonical map $i_* : K(U) \to K(M)$ induced by the inclusion $U \subset M$. Then the image of the class $[\mathcal{E}^+|_U, \sigma, \mathcal{E}^-|_U]$ is the formal difference of bundles $[\mathcal{G}(\sigma)] - [\mathcal{E}^-]$. In the particular case where σ extends to $M - F$ in an isomorphism, this virtual bundle is isomorphic to $[\mathcal{E}^+] - [\mathcal{E}^-]$ but in general it is not.

1.2 Volterra expansion formula

We will make constant use of Volterra's expansion formula for the exponential (see [9],Chapter 2, Section 4). Consider the simplex

$$\Delta_k = \{(s_0, \dots, s_k) \in \mathbf{R}^{k+1} \mid \textstyle\sum_{i=0}^k s_i = 1, s_i \geq 0\}$$

with the measure $ds_1 \dots ds_k$. Let A and Z be two elements of a Banach algebra. Then

$$e^{A+Z} - e^A = \sum_{k=1}^{\infty} I_k, \tag{4}$$

where

$$I_k = \int_{\Delta_k} e^{s_0 A} Z e^{s_1 A} Z \dots e^{s_{k-1} A} Z e^{s_k A}\, ds_1 \dots ds_k.$$

This infinite sum is convergent: the volume of the simplex Δ_k for the measure $ds_1 \dots ds_k$ is $\mathrm{vol}(\Delta_k) = \frac{1}{k!}$. Thus

$$\|I_k\| \leq e^{\|A\|} \|Z\|^k / k!. \tag{5}$$

Volterra's formula can be viewed as a Taylor's formula for the exponential map. A consequence is Duhamel's formula for the derivative of an exponential: assume that $A(z)$ depends smoothly on a parameter $z \in \mathbf{R}$. By setting $A = A(z)$ and $Z = \epsilon \frac{dA}{dz}$ we obtain

$$\frac{d}{dz}(e^{A(z)}) = \int_0^1 e^{(1-s)A(z)} \frac{dA}{dz} e^{sA(z)} ds. \tag{6}$$

We also write Volterra's expansion as

$$e^{A+Z} = \sum_{k=0}^{\infty} I_k, \text{ with } I_0 = e^A. \tag{7}$$

We will use Volterra's expansion in the case of an algebra $\mathrm{End}\, E \otimes L$, where E is a finite dimensional Hermitian vector space and $L^\bullet = \sum L^k$ is a finite dimensional

$\mathbf{Z}_+$-graded algebra, such that $L^0 = \mathbf{C}$ (for example an exterior algebra). We identify $\operatorname{End} E$ with the component of degree 0 of $\operatorname{End} E \otimes L^\bullet$. We consider L acting on itself by left multiplication. The norm on $\operatorname{End} E \otimes L$ is the operator norm on $\operatorname{End}(E \otimes L)$. Assume that A is a Hermitian element of $\operatorname{End} E$ and let $\lambda \in \mathbf{R}$ be the largest eigenvalue of A, then $\|e^{sA}\| = e^{s\lambda}$ for every $s \geq 0$, thus

$$\|I_k\| \leq e^\lambda \|Z\|^k / k!, \tag{8}$$

$$\|e^{A+Z}\| \leq e^{\lambda + \|Z\|}. \tag{9}$$

We obtain a refinement of (8) by decomposing $Z = Z_0 + Z_1$, where $Z_0 \in \operatorname{End} E$ is the degree zero component of $Z \in \operatorname{End} E \otimes L^\bullet$. The integral I_k expands in the sum of 2^k terms of the form

$$\int_{\Delta_k} e^{s_0 A} Z_{\epsilon_1} e^{s_1 A} Z_{\epsilon_2} \dots e^{s_{k-1} A} Z_{\epsilon_k} e^{s_k A} \, ds_1 \dots ds_k \, .$$

Let N be the highest degree allowed in $L^\bullet$. Then any term which contains more than N factors Z_1 is zero, therefore there are at most N factors Z_1 and k factors Z_0. If we set $z_0 = \sup(\|Z_0\|, 1)$ and $z_1 = \sup(\|Z_1\|, 1)$, we have

$$\|I_k\| \leq e^\lambda z_1^N z_0^k 2^k / k!. \tag{10}$$

1.3 Superconnections and Chern characters

From now on, M will be a smooth paracompact manifold and all vector bundles over M will be smooth. If $\mathcal{E} \to M$ is a (smooth) vector bundle over M the space of *smooth* sections of $\mathcal{E}$ is denoted by $\Gamma(M, \mathcal{E})$. Let $\Lambda T^*M = \oplus_{k=0}^{\dim M} \Lambda^k T^*M$ be the $\mathbf{Z}_+$-graded bundle of exterior algebras over M. Let $\mathcal{A}^\bullet(M) = \Gamma(M, \Lambda T^*M) = \sum_k \mathcal{A}^k(M)$ be the graded algebra of differential forms. Let

$$d : \mathcal{A}^\bullet(M) \to \mathcal{A}^{\bullet+1}(M)$$

be the exterior differential. The cohomology of the complex $(\mathcal{A}^\bullet(M), d)$ is de Rham cohomology $H^\bullet(M)$ of M. We denote by $\mathcal{A}^\bullet_{cpt}(M)$ the space of differential forms with compact support. The cohomology of $(\mathcal{A}^\bullet_{cpt}(M), d)$ is denoted by $H^\bullet_{cpt}(M)$ and is called the de Rham cohomology with compact support. If M is oriented and $\alpha \in \mathcal{A}^\bullet_{cpt}(M)$, we denote by $\int_M \alpha$ the integral of the highest degree term of α (on each connected component of M).

Let $\mathcal{E} \to M$ be a vector bundle over M and let $\mathcal{A}(M, \mathcal{E}) = \Gamma(M, \Lambda T^*M \otimes \mathcal{E})$ be the space of $\mathcal{E}$-valued differential forms. On the space $\mathcal{A}(M, \mathcal{E})$ we will consider the topology of uniform convergence on compact subsets as well as uniform convergence of derivatives . This topology is defined by the family of C_ℓ-semi-norms on compact subsets of M.

The notion of connection is a generalisation of the exterior differential: a connection ∇ on $\mathcal{E}$ is an operator

$$\nabla : \mathcal{A}^\bullet(M, \mathcal{E}) \to \mathcal{A}^{\bullet+1}(M, \mathcal{E})$$

which satisfies Leibniz's rule

$$\nabla(\alpha\nu) = (d\alpha) \wedge \nu + (-1)^k \alpha \wedge \nabla\nu$$

if $\alpha \in \mathcal{A}^k(M)$ and $\nu \in \mathcal{A}(M,\mathcal{E})$.

Let us explicitly describe a connection in a local trivialisation. If the bundle $\mathcal{E} \to M$ is trivial, $\mathcal{E} = M \times E$, then

$$\mathcal{A}(M,\mathcal{E}) = \mathcal{A}(M) \otimes E$$

and a connection on $\mathcal{E} = M \times E$ is an operator ∇ on $\mathcal{A}(M) \otimes E$ of the form

$$\nabla = d \otimes I + \sum \omega^a \otimes A_a$$

where ω^a is the multiplication by a 1-form $\omega^a \in \mathcal{A}^1(M)$ and $A_a \in \operatorname{End} E$ is a linear operator on E. We write $\omega = \sum_a \omega^a \otimes A_a$ and we say that $\omega \in \mathcal{A}^1(M) \otimes \operatorname{End} E$ is the connection form of ∇. The connection $d \otimes I$ on the trivial vector bundle $M \times E$ is called the trivial connection and is denoted simply by d. Let $M = \bigcup_i U_i$ be a covering by open sets with bundle isomorphisms $I_i : \mathcal{E}|_{U_i} \to U_i \times E$. Let ϕ_i be a partition of unity subordinate to the covering U_i. The operator

$$\nabla s = \sum_i I_i^{-1} d(I_i \phi_i s) \tag{11}$$

is a connection on $\mathcal{E}$. Thus any bundle over a (paracompact) manifold admits a connection.

Let $\mathcal{E} \to M$ be a vector bundle with an Hermitian structure. Then if $\mathcal{E}$ has a connection ∇, any subbundle $\mathcal{E}_0$ of $\mathcal{E}$ inherits a connection ∇_0. Indeed, let us consider the orthogonal projection $P_0 : \mathcal{E} \to \mathcal{E}_0$ and denote still by P_0 the projection map $P_0 : \mathcal{A}(M,\mathcal{E}) \to \mathcal{A}(M,\mathcal{E}_0)$.

Definition 6 *Let $\mathcal{E}_0$ be a subbundle of an Hermitian vector bundle $\mathcal{E}$. Let ∇ be a connection on $\mathcal{E}$. The projected connection $\nabla_0 : \mathcal{A}(M,\mathcal{E}_0) \to \mathcal{A}(M,\mathcal{E}_0)$ is defined by:*

$$\nabla_0 s = P_0 \nabla s \quad \text{for } s \in \mathcal{A}(M,\mathcal{E}_0) \subset \mathcal{A}(M,\mathcal{E}).$$

Let $\mathcal{E} = \mathcal{E}^+ \oplus \mathcal{E}^-$ be a $\mathbb{Z}/2\mathbb{Z}$-graded vector bundle on a manifold M. We will say that $\mathcal{E}$ is a superbundle. (An ungraded vector bundle is considered as a superbundle such that $\mathcal{E}^- = 0$.) The space $\mathcal{A}(M,\mathcal{E})$ inherits a total $\mathbb{Z}/2\mathbb{Z}$-grading and a $\mathbb{Z}_+$-grading. The notion of superconnection is a generalisation due to Quillen of the notion of connection. In the remaining of this subsection, we recall some of the results of Quillen [21].

Definition 7 *A superconnection $\mathbb{A}$ on $\mathcal{E}$ is an operator*

$$\mathbb{A} : \mathcal{A}(M,\mathcal{E}) \to \mathcal{A}(M,\mathcal{E})$$

which is odd (with respect to the total $\mathbb{Z}/2\mathbb{Z}$-grading) and which satisfies Leibniz's rule

$$\mathbb{A}(\alpha\nu) = (d\alpha) \wedge \nu + (-1)^k \alpha \wedge \mathbb{A}\nu$$

if $\alpha \in \mathcal{A}^k(M)$ and $\nu \in \mathcal{A}(M,\mathcal{E})$.

Let us explicitly describe a superconnection in a local trivialization. Assume that the superbundle $\mathcal{E}$ is trivial: $\mathcal{E}^{\pm} = M \times E^{\pm}$ where $E^{\pm}$ are two finite-dimensional vector spaces. Let $E = E^{+} \oplus E^{-}$. Then $\mathcal{A}(M, \mathcal{E}) = \mathcal{A}(M) \otimes E$. A superconnection on $\mathcal{E} = M \times E$ is an operator $\mathbb{A}$ on $\mathcal{A}(M) \otimes E$ of the form

$$\mathbb{A} = d \otimes I + \sum \omega^{[i],a} \otimes A_{i,a} \tag{12}$$

where $\omega^{[i],a}$ is the multiplication by a differential form of degree i and $A_{i,a}$ is a linear operator on $E = E^{+} \oplus E^{-}$ which is odd if i is even and even if i is odd.

If $K : \mathcal{A}(M, \mathcal{E}) \to \mathcal{A}(M, \mathcal{E})$ is an operator, we write

$$K = \begin{pmatrix} a & b \\ c & d \end{pmatrix}$$

where $a : \mathcal{A}(M, \mathcal{E}^{+}) \to \mathcal{A}(M, \mathcal{E}^{+})$, $b : \mathcal{A}(M, \mathcal{E}^{+}) \to \mathcal{A}(M, \mathcal{E}^{-})$, etc..

We can write a superconnection $\mathbb{A}$ on a superbundle $\mathcal{E}$ as

$$\mathbb{A} = A^{[0]} \oplus A^{[1]} \oplus \ldots$$

where

$$A^{[i]} : \mathcal{A}^{\bullet}(M, \mathcal{E}) \to \mathcal{A}^{\bullet + i}(M, \mathcal{E}).$$

In particular the component

$$A^{[1]} = \begin{pmatrix} \nabla^{+} & 0 \\ 0 & \nabla^{-} \end{pmatrix}$$

is a direct sum of connections $\nabla^{\pm}$ on each bundle $\mathcal{E}^{+}$, $\mathcal{E}^{-}$ and $A^{[0]}$ is an odd endomorphism of $\mathcal{E} = \mathcal{E}^{+} \oplus \mathcal{E}^{-}$.

Although a superconnection $\mathbb{A}$ is a first order differential operator, its square

$$F = \mathbb{A}^{2} : \mathcal{A}(M, \mathcal{E}) \to \mathcal{A}(M, \mathcal{E})$$

is a differential operator of order 0: this is easily deduced from the fact that $d^2 = 0$ on $\mathcal{A}(M)$. Thus the operator F is given by the action on $\mathcal{A}(M, \mathcal{E})$ of an even element $F \in \mathcal{A}(M, \operatorname{End} \mathcal{E})$ called the *curvature* of $\mathbb{A}$ and still denoted by F.

Let $\alpha \in \mathcal{A}(M, \operatorname{End} \mathcal{E})$, then the operator $[\mathbb{A}, \alpha] = \mathbb{A}\alpha - (-1)^{|\alpha|}\alpha\mathbb{A}$ is again given by the action of an element of $\mathcal{A}(M, \operatorname{End} \mathcal{E})$ denoted by $\mathbb{A} \cdot \alpha$.

As the space of superconnections is an affine space, the derivative of a smooth one-parameter family of superconnections $\mathbb{A}_t$ is given by the action of a differential form still denoted by

$$\frac{d}{dt}\mathbb{A}_t \in \mathcal{A}(M, \operatorname{End} \mathcal{E}).$$

This remark will be applied in particular to the segment joining two superconnections: $\mathbb{A}_t = t\mathbb{A}_0 + (1-t)\mathbb{A}_1$.

Let $K \in \Gamma(M, \operatorname{End} \mathcal{E})$, $K = \begin{pmatrix} a & b \\ c & d \end{pmatrix}$, with $a \in \Gamma(M, \operatorname{Hom}(\mathcal{E}^{+}, \mathcal{E}^{+}))$, $c \in \Gamma(M, \operatorname{Hom}(\mathcal{E}^{+}, \mathcal{E}^{-}))$, $b \in \Gamma(M, \operatorname{Hom}(\mathcal{E}^{-}, \mathcal{E}^{+}))$, $d \in \Gamma(M, \operatorname{Hom}(\mathcal{E}^{-}, \mathcal{E}^{-}))$. The *supertrace* of K is defined by

$$\operatorname{Str}(K) = \operatorname{Tr}(a) - \operatorname{Tr}(d).$$

The supertrace is extended by $\mathcal{A}(M)$-linearity to a map $\mathcal{A}(M, \mathrm{End}\,\mathcal{E}) \to \mathcal{A}(M)$: $\mathrm{Str}(\alpha K) = \alpha\,\mathrm{Str}(K)$ if $\alpha \in \mathcal{A}(M)$, $K \in \Gamma(M, \mathrm{End}\,\mathcal{E})$. The supertrace vanishes on supercommutators. Thus the following formula is easy to prove using the description (12) of $\mathbb{A}$ in local coordinates :

$$d\,\mathrm{Str}(\alpha) = \mathrm{Str}[\mathbb{A}, \alpha]. \tag{13}$$

Definition 8 *The Chern character form* $\mathrm{ch}(\mathcal{E}, \mathbb{A})$ *of the superbundle* $\mathcal{E}$ *with superconnection* $\mathbb{A}$ *is the differential form on* M *defined by:*

$$\mathrm{ch}(\mathcal{E}, \mathbb{A}) = \mathrm{Str}(e^F)$$

where $F = \mathbb{A}^2$ *is the curvature of* $\mathbb{A}$.

As F is an even element of $\mathcal{A}(M, \mathrm{End}\,\mathcal{E})$, the Chern character $\mathrm{ch}(\mathcal{E}, \mathbb{A})$ is even dimensional. The formula (13) implies that $\mathrm{ch}(\mathcal{E}, \mathbb{A})$ is a de Rham closed form on M.

Let $\mathbb{A}_t$ ($t \in \mathbf{R}$) be a smooth family of superconnections on $\mathcal{E} \to M$ with curvature F_t. Let us define the *transgression* form of the family $\mathbb{A}_t$ by

$$\alpha(t) = \mathrm{Str}\Big(\frac{d\mathbb{A}_t}{dt} e^{F_t}\Big) \in \mathcal{A}(M). \tag{14}$$

It is easy to prove that

$$\frac{d}{dt}\,\mathrm{ch}(\mathcal{E}, \mathbb{A}_t) = d\alpha(t).$$

Integrating this formula, we see that the difference $\mathrm{ch}(\mathcal{E}, \mathbb{A}_T) - \mathrm{ch}(\mathcal{E}, \mathbb{A}_{t_0})$ is an exact form. Precisely

$$\mathrm{ch}(\mathcal{E}, \mathbb{A}_T) - \mathrm{ch}(\mathcal{E}, \mathbb{A}_{t_0}) = d \int_{t_0}^{T} \alpha(t)dt. \tag{15}$$

Thus the de Rham cohomology class of the Chern character form $\mathrm{ch}(\mathcal{E}, \mathbb{A})$ is independent of the choice of the superconnection $\mathbb{A}$ on $\mathcal{E}$. We denote the de Rham cohomology class of $\mathrm{ch}(\mathcal{E}, \mathbb{A})$ by $\mathrm{ch}(\mathcal{E})$. In particular choosing $\mathbb{A}$ to be simply equal to a direct sum connection $\nabla = \nabla^+ \oplus \nabla^-$, we see that

$$\mathrm{ch}(\mathcal{E}, \mathbb{A}) \equiv \mathrm{ch}(\mathcal{E}^+) - \mathrm{ch}(\mathcal{E}^-)$$

in de Rham cohomology. However we will mainly be interested to compare cohomology classes with growth conditions.

It is clear that, when the manifold M is compact, the Chern character induces a map

$$\mathrm{ch} : K(M) \to H^\bullet(M). \tag{16}$$

If U is a locally compact manifold, there is a Chern character map

$$\mathrm{ch} : K(U) \to H^\bullet_{cpt}(U).$$

Let us describe this map using excision (Proposition 3). Choose an imbedding of U as an open subset of a compact manifold M. Let $\mathcal{F}^\pm$ be two vector bundles over M and let $\mathcal{F}^+ \xrightarrow{\nu} \mathcal{F}^-$ be a bundle map with compact support contained in U. It is possible, using partition of unity (Formula 11), to find connections $\nabla^\pm$ on $\mathcal{F}^\pm \to M$ such that ∇^+ and $\nu^*\nabla^-$ coincide near infinity. Then the differential form $\mathrm{ch}(\mathcal{F}^+, \nabla^+) - \mathrm{ch}(\mathcal{F}^-, \nabla^-)$ has compact support contained in U. It is easy to see that the cohomology class of $\mathrm{ch}(\mathcal{F}^+, \nabla^+) - \mathrm{ch}(\mathcal{F}^-, \nabla^-)$ in $H^\bullet_{cpt}(U)$ depends only of the element $[\mathcal{F}^+|_U, \nu|_U, \mathcal{F}^-|_U]$ of $K(U)$.

Definition 9 *For* $[\sigma] = [\mathcal{F}^+|_U, \nu|_U, \mathcal{F}^-|_U]$, *the cohomology class of* $\mathrm{ch}(\mathcal{F}^+, \nabla^+) - \mathrm{ch}(\mathcal{F}^-, \nabla^-)$ *in* $H^\bullet_{cpt}(U)$ *is denoted by* $\mathrm{ch}([\sigma])$.

Consider two vector bundles $\mathcal{E}^\pm$ over M with connections $\nabla^\pm$ and a bundle map $\sigma : \mathcal{E}^+|_U \to \mathcal{E}^-|_U$ with compact support (σ is defined only over U). Let ϕ be a cut-off function on U adapted to σ. We construct the subbundle $\mathcal{F}(\sigma, \phi)$ of $\mathcal{E}^+ \oplus \mathcal{E}^-$ with its projection π on $\mathcal{E}^-$ (Definition 1). Consider the connection ∇_0 on $\mathcal{F}(\sigma, \phi)$ obtained by projection of the connection $\nabla^+ \oplus \nabla^-$. As the bundles $\mathcal{F}(\sigma, \phi)$ and $\mathcal{E}^-$ coincide near infinity, the connections $\pi^*\nabla^-$ and ∇_0 coincide at infinity. From Definition 9 we have (see Lemma 4):

$$\mathrm{ch}([\sigma]) \equiv \mathrm{ch}(\mathcal{F}(\sigma, \phi), \nabla_0) - \mathrm{ch}(\mathcal{E}^-, \nabla^-) \quad \text{in } H^\bullet_{cpt}(U). \tag{17}$$

Remark 10 By means of a superconnection $\mathbb{A}(\sigma)$ with [0]-exterior degree term $i\begin{pmatrix} 0 & \sigma^* \\ \sigma & 0 \end{pmatrix}$, D.Quillen [21] associates to a bundle map σ with support contained in a compact set C a Chern character $\mathrm{ch}_Q(\sigma)$ in the *relative* cohomology group $H^\bullet(U, U - C)$. Using a transgression formula in relative cohomology, it is possible to prove that the image of $\mathrm{ch}_Q(\sigma)$ in $H^\bullet_{cpt}(U)$ is equal to $\mathrm{ch}([\sigma])$.

For some non compact manifolds U the notion of rapidly decreasing differential forms is well defined and $H^\bullet_{cpt}(U)$ coincides with the cohomology $H^\bullet_{rapid}(U)$ of rapidly decreasing forms. For such a manifold U and for a bundle map σ which satisfies some growth conditions, the Chern character of the superconnection $\mathbb{A}(\sigma)$ is rapidly decreasing. The purpose of section 1.5 is to prove that its class in $H^\bullet_{rapid}(U)$ is also equal to $\mathrm{ch}([\sigma])$.

1.4 Chern character of index bundles

Let $\mathcal{E} \to M$ be a superbundle. Suppose that $\mathcal{E}^+$ and $\mathcal{E}^-$ have Hermitian structures. We consider on the vector bundle $\mathcal{E} = \mathcal{E}^+ \oplus \mathcal{E}^-$ the Hermitian structure obtained by direct sum. Let $v \in \Gamma(M, \mathrm{End}\,\mathcal{E})$ be an odd Hermitian endomorphism of the superbundle $\mathcal{E}$:

$$v = \begin{pmatrix} 0 & u^* \\ u & 0 \end{pmatrix}.$$

Assume that $\mathrm{Ker}(v)$ has *constant rank*, so that the family of superspaces $(\mathrm{Ker}(v_z) \mid z \in M)$ forms a superbundle over M, called the **index bundle** of v:

$$\mathrm{Ker}(v) = \mathrm{Ker}(u) \oplus \mathrm{Ker}(u^*).$$

Let us denote by $\mathcal{E}_0 \subset \mathcal{E}$ the graded superbundle $\mathrm{Ker}(v)$. Likewise, let $\mathcal{E}_1 = \mathrm{Im}(v) \subset \mathcal{E}$ be the image of the operator v, and let $P_1 = 1 - P_0$ be the orthogonal projection of $\mathcal{E}$ on $\mathcal{E}_1$. If $x \in \mathcal{E}_z$ and $y \in \mathrm{Ker}(v_z)$, then by the assumption that v is self-adjoint,

$$(v_z x, y) = (x, v_z y) = 0,$$

from which we see that $\mathcal{E}_1$ is the orthogonal complement of $\mathcal{E}_0$,

$$\begin{aligned} \mathcal{E}^+ &= \mathcal{E}_0^+ \oplus \mathcal{E}_1^+, \\ \mathcal{E}^- &= \mathcal{E}_0^- \oplus \mathcal{E}_1^-, \end{aligned}$$

and hence the bundle $\mathrm{Ker}(u^*)$ is isomorphic to the bundle $\mathrm{Coker}(u) = \mathcal{E}^- / \mathrm{Im}(u)$, and the bundles $\mathcal{E}_1^+$ and $\mathcal{E}_1^-$ are isomorphic by the restriction of u. Let $\nabla^\pm$ be connections on $\mathcal{E}^\pm$. We denote by ∇ the direct sum connection on $\mathcal{E}$. Let $\mathbb{A}(v) = iv + \nabla$ be the superconnection on $\mathcal{E}$ associated to v and ∇:

$$\mathbb{A}(v) = i \begin{pmatrix} 0 & u^* \\ u & 0 \end{pmatrix} + \begin{pmatrix} \nabla^+ & 0 \\ 0 & \nabla^- \end{pmatrix}. \tag{18}$$

The curvature $F(v) = \mathbb{A}(v)^2$ of $\mathbb{A}(v)$ is

$$F(v) = -v^2 + i\nabla \cdot v + F \tag{19}$$

i.e.

$$F(v) = -\begin{pmatrix} u^*u & 0 \\ 0 & uu^* \end{pmatrix} + i \begin{pmatrix} 0 & \nabla \cdot u^* \\ \nabla \cdot u & 0 \end{pmatrix} + \begin{pmatrix} F^+ & 0 \\ 0 & F^- \end{pmatrix} \tag{20}$$

where $\nabla \cdot u^* = \nabla^+ u^* + u^* \nabla^- \in \mathcal{A}^1(M, \mathrm{Hom}(\mathcal{E}^-, \mathcal{E}^+))$, $\nabla \cdot u = \nabla^- u + u\nabla^+ \in \mathcal{A}^1(M, \mathrm{Hom}(\mathcal{E}^+, \mathcal{E}^-))$ are 1-forms with matrix coefficients, $F^+ = (\nabla^+)^2 \in \mathcal{A}^2(M, \mathrm{End}\,\mathcal{E}^+)$ is the curvature of ∇^+ and $F^- = (\nabla^-)^2 \in \mathcal{A}^2(M, \mathrm{End}\,\mathcal{E}^-)$ is the curvature of ∇^-.

We have the equalities in de Rham cohomology

$$\begin{aligned} \mathrm{ch}(\mathcal{E}, \mathbb{A}) &= \mathrm{ch}(\mathcal{E}^+) - \mathrm{ch}(\mathcal{E}^-) \\ &= \mathrm{ch}(\mathcal{E}_0^+) + \mathrm{ch}(\mathcal{E}_1^+) - \mathrm{ch}(\mathcal{E}_0^-) - \mathrm{ch}(\mathcal{E}_1^-) \\ &= \mathrm{ch}(\mathcal{E}_0^+) - \mathrm{ch}(\mathcal{E}_0^-) \\ &= \mathrm{ch}(\mathrm{Ker}(v)). \end{aligned}$$

We have obtained in [9] a refinement of this formula at the level of differential forms. Let us recall the result. Denote by ∇_0 the connection on the subbundle $\mathcal{E}_0$ given by projecting the connection ∇:

$$\nabla_0 = P_0 \nabla P_0.$$

Theorem 11 *Let $\mathcal{E} = \mathcal{E}^+ \oplus \mathcal{E}^-$ be a Hermitian superbundle with a direct sum connection ∇. Let v be a Hermitian odd endomorphism of $\mathcal{E}$ whose kernel has constant rank. Let $\mathbb{A}(v) = iv + \nabla$ be the superconnection on $\mathcal{E}$ associated to v and ∇. For $t > 0$, let*

$$\mathbb{A}_t = itv + \nabla$$

be the rescaled superconnection. Then the limit $\lim_{t\to\infty} \mathrm{ch}(\mathcal{E}, \mathbb{A}_t)$ exists in the sense of all C^ℓ-norms on compact sets of M. The limit equals the Chern character of the projected connection ∇_0 on the superbundle $\mathcal{E}_0 = \mathrm{Ker}(v)$:

$$\lim_{t\to\infty} \mathrm{ch}(\mathcal{E}, \mathbb{A}_t) = \mathrm{ch}(\mathcal{E}_0, \nabla_0).$$

We have also obtained an explicit homotopy between $\mathrm{ch}(\mathcal{E}, \mathbb{A})$ and $\mathrm{ch}(\mathcal{E}_0, \nabla_0) = \lim_{t\to\infty} \mathrm{ch}(\mathcal{E}, \mathbb{A}_t)$. Let us write

$$\alpha(t) = \mathrm{Str}\Big(\frac{d\mathbb{A}_t}{dt} e^{\mathbb{A}_t^2}\Big) \in \mathcal{A}(M).$$

The transgression formula (15) states that

$$\mathrm{ch}(\mathcal{E}, \mathbb{A}_T) - \mathrm{ch}(\mathcal{E}, \mathbb{A}(v)) = d \int_1^T \alpha(t)\, dt.$$

Theorem 12 *The differential form* $\alpha(t) = \mathrm{Str}\left(\frac{d\mathbb{A}_t}{dt} e^{\mathbb{A}_t^2}\right)$ *satisfies the estimates, for* t *large,*

$$\|\alpha(t)\|_\ell \leq c_\ell t^{-2}$$

on compact subsets of M*, for all* C^ℓ*-norms.*

The integral $\int_1^\infty \alpha(t)\, dt$ *is convergent in* $\mathcal{A}(M)$*, and*

$$\mathrm{ch}(\mathcal{E}_0, \nabla_0) - \mathrm{ch}(\mathcal{E}, \mathbb{A}(v)) = d \int_1^\infty \alpha(t)\, dt.$$

Proof. These two theorems are proved in ([9], Chapter 9, Section 1). (In [9], the rescaled superconnection is defined with $t^{1/2}$ instead of t). We give here a modified proof which is simpler and which will generalise to the equivariant case. We will make constant use of the following notation: if K is an operator acting on $\mathcal{A}(M, \mathrm{End}\, \mathcal{E})$, we write K as a 2×2 block matrix *with respect to the decomposition* $\mathcal{E} = \mathcal{E}_0 \oplus \mathcal{E}_1$. To avoid confusion with the decomposition $\mathcal{E} = \mathcal{E}^+ \oplus \mathcal{E}^-$ we use brackets instead of parentheses to denote the matrix:

$$K = \begin{bmatrix} \alpha & \beta \\ \gamma & \delta \end{bmatrix}.$$

This means that

$$\alpha = P_0 K P_0, \quad \beta = P_0 K P_1 \quad \gamma = P_1 K P_0, \quad \delta = P_1 K P_1.$$

Let $F_t \in \mathbb{C}[t] \otimes \mathcal{A}(M, \mathrm{End}\, \mathcal{E})$ be the curvature of $\mathbb{A}_t$. In the next lemma we study the expansion in powers of t of the blocks of F_t with respect to the decomposition $\mathcal{E} = \mathcal{E}_0 \oplus \mathcal{E}_1$.

Lemma 13 *Let*

$$F_t = \begin{bmatrix} A(t) & B(t) \\ C(t) & D(t) \end{bmatrix}$$

be the block decomposition of F_t *with respect to* $\mathcal{E} = \mathcal{E}_0 \oplus \mathcal{E}_1$.

1. *The expansion of* $D(t)$ *in powers of* t *is* $D(t) = t^2 D_2 + t D_1 + D_0$ *where* $D_2 = -P_1 v^2 P_1$ *so that* D_2 *is a negative definite endomorphism of* $\mathcal{E}_1$.

2. *The other blocks have power expansion*

$$A(t) = A_0, \quad B(t) = t B_1 + B_0, \quad C(t) = t C_1 + C_0\, .$$

3. *Furthermore, let* $R \in \mathcal{A}(M, \mathrm{End}\, \mathcal{E}_0)$ *be the curvature of* $\nabla_0 = P_0 \nabla P_0$*, then*

$$R = A_0 - B_1 D_2^{-1} C_1.$$

Proof.

This is proven in ([9], Chapter 9, Section 1). We recall the proof. The connection ∇ has block decomposition

$$\nabla = \begin{bmatrix} \nabla_0 & \mu \\ \nu & \nabla_1 \end{bmatrix}$$

where ∇_0, ∇_1 are the connections on $\mathcal{E}_0$ and $\mathcal{E}_1$ respectively obtained by projecting ∇. The blocks $\mu = P_0 \nabla P_1 = P_0(\nabla P_1 + P_1 \nabla) = P_0(\nabla \cdot P_1)$ and $\nu = P_1 \nabla P_0 = P_1(\nabla \cdot P_0)$

are operators given by multiplication by a matrix of differential forms which we also denote by $\mu \in \mathcal{A}^1(M, \mathrm{Hom}(\mathcal{E}_1, \mathcal{E}_0))$ and $\nu \in \mathcal{A}^1(M, \mathrm{Hom}(\mathcal{E}_0, \mathcal{E}_1))$.

Thus the block decomposition of $\mathbb{A} = iv + \nabla$ with respect to $\mathcal{E} = \mathcal{E}_0 \oplus \mathcal{E}_1$ is

$$\mathbb{A} = \begin{bmatrix} 0 & 0 \\ 0 & iv \end{bmatrix} + \begin{bmatrix} \nabla_0 & 0 \\ 0 & \nabla_1 \end{bmatrix} + \begin{bmatrix} 0 & \mu \\ \nu & 0 \end{bmatrix}.$$

Let $\tilde{\nabla} = \begin{bmatrix} \nabla_0 & 0 \\ 0 & \nabla_1 \end{bmatrix}$. Denote by R, S the curvature of ∇_0 and ∇_1 respectively. We obtain for $F = \mathbb{A}^2$ the formula:

$$F = \begin{bmatrix} R + \mu\nu & i\mu v + \tilde{\nabla} \cdot \mu \\ iv\nu + \tilde{\nabla} \cdot \nu & -v^2 + i\nabla_1 \cdot v + S + \nu\mu \end{bmatrix} \tag{21}$$

Here $\tilde{\nabla} \cdot \mu = \nabla_0 \mu + \mu \nabla_1$ and $\tilde{\nabla} \cdot \nu = \nabla_1 \nu + \nu \nabla_0$ have exterior degree 2 while $\nabla_1 \cdot v = \nabla_1 v + v \nabla_1$ has exterior degree 1. Now replacing v by tv we obtain

$$F_t = \begin{bmatrix} R + \mu\nu & it\mu v + \tilde{\nabla} \cdot \mu \\ itv\nu + \tilde{\nabla} \cdot \nu & -t^2 v^2 + it\nabla_1 \cdot v + S + \nu\mu \end{bmatrix}.$$

This clearly implies the lemma.

The following lemma is the key step in the proof.

Lemma 14 *Let $E = E_0 \oplus E_1$ be an orthogonal decomposition of a Hermitian space E. Consider a Laurent polynomial $L(t) \in \mathbb{C}[t, t^{-1}] \otimes \mathrm{End}\, E$. Assume the blocks of $L(t)$*

$$L(t) = \begin{bmatrix} A(t) & B(t) \\ C(t) & D(t) \end{bmatrix}$$

with respect to the decomposition $E = E_0 \oplus E_1$ satisfy

$$\begin{aligned} A(t) &= \sum_{k \le 0} t^k A_k \qquad B(t) = \sum_{k \le 1} t^k B_k \\ C(t) &= \sum_{k \le 1} t^k C_k \qquad D(t) = \sum_{k \le 2} t^k D_k \end{aligned}$$

Assume further that D_2 is a negative definite Hermitian endomorphism of E_1. Let $R = A_0 - B_1 D_2^{-1} C_1$. Then when $t \to +\infty$

$$e^{L(t)} = \begin{bmatrix} e^R + O(t^{-1}) & O(t^{-1}) \\ O(t^{-1}) & O(t^{-2}) \end{bmatrix}.$$

Here $O(t^{-j})$ is the usual Landau notation: $M(t) = O(t^{-j})$ means that $t^j \|M(t)\|$ is bounded.

Proof. We define $\mathcal{L}_i$ as the space of Laurent polynomials of the form

$$\sum_{k \le i} t^k X_k.$$

We first prove that there exists an invertible matrix $g_1(t)$ of the form

$$g_1(t) = \begin{bmatrix} 1 & U(t) \\ 0 & 1 \end{bmatrix}$$

with $U(t) = t^{-1}U_{-1} + t^{-2}U_{-2}$ such that

$$g_1(t)L(t) = \begin{bmatrix} R & 0 \\ C(t) & D(t) \end{bmatrix} + s_1(t)$$

with $s_1(t) \in \mathcal{L}_{-1}$. Indeed

$$g_1(t)L(t) = \begin{bmatrix} A(t) + U(t)C(t) & B(t) + U(t)D(t) \\ C(t) & D(t) \end{bmatrix}.$$

By hypothesis on $B(t), D(t)$, the Laurent polynomial $B(t)+U(t)D(t)$ belongs to $\mathcal{L}_1$. The coefficient of t in $B(t) + U(t)D(t)$ is $B_1 + U_{-1}D_2$, while the constant coefficient is $B_0 + U_{-1}D_1 + U_{-2}D_2$. As D_2 is invertible we can choose U_i such that both of these terms are 0. In particular, $U_{-1} = -B_1D_2^{-1}$. By hypothesis on $A(t), C(t)$, the Laurent polynomial $A(t) + U(t)C(t)$ belongs to $\mathcal{L}_0$. Its constant term is

$$R = A_0 - B_1D_2^{-1}C_1.$$

We have

$$g_1(t)^{-1} = \begin{bmatrix} 1 & -U(t) \\ 0 & 1 \end{bmatrix}.$$

Let $L_1(t) = g_1(t)L(t)g_1(t)^{-1}$. Then

$$L_1(t) = \begin{bmatrix} R & 0 \\ C(t) & D'(t) \end{bmatrix} + s_1'(t)$$

with $s_1'(t) \in \mathcal{L}_{-1}$ and

$$D'(t) = t^2D_2 + tD_1 + D_0'.$$

Similarly we can find a matrix $g_2(t)$ of the form $g_2(t) = \begin{bmatrix} 1 & 0 \\ V(t) & 1 \end{bmatrix}$ with $V(t) = t^{-1}V_{-1} + t^{-2}V_{-2}$ such that $L_1(t)g_2(t)^{-1}$ is of the form

$$L_1(t)g_2(t)^{-1} = \begin{bmatrix} R & 0 \\ 0 & D'(t) \end{bmatrix} + s_2(t)$$

with $s_2(t) \in \mathcal{L}_{-1}$. Finally if $g(t) = g_2(t)g_1(t)$ we see that

$$g(t)L(t)g(t)^{-1} = \begin{bmatrix} R & 0 \\ 0 & D'(t) \end{bmatrix} + Z(t) \tag{22}$$

with $Z(t) \in \mathcal{L}_{-1}$. Let

$$H(t) = \begin{bmatrix} R & 0 \\ 0 & D'(t) \end{bmatrix},$$

We write the Volterra expansion (4) $e^{H(t)+Z(t)} - e^{H(t)} = \sum_{k\geq 1} I_k(t)$. Let λ be the largest eigenvalue of $-D_2$ so that $\lambda > 0$. From the inequality (9), we obtain

$$\|e^{sD'(t)}\| \leq e^{-st^2\lambda + st\|D_1\| + s\|D_0'\|}. \tag{23}$$

In particular, $\|e^{sH(t)}\|$ is uniformly bounded for $s \in [0,1]$. Thus, as $Z(t) = O(t^{-1})$, it follows from the formula (8) that

$$e^{H(t)+Z(t)} - e^{H(t)} - I_1(t) = O(t^{-2}).$$

Let us see that the block decomposition of $I_1(t)$ has the form $I_1(t) = \begin{bmatrix} O(t^{-1}) & O(t^{-1}) \\ O(t^{-1}) & W(t) \end{bmatrix}$, where $W(t)$ is rapidly decreasing. Indeed if we write $Z(t) = \begin{bmatrix} Z_0(t) & Z_{1,0}(t) \\ Z_{1,0}(t) & Z_1(t) \end{bmatrix}$, then

$$W(t) = \int_{\Delta_1} e^{s_0 D'(t)} Z_1(t) e^{s_1 D'(t)}.$$

On the simplex Δ_1 one of the two numbers s_0, s_1 is greater than 1/2. Thus $\|W(t)\|$ is rapidly decreasing since by (23) $\|e^{sD'(t)}\|$ remains bounded for all s, t and is rapidly decreasing in t for $s \geq 1/2$, while $\|Z_1(t)\| = O(t^{-1})$.

Thus we obtain

$$e^{H(t)+Z(t)} = \begin{bmatrix} e^R + O(t^{-1}) & O(t^{-1}) \\ O(t^{-1}) & O(t^{-2}) \end{bmatrix}.$$

Now we have

$$e^{L(t)} = g(t)^{-1} e^{H(t)+Z(t)} g(t).$$

As $g(t) - 1 = O(t^{-1})$ as well as $g(t)^{-1} - 1$ we obtain easily our lemma from the estimate above.

Remark 15 The estimates of Lemma 14 cannot be improved. Indeed if

$$L(t) = \begin{bmatrix} A & tB \\ tC & t^2 D \end{bmatrix}$$

with D a negative definite endomorphism of V_1, then, following the same method, it is easy to prove that, when t tends to ∞

$$e^{L(t)} = \begin{bmatrix} e^R + O(t^{-2}) & -t^{-1} e^R B D^{-1} + O(t^{-3}) \\ -t^{-1} D^{-1} C e^R + O(t^{-3}) & t^{-2} D^{-1} C e^R B D^{-1} + O(t^{-4}) \end{bmatrix}.$$

The limit of $\mathrm{ch}(\mathcal{E}, \mathbb{A}_t)$ in Theorem 11 is then a consequence of the two Lemmas 13 and 14. We apply Lemma 14 to the curvature F_t. It is clear that all the preceeding estimates are valid uniformly on compact subsets of M. By Lemma 13 we obtain

$$e^{F_t} = \begin{bmatrix} e^R + O(t^{-1}) & O(t^{-1}) \\ O(t^{-1}) & O(t^{-2}) \end{bmatrix} \tag{24}$$

where the estimates $O(t^{-j})$ are uniform on compact subsets of M. We show the convergence of the derivatives(with respect to the coordinates in M) by similar arguments, using Duhamel's formula (6). This completes the proof of Theorem 11.

Let us now prove Theorem 12.

We have $\alpha(t) = \mathrm{Str}\left(\frac{d\mathbb{A}_t}{dt} e^{F_t}\right)$, and

$$\frac{d\mathbb{A}_t}{dt} e^{F_t} = \begin{bmatrix} 0 & 0 \\ 0 & iv \end{bmatrix} \begin{bmatrix} e^R + O(t^{-1}) & O(t^{-1}) \\ O(t^{-1}) & O(t^{-2}) \end{bmatrix} = \begin{bmatrix} 0 & 0 \\ O(t^{-1}) & O(t^{-2}) \end{bmatrix}$$

Since only the diagonal blocks contribute to the supertrace, this proves Theorem 12.

1.5 Good bundle maps and their Chern character

Let $\mathcal{V}$ be the total space of a real vector bundle over a compact base B. A point of $\mathcal{V}$ will be denoted by (z,x) with $z \in B$ and $x \in \mathcal{V}_z$. In a local trivialisation of $\mathcal{V}$ and in local coordinates z on B a differential form α on $\mathcal{V}$ is witten as $\sum_{I,J} a_{I,J}(z,x)dz_I dx_J$. Then α is said to be rapidly decreasing if its coefficients $a_{I,J}(z,x)$ and all their derivatives in z,x are rapidly decreasing with respect to x. The space of rapidly decreasing differential forms is denoted by $\mathcal{A}^\bullet_{rapid}(\mathcal{V})$. The cohomology ring of the complex $(\mathcal{A}^\bullet_{rapid}(\mathcal{V}),d)$ is denoted by $H^\bullet_{rapid}(\mathcal{V})$. We will now see that using superconnections we can define a Chern character map

$$\mathrm{ch}: K(\mathcal{V}) \to H^\bullet_{rapid}(\mathcal{V}).$$

(The same construction would be valid for manifolds with a notion of rapidly decreasing differential forms, for example a real algebraic submanifold of $\mathbf{R}^n$).

We need to define the set of *good bundle maps* on $\mathcal{V}$. Every complex vector bundle $\mathcal{F}$ over $\mathcal{V}$ is isomorphic to the pull back $p^*\mathcal{F}|_B$. This may be seen as follows: choose a connection ∇ on the bundle $\mathcal{F} \to \mathcal{V}$ and use parallel transport along the curve (z,tx) to identify $\mathcal{F}_{z,0}$ to $\mathcal{F}_{z,tx}$. Let $\mathcal{E}^\pm$ be two complex vector bundles on B. Let $\sigma : p^*\mathcal{E}^+ \to p^*\mathcal{E}^-$ be a bundle map. We consider the superbundle $\mathcal{E}^+ \oplus \mathcal{E}^-$. We choose Hermitian metrics on $\mathcal{E}^\pm$ and we associate to σ the odd endomorphism $v = v(\sigma)$ of $p^*\mathcal{E}$ defined by

$$v = \begin{pmatrix} 0 & \sigma^* \\ \sigma & 0 \end{pmatrix}. \tag{25}$$

Note that the square $v^2 = \begin{pmatrix} \sigma^*\sigma & 0 \\ 0 & \sigma\sigma^* \end{pmatrix}$ is a Hermitian positive endomorphism of the Hermitian bundle $p^*\mathcal{E}$.

If the bundle map σ is invertible outside of a compact subset of $\mathcal{V}$, it represents an element of $K(\mathcal{V})$. We will use representatives which satisfy further conditions. Let us choose an Euclidean structure on $\mathcal{V}$.

Definition 16 *The bundle map $\sigma : p^*\mathcal{E}^+ \to p^*\mathcal{E}^-$ is called a* good bundle map *if*

1. *The bundle map σ is a smooth section of* $\mathrm{Hom}(p^*\mathcal{E}^+, p^*\mathcal{E}^-)$.

2. *The bundle map σ and all its derivatives have at most polynomial growth in the fiber direction.*

3. *There exist $r > 0$ and $c > 0$ such that*

$$v(\sigma)(z,x)^2 \geq c\|x\|^2 I_{\mathcal{E}_z}$$

for all $z \in B, x \in \mathcal{V}_z$ such that $\|x\| \geq r$.

Clearly the notion of good bundle map is independent of the choice of Hermitian structures on $\mathcal{E}^\pm \to B$ and of the Euclidean structure on $\mathcal{V}$. If $\sigma(z,x)$ is homogeneous of degree 1 in x and is invertible outside the zero section of $\mathcal{V}$, then clearly σ is a good bundle map.

Lemma 17 *(i)Every element of $K(\mathcal{V})$ has a good representative.*

(ii) Let σ_0, $\sigma_1 : p^*\mathcal{E}^+ \to p^*\mathcal{E}^-$ be good bundle maps and assume that there exists a homotopy of bundle maps with compact support $\sigma_t : p^*\mathcal{E}^+ \to p^*\mathcal{E}^-$. Then there exists a homotopy $\tau_t : p^*\mathcal{E}^+ \to p^*\mathcal{E}^-$ between τ_0 and τ_1 such that each τ_t is a good bundle map and furthermore such that $\frac{d}{dt}\tau_t$ and all its derivatives have at most polynomial growth in the fiber direction.

Proof. We will indeed construct a representative ρ which is smooth and homogeneous "away from the zero section" in the sense that $\rho(z, \lambda x) = \lambda\rho(z, x)$ for $\lambda \geq 1$ and $\|x\| \geq 1$. Such a representative will automatically be good. In the case when $\mathcal{V}$ is the cotangent bundle T^*B this amounts to the fact that $K(T^*B)$ is represented by the principal symbols of pseudodifferential operators of order 1 on B. The reason for using representatives which are not principal symbols is that the tensor product of pseudodifferential operators is not pseudodifferential in general [7]. In contrast, the tensor product rule in $K(\mathcal{V})$ [7] lifts in a straightforward way on good representatives.

Clearly a given element of $K(\mathcal{V})$ has a representative σ which is a smooth bundle map with $\sigma(z, x)$ invertible for $\|x\| \geq 1$. Let $\phi : \mathbf{R} \to \mathbf{R}$ be a smooth function such that $\phi(t) = 0$ for $t \leq \frac{1}{2}$ and $\phi(t) = 1$ for $t \geq 1$. Define $\sigma_t(z, x) = \phi(\|x\|)(t\|x\| + 1 - t)\sigma(z, t\frac{x}{\|x\|} + (1-t)x)$. Then σ_t is invertible for $|x| \geq 1$, $[\sigma_0] = [\sigma]$, and $\sigma_1(z, x) = \phi(\|x\|)\|x\|\sigma(x/\|x\|)$ satisfies $\sigma_1(z, tx) = t\sigma_1(z, x)$ for $t \geq 1$ and $\|x\| \geq 1$. Let us call "good" a homotopy which satisfies the conditions (ii) of the lemma: Let us remark that if σ is good, so is the homotopy σ_t between σ and $\sigma_1(z, x) = \phi(\|x\|)\|x\|\sigma(x/\|x\|)$.

We obtain the second part of the lemma with the same method applied to $B \times [0, 1]$. From the discussion above, we may assume that σ_0 and σ_1 are good bundles maps of the form $\sigma_i(z, x) = \phi(\|x\|)\|x\|\sigma_i(x/\|x\|)$. We may assume that σ_t is invertible for $\|x\| \geq 1$. Define $\rho_t = \phi(\|x\|)\|x\|\sigma_t(z, \frac{x}{\|x\|})$. Then ρ_t is a good homotopy between σ_0 and σ_1.

Let $\nabla = \nabla^+ \oplus \nabla^-$ be a connection on $\mathcal{E} = \mathcal{E}^+ \oplus \mathcal{E}^-$. With a slight change of notations with respect to the notation 18, we now denote by $\mathbb{A}(\sigma, \nabla)$ or $\mathbb{A}(\sigma)$ the superconnection on $p^*\mathcal{E}$ defined by

$$\mathbb{A}(\sigma) = iv(\sigma) + p^*\nabla = i\begin{pmatrix} 0 & \sigma^* \\ \sigma & 0 \end{pmatrix} + \begin{pmatrix} p^*\nabla^+ & 0 \\ 0 & p^*\nabla^- \end{pmatrix}. \tag{26}$$

We write simply $\mathrm{ch}(\mathbb{A}(\sigma))$ for $\mathrm{ch}(p^*\mathcal{E}, \mathbb{A}(\sigma))$.

Proposition 18 *Let σ be a good bundle map. Then the Chern character form $\mathrm{ch}(\mathbb{A}(\sigma))$ is a rapidly decreasing differential form on $\mathcal{V}$. Furthermore the cohomology class of $\mathrm{ch}(\mathbb{A}(\sigma))$ in $H^\bullet_{rapid}(\mathcal{V})$ depends only on the class $[p^*\mathcal{E}^+, \sigma, p^*\mathcal{E}^-] \in K(\mathcal{V})$.*

Proof.

Denote by F the curvature of ∇. The curvature of $\mathbb{A}(\sigma)$ is

$$F(\sigma) = -v^2 + i(p^*\nabla) \cdot v + p^*F.$$

We fix a point $z \in B$. We compute the Chern character $\mathrm{ch}(\mathbb{A}(\sigma)) = \mathrm{Str}(e^{F(\sigma)})$ at the point (z, x). Using a local trivialisation of $\mathcal{V}$, we identify $F(\sigma)_{(z,x)}$ to an element of the fixed algebra $\operatorname{End}\mathcal{E}_z \otimes \Lambda T_z^*B \otimes \Lambda \mathcal{V}_z^*$. Let us study the behaviour in x of $e^{F(\sigma)}$. We use the Volterra expansion (7) with $A(z,x) = -v^2(z,x)$ and $Z(z,x) = (i(p^*\nabla)\cdot v + p^*F)_{(z,x)}$. As $Z(z,x)$ has no term of exterior degree 0, the Volterra expansion is a finite sum:

$$e^{F(\sigma)(z,x)} = \sum_0^N I_k \tag{27}$$

with $N = \dim B + \dim \mathcal{V}_z$.

The connection $p^*\nabla$ being the pull back of a connection on the basis B, the form p^*F is constant in x and the covariant derivative $(p^*\nabla)\cdot v$ is of at most polynomial growth on x by our assumption on σ. Thus $\|Z(z,x)\|$ is bounded by a polynomial in x. As the largest eigenvalue of $-v^2(z,x)$ is smaller than $-c\|x\|^2$ for $\|x\| \geq r$, the inequality (8) shows that we have

$$\|I_k\| \leq e^{-c\|x\|^2} \frac{\|Z(z,x)\|^k}{k!} \tag{28}$$

so that summing up these inequalities from 0 to N, we obtain that $e^{F(\sigma)(z,x)}$ is rapidly decreasing in x.

We study the derivatives of $e^{F(\sigma)}$ with respect to the variables (z,x) in a similar manner, using Duhamel's formula (6), the polynomial growth of the derivatives of the bundle map σ and again (8).

In order to prove the second claim we take two good bundle maps σ_0 , σ_1 of the same class in $K(\mathcal{V})$. We must show that

$$\operatorname{ch}(\mathbb{A}(\sigma_1, \nabla_1)) - \operatorname{ch}(\mathbb{A}(\sigma_0, \nabla_0)) = d\beta, \tag{29}$$

with a rapidly decreasing differential form β. The equality $[p^*\mathcal{E}_0^+, \sigma_0, p^*\mathcal{E}_0^-] = [p^*\mathcal{E}_1^+, \sigma_1, p^*\mathcal{E}_1^-]$ means that there exists two bundles $\mathcal{F}_0$ and $\mathcal{F}_1$ over B , isomorphisms

$$\begin{aligned} \mathcal{E}_0^+ \oplus \mathcal{F}_0 &\simeq \mathcal{E}_1^+ \oplus \mathcal{F}_1, \\ \mathcal{E}_0^- \oplus \mathcal{F}_0 &\simeq \mathcal{E}_1^- \oplus \mathcal{F}_1, \end{aligned}$$

and a homotopy between

$$\begin{aligned} \sigma_0 \oplus I_{\mathcal{F}_0} : p^*(\mathcal{E}_0^+ \oplus \mathcal{F}_0) &\to p^*(\mathcal{E}_0^- \oplus \mathcal{F}_0) \\ \sigma_1 \oplus I_{\mathcal{F}_1} : p^*(\mathcal{E}_1^+ \oplus \mathcal{F}_1) &\to p^*(\mathcal{E}_1^- \oplus \mathcal{F}_1). \end{aligned}$$

In order to deal only with good bundle maps we replace $\sigma_0 \oplus I_{\mathcal{F}_0}$ with the homotopic $\sigma_0 \oplus \phi(\|x\|)\|x\| I_{\mathcal{F}_0}$, and do the same with σ_1. Thus, changing notations, we need to prove (29) in the following three cases :

1. $\sigma_0 = \sigma_1$, but the Hermitian metrics and the connections on $\mathcal{E}_0$ and $\mathcal{E}_1$ may be different.

2. $\sigma_1 = \sigma_0 \oplus (\phi(\|x\|)\|x\| I_F$.

3. σ_0 and σ_1 are homotopic good bundle maps.

Case 2 is immediate: the Chern character is additive, and obviously the supertrace $\mathrm{Str}(e^{\mathbb{A}(\sigma,\nabla)^2})$ is equal to 0 when $\mathcal{E}^+ = \mathcal{E}^-$ with the same connection and σ is self-adjoint.

We treat together case 1 and case 3. Let σ_t be a good homotopy given by Lemma 17. Let g_t be the segment joining the two Hermitian metrics on $\mathcal{E}$ and let $v_t = \begin{pmatrix} 0 & \sigma_t' \\ \sigma_t & 0 \end{pmatrix}$, where σ_t' is the adjoint of σ_t with respect to the metric g_t. Let $\mathbb{A}_t$ be the superconnection $\mathbb{A}_t = iv_t + p^*\nabla_t$, where $\nabla_t = \nabla_0 + t(\nabla_1 - \nabla_0)$. Let F_t the curvature of $\mathbb{A}_t$. The transgression formula 15 gives $\mathrm{ch}(\mathbb{A}(\sigma_1, \nabla_1)) - \mathrm{ch}(\mathbb{A}(\sigma_0, \nabla_0)) = d\beta$, with

$$\beta = \int_0^1 \mathrm{Str}((\frac{dv_t}{dt} + p^*(\nabla_1 - \nabla_0))e^{F_t})\,dt.$$

Let us study the behaviour of the terms under the integral. To study e^{F_t}, we apply again Volterra expansion formula with $A(t,z,x) = -v_t^2(z,x)$ and $Z(t,z,x) = (i(p^*\nabla_t)\cdot v_t + p^*\nabla_t^2)_{(z,x)}$. Again $Z(t,z,x)$ has no term of exterior degree 0 and Volterra's expansion is finite. By our assumption on σ_t we see that $\|Z(t,z,x)\|$ is bounded by a polynomial in x, when $t \in [0,1]$. Thus $\|e^{F_t}\| \leq e^{-c\|x\|^2} I(x)$ where $I(x)$ is polynomial in x. Now $\frac{dv_t}{dt} = \begin{pmatrix} 0 & \frac{d\sigma_t'}{dt} \\ \frac{d\sigma_t}{dt} & 0 \end{pmatrix}$. By our assumptions, the derivative $\frac{d\sigma_t}{dt}$ of a good homotopy has at most polynomial growth in x. Moreover the change of metric depends only of the variable z so that $\frac{dv_t}{dt}$ has also at most polynomial growth in x, while the 1-form $p^*(\nabla_1 - \nabla_0)$ is constant in x. Thus the form β is rapidly decreasing. This completes the proof of Proposition 18.

Recall the Chern character map with compact support

$$\mathrm{ch} : K(\mathcal{V}) \to H^\bullet_{cpt}(\mathcal{V}).$$

The class of $\mathrm{ch}([\sigma])$ in $H^\bullet_{cpt}(\mathcal{V})$ determines a fortiori a class in $H^\bullet_{rapid}(\mathcal{V})$. We now prove the main theorem of this section.

Theorem 19 *(Quillen) Let σ be a good bundle map. Then we have the following equality in $H^\bullet_{rapid}(\mathcal{V})$:*

$$\mathrm{ch}([\sigma]) \equiv \mathrm{ch}(\mathbb{A}(\sigma)).$$

Proof. This theorem can be proved using Quillen's construction of $\mathrm{ch}_Q(\sigma)$ as a relative class (Remark 10).

We will give in full details a proof using the construction of $\mathrm{ch}([\sigma])$ as difference of Chern characters of specific bundles over a specific compactification of $\mathcal{V}$. This last proof leads to explicit formulas at the level of differential forms and extends (without further work) to the equivariant case.

To describe $\mathrm{ch}([\sigma])$, we need to realize $\mathcal{V}$ as an open subset of a compact manifold. Consider the vector space $\mathbf{R}^n$ of dimension n. We identify $\mathbf{R}^n$ as an open subset of the sphere S_n of dimension n by adding a point at infinity to $\mathbf{R}^n$. If the vector bundle $\mathcal{V} \to B$ has typical fiber $\mathbf{R}^n$, we denote by $\Sigma(\mathcal{V}) \to B$ the bundle with typical fiber S_n obtained by adding a point at infinity in each fiber. More precisely, the manifold $\Sigma(\mathcal{V})$ contains $\mathcal{V}$ as an open subset and $\Sigma(\mathcal{V}) - \mathcal{V}$ (called the section at ∞) is isomorphic to our base B. The compact manifold $\Sigma(\mathcal{V})$ is covered by two open sets $U_0 = \mathcal{V}$ and U_1, where U_1 is a neighborhood of the section at infinity with diffeomorphisms

$$I_0 : U_0 \to \mathcal{V} \qquad I_\infty : U_1 \to \mathcal{V}$$

such that I_0 carries $U_0 \cap U_1$ to $\mathcal{V} - \{0\}$ and such that for $(z,x) \in \mathcal{V} - \{0\}$

$$I_\infty I_0^{-1}(z,x) = (z, \frac{x}{\|x\|^2}).$$

We still denote by $p : \Sigma(\mathcal{V}) \to B$ the fibration $\Sigma(\mathcal{V}) \to B$. Thus $p^*\mathcal{E}^\pm$ are bundles on $\Sigma(\mathcal{V})$ and $p^*\nabla^\pm$ are connections defined on $\Sigma(\mathcal{V})$. Consider the bundle map $\sigma : p^*\mathcal{E}^+|_\mathcal{V} \to p^*\mathcal{E}^-|_\mathcal{V}$. Let $\phi(z,x)$ be a cut-off function on $\mathcal{V}$ adapted to σ. Consider the map

$$u(\sigma,\phi) : (p^*\mathcal{E}^+ \oplus p^*\mathcal{E}^-)|_\mathcal{V} \to p^*\mathcal{E}^-|_\mathcal{V}$$

given for $(v_0, v_1) \in \mathcal{E}_x^+ \oplus \mathcal{E}_x^-$ by

$$u(\sigma,\phi)(v_0,v_1) = \sigma(z,x)v_0 - \phi(z,x)v_1.$$

To $u(\sigma,\phi)$ we associate the subbundle $\mathcal{F} = \mathcal{F}(\sigma,\phi)$ of $p^*\mathcal{E}^+ \oplus p^*\mathcal{E}^-$ (Definition 1). Let ∇_0 be the projected connection of $p^*\nabla$ on $\mathcal{F}$. In view of Formula 17 for $\mathrm{ch}([\sigma])$, we need to prove:

$$\mathrm{ch}(\mathcal{F},\nabla_0) - \mathrm{ch}(p^*\mathcal{E}^-, p^*\nabla^-) \equiv \mathrm{ch}(\mathbb{A}(\sigma))$$

in $H^\bullet_{rapid}(\mathcal{V})$.

Consider the bundles $\mathcal{D}^+ = (p^*\mathcal{E}^+ \oplus p^*\mathcal{E}^-)|_\mathcal{V}$ and $\mathcal{D}^- = p^*\mathcal{E}^-|_\mathcal{V}$, so that $u = u(\sigma,\phi)$ is a surjective bundle map $u : \mathcal{D}^+ \to \mathcal{D}^-$. Recall that $\mathcal{F}|_\mathcal{V}$ is isomorphic to $\mathrm{Ker}(u)$. Consider the odd endomorphism of $\mathcal{D} = \mathcal{D}^+ \oplus \mathcal{D}^-$ associated to u:

$$L = \begin{pmatrix} 0 & u^* \\ u & 0 \end{pmatrix}.$$

As u is surjective, u^* is injective and $\mathrm{Ker}(L)^+ = \mathcal{F}|_\mathcal{V}$ while $\mathrm{Ker}(L)^- = \{0\}$.

In the decomposition $\mathcal{D} = (p^*\mathcal{E}^- \oplus p^*\mathcal{E}^+) \oplus p^*\mathcal{E}^-$

$$L = \begin{pmatrix} 0 & 0 & -\phi \\ 0 & 0 & \sigma^* \\ -\phi & \sigma & 0 \end{pmatrix}.$$

Consider the superconnection

$$\mathbb{A}(L) = i\begin{pmatrix} 0 & 0 & -\phi \\ 0 & 0 & \sigma^* \\ -\phi & \sigma & 0 \end{pmatrix} + \begin{pmatrix} p^*\nabla^- & 0 & 0 \\ 0 & p^*\nabla^+ & 0 \\ 0 & 0 & p^*\nabla^- \end{pmatrix}$$

on $\mathcal{D} = (p^*\mathcal{E}^- \oplus p^*\mathcal{E}^+) \oplus p^*\mathcal{E}^-$.

Recall that

$$\mathbb{A}(\sigma) = iv(\sigma) + p^*\nabla = i\begin{pmatrix} 0 & \sigma^* \\ \sigma & 0 \end{pmatrix} + \begin{pmatrix} p^*\nabla^+ & 0 \\ 0 & p^*\nabla^- \end{pmatrix}.$$

Let $F(t,L)$ be the curvature of the superconnection $\mathbb{A}(tL)$ and let $F(t,\sigma)$ be the curvature of $\mathbb{A}(t\sigma)$. Let

$$\alpha(t) = \mathrm{Str}(iLe^{F(t,L)})$$

be the transgression form of the family $\mathbb{A}(tL)$. By Theorem 12

$$\mathrm{ch}(\mathcal{F},\nabla_0)-\mathrm{ch}(\mathcal{D},\mathbb{A}(L))=d\Big(\int_1^\infty \alpha(t)dt\Big).$$

Let us prove that

$$\beta=\int_1^\infty \alpha(t)dt$$

is an element of $\mathcal{A}^\bullet_{rapid}(\mathcal{V})$. Outside the compact support of ϕ,

$$\alpha(t)=\mathrm{Str}(i\sigma e^{F(t,\sigma)}).$$

To study the behaviour of $e^{F(t,\sigma)}$ in (t,x), we use again Volterra expansion(4) with $A(t,z,x)=-t^2v^2(z,x)$ and $Z(t,z,x)=(it(p^*\nabla)\cdot v+p^*F)_{(z,x)}$. The term Z has no term of exterior degree 0 and Volterra's expansion is finite. The function $\|Z(t,z,x)\|$ is bounded by a polynomial in (t,x). As the largest eigenvalue of $-t^2v^2(z,x)$ is smaller than $-ct^2\|x\|^2$ for $\|x\|\geq r$, summing up the inequalities (8) we obtain

$$\|e^{F(t,\sigma)}\|_{(z,x)}\leq e^{-ct^2\|x\|^2}I(t,x)$$

where $I(t,x)$ is polynomial in (t,x). The function $e^{-ct^2\|x\|^2/2}I(t,x)$ is uniformly bounded for all $\|x\|\geq r, t\geq 1$ by a constant K. Thus

$$\|e^{F(t,\sigma)(z,x)}\|\leq Ke^{-ct^2\|x\|^2/2}\leq Ke^{-c\|x\|^2/4}e^{-ct^2\|x\|^2/4}\quad \text{for } t\geq 1.$$

Similar estimates can be made for the derivatives. Thus the form $\int_1^\infty \alpha(t)dt$ is rapidly decreasing on $\mathcal{V}$ and we obtain the equality

$$\mathrm{ch}(\mathcal{F},\nabla_0)-\mathrm{ch}(\mathcal{D},\mathbb{A}(L))=d\beta \quad \text{with } \beta\in\mathcal{A}^\bullet_{rap}(\mathcal{V}). \tag{30}$$

Now consider the superconnection

$$\mathbb{A}(L'_t)=i\begin{pmatrix}0&0&-t\phi\\0&0&\sigma^*\\-t\phi&\sigma&0\end{pmatrix}+\begin{pmatrix}p^*\nabla^-&0&0\\0&p^*\nabla^+&0\\0&0&p^*\nabla^-\end{pmatrix}$$

on $\mathcal{D}=(p^*\mathcal{E}^-\oplus p^*\mathcal{E}^+)\oplus p^*\mathcal{E}^-$. Let F'_t be the curvature of $\mathbb{A}(L'_t)$. Clearly for $t=0$,

$$\mathrm{ch}(\mathcal{D},\mathbb{A}(L'_0))=\mathrm{ch}(p^*\mathcal{E}^-,p^*\nabla^-)+\mathrm{ch}(p^*\mathcal{E},\mathbb{A}(\sigma)).$$

Furthermore as $\frac{d}{dt}\mathbb{A}(L'_t)$ has compact support, it follows from the transgression formula for $\mathrm{ch}(\mathcal{D},\mathbb{A}(L'_t))$ between $t=0$ and $t=1$ that

$$\mathrm{ch}(p^*\mathcal{E}^-,\nabla^-)+\mathrm{ch}(p^*\mathcal{E},\mathbb{A}(\sigma))-\mathrm{ch}(\mathcal{D},\mathbb{A}(L))=d\mu \tag{31}$$

with $\mu\in\mathcal{A}^\bullet_{cpt}(\mathcal{V})$. Substracting the two equalities (30) and (31), we obtain our proposition.

1.6 Index of elliptic operators

In order to state Atiyah-Singer formula for the index of elliptic operators, we need to define another cohomology class associated to the tangent bundle. Let $\mathcal{V} \to B$ be a real vector bundle of rank N. Consider a connection ∇ on $\mathcal{V}$. Let $R = \nabla^2$ be the curvature of ∇. Then in a local frame $e_i, 1 \leq i \leq N$, R is a matrix of 2-forms $R = R_{i,j}, 1 \leq i,j \leq N$. Define

$$J(B, \mathcal{V}, \nabla) = \det\left(\frac{e^{R/2} - e^{-R/2}}{R}\right) = 1 + \frac{1}{24}\operatorname{Tr}(R^2) + \dots \tag{32}$$

Then $J(B, \mathcal{V}, \nabla)$ is a de Rham closed form on B and the cohomology class $J(B, \mathcal{V})$ of $J(B, \mathcal{V}, \nabla)$ is independent of the choice of ∇. If $TB \to B$ is the tangent bundle, we denote $J(B, TB)$ simply by $J(B)$. The form $J(B)$ is invertible.

With this definition we are now able to state the index theorem of Atiyah-Singer in its cohomological formulation. Let B be a compact manifold. Let

$$P : \Gamma(B, \mathcal{E}^+) \to \Gamma(B, \mathcal{E}^-)$$

be an elliptic pseudodifferential operator on B. Let $\operatorname{Ker} P$ be the space of solutions of P and let $\operatorname{Coker} P = \Gamma(M, \mathcal{E}^-)/P(\Gamma(M, \mathcal{E}^+))$ be the cokernel of the map P. As P is elliptic, the spaces $\operatorname{Ker} P$ and $\operatorname{Coker} P$ are finite dimensional vector spaces. We have also the following description of $\operatorname{Coker} P$: let us choose a smooth positive density on B and Hermitian structures on $\mathcal{E}^\pm$. This gives to $\Gamma(B, \mathcal{E}^\pm)$ a pre-Hilbert space structure. Let P^* be the adjoint of P. Then $\operatorname{Coker} P$ is isomorphic to $\operatorname{Ker} P^*$. Let T^*B be the cotangent bundle to B. Let $p : T^*B \to B$ be the projection. The principal symbol of P defines a bundle map $\sigma(P) : p^*\mathcal{E}^+ \to p^*\mathcal{E}^-$ (defined outside the zero section). The operator P being elliptic, the bundle map $\sigma(P)$ is defined and invertible outside the zero section of T^*B. In particular it defines an element $[\sigma(P)]$ of $K(T^*B)$. Recall that $\operatorname{ch}([\sigma(P)]) \in H_{cpt}(T^*B)$. Consider on T^*B the orientation given by its canonical symplectic form. Atiyah-Singer index formula for $\operatorname{index}(P) = \dim \operatorname{Ker}(P) - \dim \operatorname{Coker}(P)$ is

$$\operatorname{index} P = (2i\pi)^{-\dim B} \int_{T^*B} \frac{\operatorname{ch}([\sigma(P)])}{J(B)}.$$

Let us choose a good bundle map σ such that $[\sigma] = [\sigma(P)]$. For example if P is a differential operator of strictly positive order, then $\sigma(P)$ is a homogeneous polynomial in the fiber variables, so that $\sigma(P)$ itself is a good bundle map.

If P is a pseudo-differential operator we may without changing $[\sigma(P)]$ assume that P has order ≥ 1. Let ϕ be a smooth function with compact support on T^*B which is identically 0 near the zero section, then

$$\sigma = \phi\sigma(P) \tag{33}$$

extends to a smooth section on T^*B and is homogeneous of degree ≥ 1 when $\|x\|$ is sufficiently large, so that σ is a good bundle map such that $[\sigma] = [\sigma(P)]$ in $K(T^*B)$. Let us choose connections $\nabla^\pm$ on $\mathcal{E}^\pm$. Let

$$\mathbb{A}(\sigma) = i\begin{pmatrix} 0 & \sigma^* \\ \sigma & 0 \end{pmatrix} + \begin{pmatrix} p^*\nabla^+ & 0 \\ 0 & p^*\nabla^- \end{pmatrix}.$$

The form $\mathrm{ch}(\mathbb{A}(\sigma))$ is a rapidly decreasing form on T^*B. In view of the equality $\mathrm{ch}([\sigma(P)]) \equiv \mathrm{ch}(\mathbb{A}(\sigma))$ in $H_{rapid}(T^*B)$, Atiyah-Singer formula can be reformulated as

Theorem 20 *Let P be an elliptic operator on B. Let σ be a good bundle map such that $[\sigma] = [\sigma(P)]$. Then*

$$\mathrm{index}(P) = (2i\pi)^{-\dim B} \int_{T^*B} \frac{\mathrm{ch}(\mathbb{A}(\sigma))}{J(B)}.$$

We will indicate in the next section the generalisation of this formula to the equivariant situation.

2 Equivariant Chern character

2.1 Equivariant vector bundles

Let G be a Lie group acting on a manifold M. We will now consider G-equivariant vector bundles over M. An important example arise as described in the first section:

Let $\mathcal{E}^{\pm}$ be two G-equivariant vector bundles over the space M. Let U be an open G-invariant subset of M. Let $\mathcal{E}^+|_U \xrightarrow{\sigma} \mathcal{E}^-|_U$ be a bundle map (only defined over U) of G-equivariant vector bundles. Assume that σ has compact support $F \subset U$. Then the bundle $\mathcal{G}(\sigma)$ defined by Formula 2 of Section 1 is a G-equivariant vector bundle. Furthermore, if the group G is compact, we can choose a G-invariant cut-off function ϕ adapted to σ by averaging a cut-off function adapted to σ. Then the map $u(\phi, \sigma)$ given by Formula 3 is a map of G-equivariant vector bundles, the subbundle $\mathcal{F}(\sigma, \phi)$ of $\mathcal{E}^+ \oplus \mathcal{E}^-$ (Definition 1 of Section 1) is a G-equivariant subbundle and clearly we have

Lemma 21 *The bundles $\mathcal{G}(\sigma)$ and $\mathcal{F}(\sigma, \phi)$ are isomorphic G-equivariant vector bundles.*

Let G be a compact Lie group acting on a compact manifold M. The isomorphism classes of G-equivariant vector bundles over M form an abelian semigroup under direct sums. The associated abelian group is denoted by $K_G(M)$. If $\mathcal{E} \to M$ is a G-equivariant vector bundle over M, we denote by $[\mathcal{E}]$ the corresponding element of $K_G(M)$. If M is a point, the group $K_G(point)$ is the group of virtual finite dimensional representations T of G. The function $g \to \mathrm{Tr}\, T(g)$ determines the isomorphism class of a representation T of G. Thus we identify $K_G(point)$ with the space $R(G)$ of virtual characters of finite dimensional representations of G. Clearly $R(G)$ is a subring of the ring $C^\infty(G)^G$ of G-invariant C^∞-functions on G.

If G acts on a locally compact manifold U, the group $K_G(U)$ has a description similar to the one given in section 1. A morphism $\mathcal{E}^+ \xrightarrow{\sigma} \mathcal{E}^-$ of G-equivariant vector bundles such that σ has compact support is a representative of an element $[\mathcal{E}^+, \sigma, \mathcal{E}^-]$ of $K_G(U)$. We sometimes denote it simply by $[\sigma]$.

Consider the case where U is an open subset of a compact G-manifold M. Let $\mathcal{E}^{\pm}$ be two G-equivariant vector bundles defined over M and consider a morphism $\mathcal{F}^+ \xrightarrow{\nu} \mathcal{F}^-$ of G-equivariant vector bundles. Assume ν has its support contained in U. Thus ν defines an element $[\mathcal{F}^+|_U, \nu|_U, \mathcal{F}^-|_U]$ of $K_G(U)$. The equivariant excision theorem (see [22]) is the analogue of Proposition 3. It asserts in particular that every element of $K_G(U)$ has a representative of the form $[\mathcal{F}^+|_U, \nu|_U, \mathcal{F}^-|_U]$.

Many elements of $K_G(U)$ will arise as follows :$[\sigma] = [\mathcal{E}^+|_U, \sigma, \mathcal{E}^-|_U]$, where the vector bundles $\mathcal{E}^\pm$ are G-equivariant vector bundles defined over M, but $\mathcal{E}^+|_U \xrightarrow{\sigma} \mathcal{E}^-|_U$ is a morphism with compact support defined only over U. Consider the sub-bundle $\mathcal{F} = \mathcal{F}(\sigma, \phi)$ of $\mathcal{E}^+ \oplus \mathcal{E}^-$ and let $\mathcal{F} \xrightarrow{\pi} \mathcal{E}^-$ be the second projection. The proof of the next lemma is identical to the proof of Lemma 4.

Lemma 22 *In $K_G(U)$ we have the equality*

$$[\mathcal{E}^+|_U, \sigma, \mathcal{E}^-|_U] = [\mathcal{F}|_U, \pi|_U, \mathcal{E}^-|_U].$$

2.2 Equivariant Chern character

Let G be a real Lie group acting on a manifold M. Let $\mathfrak{g}$ be the Lie algebra of G. If G acts on a vector space E, we denote by E^G the subspace of invariants.

Let $X \in \mathfrak{g}$. We denote by X_M the vector field produced by the action of $\exp(-tX)$ on M. Let $\iota(X_M) : \mathcal{A}^\bullet(M) \to \mathcal{A}^{\bullet-1}(M)$ be the contraction by X_M.

Consider the space:

$$\mathcal{A}_G^\infty(\mathfrak{g}, M) = C^\infty(\mathfrak{g}, \mathcal{A}(M))^G.$$

of C^∞-maps from $\mathfrak{g}$ to $\mathcal{A}(M)$. An element $\alpha \in \mathcal{A}_G^\infty(\mathfrak{g}, M)$ will be called an equivariant form on M. Define $d_\mathfrak{g} : \mathcal{A}_G^\infty(\mathfrak{g}, M) \to \mathcal{A}_G^\infty(\mathfrak{g}, M)$ by

$$(d_\mathfrak{g}\alpha)(X) = d_M(\alpha(X)) - \iota(X_M)(\alpha(X)).$$

The operator $d_\mathfrak{g}$ satisfies $d_\mathfrak{g}^2 = 0$. We say that an equivariant form α is closed if $d_\mathfrak{g}\alpha = 0$, exact if $\alpha = d_\mathfrak{g}\beta$ for some equivariant form β. We denote by $\mathcal{H}_G^\infty(\mathfrak{g}, M)$ the space $\operatorname{Ker} d_\mathfrak{g} / \operatorname{Im} d_\mathfrak{g}$ for $d_\mathfrak{g} : \mathcal{A}_G^\infty(\mathfrak{g}, M) \to \mathcal{A}_G^\infty(\mathfrak{g}, M)$. If G is reduced to the identity transformation, then $d_\mathfrak{g} = d$ and $\mathcal{H}_G^\infty(\mathfrak{g}, M)$ is just de Rham cohomology. If $M = \bullet$ is a point,

$$\mathcal{H}_G^\infty(\mathfrak{g}, \bullet) = C^\infty(\mathfrak{g})^G.$$

We introduce also the space

$$\mathcal{A}_{G,cpt}^\infty(\mathfrak{g}, M) = C^\infty(\mathfrak{g}, \mathcal{A}_{cpt}(M))^G$$

and the group $\mathcal{H}_{G,cpt}^\infty(\mathfrak{g}, M)$ of equivariant cohomology with compact support.

We say that M is G-oriented if M is oriented and if the G-action preserves the orientation of M. If M is G-oriented the integration over M of compactly supported equivariant forms is defined by $(\int_M \alpha)(X) = \int_M \alpha(X)$. The integration map sends $\mathcal{A}_{G,cpt}^\infty(\mathfrak{g}, M)$ to $C^\infty(\mathfrak{g})^G$. Furthermore, if α is closed, the integral of α is a function on $\mathfrak{g}$ depending only of the cohomology class of α. Thus we can define a map

$$\int_M : \mathcal{H}_{G,cpt}^\infty(\mathfrak{g}, M) \to C^\infty(\mathfrak{g})^G.$$

Let $p : \mathcal{V} \to B$ be a real G-equivariant vector bundle over a compact base B. We define the space:

$$\mathcal{A}_{G,rapid}^\infty(\mathfrak{g}, \mathcal{V}) = C^\infty(\mathfrak{g}, \mathcal{A}_{rapid}(\mathcal{V}))^G.$$

The operator $d_{\mathfrak{g}}$ preserves this space and we denote by $\mathcal{H}^{\infty}_{G,rapid}(\mathfrak{g}, M)$ the space $\operatorname{Ker} d_{\mathfrak{g}} / \operatorname{Im} d_{\mathfrak{g}}$ for the operator $d_{\mathfrak{g}} : \mathcal{A}^{\infty}_{G,rapid}(\mathfrak{g}, \mathcal{V}) \to \mathcal{A}^{\infty}_{G,rapid}(\mathfrak{g}, \mathcal{V})$. If G is compact, it is easy to prove using an equivariant Thom form with compact support (see [18]) that the natural map $\mathcal{H}^{\infty}_{G,cpt}(\mathfrak{g}, \mathcal{V}) \to \mathcal{H}^{\infty}_{G,rapid}(\mathfrak{g}, \mathcal{V})$ is an isomorphism.

Recall the definition of the equivariant Chern character of a superconnection. Let $\mathcal{E}^{\pm}$ be two G-equivariant vector bundles over M. Let $\mathcal{E} = \mathcal{E}^{+} \oplus \mathcal{E}^{-}$. A superconnection $\mathbb{A}$ is G-invariant, if the operator $\mathbb{A} : \mathcal{A}(M, \mathcal{E}) \to \mathcal{A}(M, \mathcal{E})$ commutes with the natural action of G on $\mathcal{A}(M, \mathcal{E})$. Let $\mathbb{A}$ be a G-invariant superconnection on $\mathcal{E}$. Consider for $X \in \mathfrak{g}$ the operator $F(X)$ on $\mathcal{A}(M, \mathcal{E})$ given by

$$F(X) = (\mathbb{A} - \iota(X_M))^2 + \mathcal{L}^{\mathcal{E}}(X)$$

where $\mathcal{L}^{\mathcal{E}}(X)$ is the first-order differential operator given by the Lie derivative of the action of G on $\mathcal{A}(M, \mathcal{E})$. The operator $F(X)$ is given by the action on $\mathcal{A}(M, \mathcal{E})$ of an element of $\mathcal{A}(M, \operatorname{End} \mathcal{E})$ that we still denote by $F(X)$ and that we call the equivariant curvature of $\mathbb{A}$.

In particular if ∇ is a G-invariant connection, then the equivariant curvature of ∇ is

$$F(X) = \mu(X) + F \tag{34}$$

where $F \in \mathcal{A}^2(M, \operatorname{End} \mathcal{E})$ is the curvature of ∇ and $\mu(X) \in \Gamma(M, \operatorname{End} \mathcal{E})$ is the vertical action of X determined by the connections ∇: for $m \in M$, $v \in \mathcal{E}_m$, $(\mu(X)v)_m$ is the vertical projection on $\mathcal{E}_m$ (determined by ∇) of the vector $(X_{\mathcal{E}})_{(m,v)}$. In [9], the endomorphism $\mu(X)$ is called the moment of X.

Definition 23 *Let $(\mathcal{E}, \mathbb{A})$ be a G-equivariant superbundle with a G-invariant superconnection $\mathbb{A}$. Let $F(X)$ be the equivariant curvature of $\mathbb{A}$. The equivariant Chern character* $\operatorname{ch}(\mathcal{E}, \mathbb{A}) \in \mathcal{A}^{\infty}_G(\mathfrak{g}, M)$ *is defined, for $X \in \mathfrak{g}$, by:*

$$\operatorname{ch}(\mathcal{E}, \mathbb{A})(X) = \operatorname{Str}(e^{F(X)}).$$

The equivariant Chern character is a closed equivariant form. Clearly by evaluating $\operatorname{ch}(\mathcal{E}, \mathbb{A})(X)$ at $X = 0$, we obtain the de Rham closed form $\operatorname{ch}(\mathcal{E}, \mathbb{A})$ defined in Definition 8.

Let $\mathbb{A}_t$ ($t \in \mathbb{R}$) be a smooth family of G-invariant superconnections on $\mathcal{E}$ with equivariant curvature $F_t(X)$. Let us define the *transgression* form $\alpha(t) \in \mathcal{A}^{\infty}_G(\mathfrak{g}, M)$ of the family $\mathbb{A}_t$ by the formula: for $X \in \mathfrak{g}$,

$$\alpha(t)(X) = \operatorname{Str}\Big(\frac{d\mathbb{A}_t}{dt} e^{F_t(X)}\Big). \tag{35}$$

It is easy to prove that

$$\frac{d}{dt} \operatorname{ch}(\mathbb{A}_t) = d_{\mathfrak{g}} \alpha(t)$$

so that the following relation holds in $\mathcal{A}^{\infty}_G(\mathfrak{g}, M)$

$$\operatorname{ch}(\mathcal{E}, \mathbb{A}_T) - \operatorname{ch}(\mathcal{E}, \mathbb{A}_{t_0}) = d_{\mathfrak{g}} \int_{t_0}^{T} \alpha(t)\, dt. \tag{36}$$

Thus the equivariant cohomology class of the Chern character form $\operatorname{ch}(\mathcal{E}, \mathbb{A})$ is independent of the choice of the invariant superconnection $\mathbb{A}$ on $\mathcal{E}$. We denote it by $\operatorname{ch}(\mathcal{E})$.

If G is a compact group acting on a compact manifold M, every vector bundle $\mathcal{E}$ is equipped with a G-invariant connection. Thus the Chern character induces a map

$$\mathrm{ch} : K_G(M) \to \mathcal{H}^\infty_G(\mathfrak{g}, M).$$

If U is a locally compact G-manifold, there is a Chern character map with compact support

$$\mathrm{ch} : K_G(U) \to \mathcal{H}^\infty_{G,cpt}(U).$$

Let us describe this map using excision . Let U be an open subset of a compact G-manifold M. Let $\mathcal{F}^\pm$ be two G-equivariant vector bundles over M and let $\mathcal{F}^+ \xrightarrow{\nu} \mathcal{F}^-$ be a bundle map with support contained in U. It is possible to find (using G-invariant partitions of unity) G-invariant connections $\nabla^\pm$ on $\mathcal{F}^\pm \to M$ such that ∇^+ and $\nu^*\nabla^-$ coincide near infinity. Then for every $X \in \mathfrak{g}$ the differential form $\mathrm{ch}(\mathcal{F}^+, \nabla^+)(X) - \mathrm{ch}(\mathcal{F}^-, \nabla^-)(X)$ is a compactly supported form on U. Its equivariant cohomology class in $\mathcal{H}^\infty_{G,cpt}(\mathfrak{g}, U)$ depends only of the element $[\mathcal{F}^+|_U, \nu|_U, \mathcal{E}^-|_U]$ of $K_G(U)$.

Definition 24 *The equivariant Chern character with compact support of the element* $[\sigma] = [\mathcal{F}^+|_U, \nu|_U, \mathcal{E}^-|_U]$ *of* $K_G(U)$ *is defined by*

$$\mathrm{ch}([\sigma]) \equiv \mathrm{ch}(\mathcal{F}^+, \nabla^+) - \mathrm{ch}(\mathcal{F}^-, \nabla^-)$$

in $\mathcal{H}^\infty_{G,cpt}(\mathfrak{g}, U)$.

Let $\mathcal{E}^\pm$ be two G-equivariant vector bundles over M with G-invariant connections $\nabla^\pm$ and let $\sigma : \mathcal{E}^+|_U \to \mathcal{E}^-|_U$ be a morphism with compact support of G-equivariant vector bundles. Let $[\sigma]$ be the corresponding element of $K_G(U)$. Let ϕ be a G-invariant cut-off function adapted to σ. Consider the G-equivariant subbundle $\mathcal{F}(\sigma, \phi)$ of $\mathcal{E}^+ \oplus \mathcal{E}^-$ (Definition 1) and let π be the second projection. Consider the G-invariant connection ∇_0 on $\mathcal{F}$ obtained by projection of the connection ∇. As near infinity, the subbundle $\mathcal{F}(\sigma, \phi)$ coincide with $\mathcal{E}^-$, the connections ∇_0 and $\pi^*\nabla^-$ coincide near infinity. In view of Lemma 22 we have

$$\mathrm{ch}([\sigma]) \equiv \mathrm{ch}(\mathcal{F}(\sigma, \phi), \nabla_0) - \mathrm{ch}(\mathcal{E}^-, \nabla^-) \tag{37}$$

in $\mathcal{H}^\infty_{G,cpt}(\mathfrak{g}, U)$.

2.3 Bouquets of Chern characters

Let G be a compact Lie group. Let us recall some definitions. Let $s \in G$. Denote by $G(s)$ the centralizer of $s \in G$ and by $\mathfrak{g}(s) = \{X \in \mathfrak{g}, s \cdot X = X\}$ the Lie algebra of $G(s)$. Consider the action of G on a manifold M. The fixed point set $M(s)$ of $s \in G$ is a submanifold of M. The group $G(s)$ acts on $M(s)$. There is a canonical decomposition

$$TM|_{M(s)} = TM(s) \oplus \mathcal{N}(s) \tag{38}$$

where for $m \in M(s)$ the space $T_mM(s)$ is the fixed subspace for the action of s on T_mM, while $\mathcal{N}_m(s) = (I - s)T_mM$. The bundle $\mathcal{N}(s) \to M(s)$ is isomorphic to the normal bundle $\mathcal{N}(M/M(s))$ of the imbedding $M(s) \subset M$.

If $S \in \mathfrak{g}$, we denote by $M(S) = \{m \in M; (S_M)_m = 0\}$ the manifold of zeros of the vector field S_M.

If $S \in \mathfrak{g}(s)$ is sufficiently small, then

$$G(se^S) = G(s) \cap G(S), \quad \mathfrak{g}(se^S) = \mathfrak{g}(s) \cap \mathfrak{g}(S)$$

and

$$M(se^S) = M(s) \cap M(S).$$

Definition 25 *A bouquet of equivariant forms is a family $(\alpha_s)_{s\in G}$ where each $\alpha_s \in \mathcal{A}^\infty_{G(s)}(\mathfrak{g}(s), M(s))$ is a closed equivariant form. Furthermore the family α_s must satisfy the following conditions:*

1. *Invariance:*

$$\alpha_{gsg^{-1}} = g \cdot \alpha_s$$

for all $g \in G$ and $s \in G_{ell}$.

2. *Compatibility: for all $s \in G$ and all $S \in \mathfrak{g}(s)$*

$$\alpha_{se^S}(Y)|_{M(s)\cap M(S)} = \alpha_s(S+Y)|_{M(s)\cap M(S)}$$

for all $Y \in \mathfrak{g}(s) \cap \mathfrak{g}(S)$.

The definition of the algebra $\mathcal{K}_G(M)$ has been given in [19] [20] (for an almost-algebraic group G). It is a global version of the algebra $\mathcal{H}^\infty_G(\mathfrak{g}, M)$. In particular $\mathcal{K}_G(point) = C^\infty(G)^G$ and $\mathcal{K}_G(M)$ is a module over $\mathcal{K}_G(point)$. A similar algebra is defined by Block and Getzler in [15]. An element $[\alpha]$ of $\mathcal{K}_G(M)$ is a family $(\alpha_s)_{s\in G}$ of elements of $\mathcal{H}^\infty_{G(s)}(\mathfrak{g}(s), M(s))$. This family of equivariant cohomology classes must satisfy compatibility conditions similar to the condition 2 of Definition 25 above. It is clear that a bouquet of equivariant forms determines an element of $\mathcal{K}_G(M)$. We have also defined in [19] [20] the space $\mathcal{K}_{G,cpt}(M)$. An element $[\alpha]$ of $\mathcal{K}_{G,cpt}(M)$ is a family $(\alpha_s)_{s\in G}$ of equivariant cohomology classes with compact support satisfying compatibility conditions.

Let $\mathcal{V} \to B$ be a real G-equivariant vector bundle over a base B. Each fixed point set $\mathcal{V}(s)$ is a vector bundle over $B(s)$. A bouquet $\alpha = (\alpha_s)_{s\in G}$ of rapidly decreasing equivariant differential forms is a bouquet such that for each $s \in G$, the form $\alpha_s \in \mathcal{A}^\infty_{G(s),rapid}(\mathfrak{g}(s), \mathcal{V}(s))$ is rapidly decreasing over the fiber of the projection $\mathcal{V}(s) \to B(s)$. We can define similarly the space $\mathcal{K}_{G,rapid}(\mathcal{V})$ as families $[\alpha] = (\alpha_s)_{s\in G}$ with $\alpha_s \in \mathcal{H}^\infty_{G(s),rapid}(\mathfrak{g}(s), M(s))$ satisfying compatibility conditions. Thus a bouquet of of rapidly decreasing equivariant differential forms determines an element of $\mathcal{K}_{G,rapid}(\mathcal{V})$. Furthermore the natural map $\mathcal{K}_{G,cpt}(\mathcal{V}) \to \mathcal{K}_{G,rapid}(\mathcal{V})$ is an isomorphism.

Let $\mathcal{E}$ be a G-equivariant superbundle with a G-invariant superconnection $\mathbb{A}$ of equivariant curvature $F(X)$. Over $M(s)$ the action of s on $\mathcal{E}|_{M(s)}$ is a transformation (still denoted by s) acting fiberwise. The $G(s)$-equivariant differential form on $M(s)$ defined by

$$\mathrm{ch}_s(\mathcal{E}, \mathbb{A})(X) = \mathrm{Tr}(se^{F(X)|M(s)}) \quad \text{for } X \in \mathfrak{g}(s) \tag{39}$$

is closed. Furthermore the family

$$\mathrm{bch}(\mathcal{E}, \mathbb{A}) = (\mathrm{ch}_s(\mathcal{E}, \mathbb{A}))_{s\in G}$$

is a bouquet of equivariant differential forms. We call it the bouquet of Chern characters. If $\mathcal{E}$ is understood, we will denote it simply by $\mathrm{bch}(\mathbb{A}) = (\mathrm{ch}_s(\mathbb{A}))_{s\in G}$. The cohomology class of $\mathrm{ch}_s(\mathcal{E},\mathbb{A})$ is independent of $\mathbb{A}$. We denote it by $\mathrm{ch}_s(\mathcal{E})$.

If G is a compact Lie group acting on a compact manifold M, we can thus define a map:

$$\mathrm{bch} : K_G(M) \to \mathcal{K}_G(M)$$

by taking bouquets of Chern characters

$$\mathrm{bch}(\mathcal{E}) = (\mathrm{ch}_s(\mathcal{E}))_{s\in G}.$$

Similarly if U is a locally compact G-manifold, there is a map

$$\mathrm{bch} : K_G(U) \to \mathcal{K}_{G,cpt}(U).$$

Indeed using excision we need to describe $\mathrm{ch}([\sigma])$ when $[\sigma] = [\mathcal{F}^+|_U, \nu|_U, \mathcal{F}^-|_U]$. We define $\mathrm{bch}(\sigma)$ as the family $(ch_s[\sigma])_{s\in G}$ where

$$\mathrm{ch}_s([\sigma]) \equiv \mathrm{ch}_s(\mathcal{F}^+,\nabla^+) - \mathrm{ch}_s(\mathcal{F}^-,\nabla^-) \quad \text{in } \mathcal{H}^\infty_{G(s),cpt}(\mathfrak{g}(s), M(s)),$$

for a choice of G-invariant connections $\nabla^\pm$ such that ∇^+ coincides with $\nu^*\nabla^-$ near infinity.

Let $\mathcal{E}^\pm$ be two G-equivariant vector bundles over M with G-invariant connections and let $\sigma : \mathcal{E}^+|_U \to \mathcal{E}^-|_U$ be a morphism with compact support of G-equivariant vector bundles. Let $[\sigma]$ be the corresponding element of $K_G(U)$. We have then

$$\mathrm{ch}_s([\sigma]) \equiv (\mathrm{ch}_s(\mathcal{F}(\sigma,\phi),\nabla_0) - \mathrm{ch}_s(\mathcal{E}^-,\nabla^-))_{s\in G}.$$

2.4 Equivariant Chern characters of index bundles

Let $\mathcal{E}^\pm \to M$ be two G-equivariant bundles with G-invariant Hermitian structures and consider on $\mathcal{E} = \mathcal{E}^+ \oplus \mathcal{E}^-$ the Hermitian structure obtained by direct sum. Let $v \in \Gamma(M, \mathrm{End}\,\mathcal{E})$ be an odd Hermitian G-invariant bundle endomorphism of the superbundle $\mathcal{E}$: $v = \begin{pmatrix} 0 & u^* \\ u & 0 \end{pmatrix}$, and assume that $\mathrm{Ker}(v)$ has *constant rank.* Let us denote by $\mathcal{E}_0 \subset \mathcal{E}$ the superbundle $\mathrm{Ker}(v)$. Then $\mathcal{E}_0$ is a G-equivariant vector bundle.

Let $\nabla^\pm$ be G-invariant connections on $\mathcal{E}^\pm$. We denote by ∇ the direct sum connection on $\mathcal{E}$. Let

$$\mathbb{A}(v) = i\begin{pmatrix} 0 & u^* \\ u & 0 \end{pmatrix} + \begin{pmatrix} \nabla^+ & 0 \\ 0 & \nabla^- \end{pmatrix} \tag{40}$$

be the G-invariant superconnection on $\mathcal{E}$ associated to v and ∇. The equivariant curvature of $\mathbb{A}(v)$ is

$$F(v)(X) = -\begin{pmatrix} u^*u & 0 \\ 0 & uu^* \end{pmatrix} + \tag{41}$$
$$i\begin{pmatrix} 0 & \nabla\cdot u^* \\ \nabla\cdot u & 0 \end{pmatrix} + \begin{pmatrix} F^+(X) & 0 \\ 0 & F^-(X) \end{pmatrix}$$

where $F^{\pm}(X)$ is the equivariant curvature (Formula 34) of $\nabla^{\pm}$.

Denote by ∇_0 the G-invariant connection on the subbundle $\mathcal{E}_0$ given by projecting the connection ∇:

$$\nabla_0 = P_0 \nabla P_0.$$

Theorem 26 *Let $\mathcal{E} = \mathcal{E}^+ \oplus \mathcal{E}^-$ be a G-equivariant Hermitian superbundle with a G-invariant direct sum connection ∇. Let v be a G-invariant Hermitian odd endomorphism of $\mathcal{E}$ whose kernel has constant rank. Consider the superconnection $\mathbb{A}(v) = iv + \nabla$. For $t > 0$, let*

$$\mathbb{A}_t = itv + \nabla$$

be the rescaled superconnection. Then the limit $\lim_{t\to\infty} \mathrm{ch}(\mathcal{E}, \mathbb{A}_t)$ exists in $\mathcal{A}^{\infty}_G(\mathfrak{g}, M)$ and equals the equivariant Chern character of the projected connection ∇_0 on the superbundle $\mathcal{E}_0 = \mathrm{Ker}(v)$:

$$\lim_{t\to\infty} \mathrm{ch}(\mathcal{E}, \mathbb{A}_t) = \mathrm{ch}(\mathcal{E}_0, \nabla_0).$$

Furthermore, the transgression form $\alpha(t)(X) = \mathrm{Str}\left(\frac{d\mathbb{A}_t}{dt} e^{F_t(X)}\right)$ of the family $\mathbb{A}_t$ satisfies the estimates, for t large

$$\|\alpha(t)\|_\ell \leq c_\ell t^{-2}$$

on compact subsets of $\mathfrak{g} \times M$, for all C^ℓ-norms and we have the transgression formula

$$\mathrm{ch}(\mathcal{E}_0, \nabla_0) - \mathrm{ch}(\mathcal{E}, \mathbb{A}(v)) = d_{\mathfrak{g}} \int_1^{\infty} \alpha(t)\, dt$$

in $\mathcal{A}^{\infty}_G(\mathfrak{g}, M)$.

Proof. We use the same notations as in the proof of Theorems 11 and 12. For $X \in \mathfrak{g}$, let $F_t(X) \in \mathbb{C}[t] \otimes \mathcal{A}(M, \mathrm{End}\,\mathcal{E})$ be the equivariant curvature of $\mathbb{A}_t$. Thus $F_t(X) = F_t + \mu(X)$ where F_t is the curvature of $\mathbb{A}_t$ and $\mu(X)$ is the vertical action of X with respect to ∇. Thus $\mu(X)$ is independent of t. In the block description with respect to the decomposition $\mathcal{E} = \mathcal{E}_0 \oplus \mathcal{E}_1$, the map $\mu(X)$ is written as

$$\mu(X) = \begin{bmatrix} \mu_0(X) & \mu_{0,1}(X) \\ \mu_{1,0}(X) & \mu_1(X) \end{bmatrix}$$

and it is easy to see that $\mu_0(X)$ and $\mu_1(X)$ are the moment maps of the projected connections ∇_0 and ∇_1.

Lemma 13 shows that

$$F_t(X) = \begin{bmatrix} A(t,X) & B(t,X) \\ C(t,X) & D(t,X) \end{bmatrix}$$

where:

(1) The expansion of $D(t,X)$ in powers of t is $D(t,X) = t^2 D_2 + t D_1 + D_0(X)$. The endomorphism D_2 is equal to $-P_1 v^2 P_1$ so that D_2 is a negative definite endomorphism of $\mathcal{E}_1$.

(2) The other blocks have power expansion

$$A(t,X) = A_0(X), \quad B(t,X) = tB_1 + B_0(X), \quad C(t,X) = tC_1 + C_0(X).$$

(3) Furthermore, if $R(X) \in \mathcal{A}(M, \mathrm{End}\,\mathcal{E}_0)$ is the *equivariant* curvature of $\nabla_0 = P_0 \nabla P_0$, then

$$R(X) = A_0(X) - B_1 D_2^{-1} C_1.$$

Thus Lemma14 allows us to conclude the proof of Theorem 26.

2.5 Equivariant good bundle maps

Let G be a compact Lie group acting on a compact manifold B. Let $\mathcal{V}$ be the total space of a real G-equivariant vector bundle over the base B. Every G-equivariant complex vector bundle $\mathcal{F}$ on $\mathcal{V}$ is isomorphic to the pullback $p^*\mathcal{F}|_B$. This may be seen as in subsection 1.5 choosing a G-invariant connection on $\mathcal{F} \to \mathcal{V}$.

Let $\mathcal{E}^\pm$ be G-equivariant complex vector bundles over B. Let $\sigma : p^*\mathcal{E}^+ \to p^*\mathcal{E}^-$ be a map of G-equivariant vector bundles. We consider the superbundle $\mathcal{E}^+ \oplus \mathcal{E}^-$. We choose G-invariant Hermitian metrics on $\mathcal{E}^\pm$ and we associate to σ the odd G-invariant endomorphism $v = v(\sigma)$ of $p^*\mathcal{E}$ defined by

$$v = \begin{pmatrix} 0 & \sigma^* \\ \sigma & 0 \end{pmatrix}. \tag{42}$$

If the bundle map σ has compact support, it represents an element $[\sigma]$ of $K_G(\mathcal{V})$.

Let us choose a G-invariant Euclidean structure on $\mathcal{V}$. We have defined (Definition 16) the notion of good bundle maps $\sigma : p^*\mathcal{E}^+ \to p^*\mathcal{E}^-$. If σ is a morphism of G-equivariant vector bundles and is good, we will say that σ is a good representative of $[\sigma]$. As Lemma 17, it is possible to prove

Lemma 27 *(i)Every element of $K_G(\mathcal{V})$ has a good representative σ.*

(ii) Let $\sigma_0, \sigma_1 : p^*\mathcal{E}^+ \to p^*\mathcal{E}^-$ be good bundle maps of G-equivariant vector bundles and assume that there exists a homotopy of G-equivariant bundle maps with compact support $\sigma_t : p^*\mathcal{E}^+ \to p^*\mathcal{E}^-$. Then there exists a homotopy $\tau_t : p^*\mathcal{E}^+ \to p^*\mathcal{E}^-$ between τ_0 and τ_1 such that each τ_t is a G-equivariant good bundle map and furthermore such that $\frac{d}{dt}\tau_t$ and all its derivatives have at most polynomial growth in the fiber direction.

Let $\nabla = \nabla^+ \oplus \nabla^-$ be a G-invariant connection on $\mathcal{E} = \mathcal{E}^+ \oplus \mathcal{E}^-$. We denote by $\mathbb{A}(\sigma, \nabla)$ or $\mathbb{A}(\sigma)$ the superconnection on $p^*\mathcal{E}$ defined by

$$\mathbb{A}(\sigma) = iv + p^*\nabla = i\begin{pmatrix} 0 & \sigma^* \\ \sigma & 0 \end{pmatrix} + \begin{pmatrix} p^*\nabla^+ & 0 \\ 0 & p^*\nabla^- \end{pmatrix}.$$

We denote now by $\mathrm{ch}(\mathbb{A}(\sigma))$ the equivariant Chern character of the superbundle $p^*\mathcal{E}$ with superconnection $\mathbb{A}(\sigma)$.

Proposition 28 *Let $\sigma : p^*\mathcal{E}^+ \to p^*\mathcal{E}^-$ be a good bundle map (of G-equivariant vector bundles). Then the equivariant Chern character form $\mathrm{ch}(\mathbb{A}(\sigma))$ belongs to the space $\mathcal{A}^\infty_{G,rapid}(\mathfrak{g}, \mathcal{V})$. Furthermore the equivariant cohomology class of $\mathrm{ch}(\mathbb{A}(\sigma))$ in $\mathcal{H}^\infty_{G,rapid}(\mathfrak{g}, \mathcal{V})$ depends only of the element $[p^*\mathcal{E}^+, \sigma, p^*\mathcal{E}^-]$ of $K_G(\mathcal{V})$.*

Proof. We proceed as in the proof of Proposition 18. Denote by $F(X) = \mu(X) + F$ the equivariant curvature of ∇ (Formula 34). The equivariant curvature of $\mathbb{A}(\sigma)$ is

$$F(\sigma)(X) = -v^2 + i(p^*\nabla)\cdot v + p^*(F(X)).$$

For $X \in \mathfrak{g}$, we compute $\mathrm{ch}(\mathbb{A}(\sigma))(X) = \mathrm{Str}(e^{F(\sigma)(X)})$ using the Volterra expansion (7)

$$e^{A+Z} = \sum_{k=0}^{\infty} I_k$$

with

$$A = -v^2,$$

$$Z = i(p^*\nabla)\cdot v + p^*(F(X)) = p^*(\mu(X)) + i(p^*\nabla)\cdot v + p^*F.$$

The term Z has a component of exterior degree 0 so there is (in contrast to the non-equivariant case of section 1) indeed an infinite number of non zero terms I_k in the sum above. We will use inequality (10) instead of (8). We write $Z = Z_0 + Z_1$ with $Z_0 = p^*(\mu(X))$, $Z_1 = i(p^*\nabla)\cdot v + p^*F$. The term Z_0 is the pullback of a function on the base while the term Z_1 is of at most polynomial growth on the fiber. Thus there exists a constant z_0 (that we may take greater than 1) so that when X varies in a compact subset of $\mathfrak{g}$, we have $\|Z_0(z,x,X)\| \leq z_0$ uniformly on $\mathcal{V}$.

Consider $z_1(z,x) = \sup(\|Z_1(z,x)\|, 1)$. There exists a polynomial $I(x)$ such that $(z_1(z,x))^N \leq I(x)$. Summing up the inequalities (10) from $k = 0$ to ∞, we obtain

$$\|e^{F(\sigma)(X)}\|_{(z,x)} \leq e^{-c\|x\|^2} e^{2z_0} I(x).$$

Similar estimates can be made for the derivatives. Thus $\mathrm{ch}(\mathbb{A}(\sigma))(X) = \mathrm{Str}(e^{F(\sigma)(X)})$ is a rapidly decreasing differential form on $\mathcal{V}$. The fact that the class of $\mathrm{ch}(\mathbb{A}(\sigma))$ in $\mathcal{H}^\infty_{G,rapid}(\mathfrak{g},\mathcal{V})$ remains constant when σ_0 and σ_1 are stably homotopic is again a consequence of estimates in transgression formulas obtained by the same method.

Let $[\sigma] \in K_G(\mathcal{V})$. By definition, the class $\mathrm{ch}([\sigma])$ belongs to $\mathcal{H}^\infty_{G,cpt}(\mathfrak{g},\mathcal{V})$. A fortiori, $\mathrm{ch}([\sigma])$ determines a class in $\mathcal{H}^\infty_{G,rapid}(\mathfrak{g},\mathcal{V})$.

Theorem 29 *Let σ be a good bundle map of G-equivariant vector bundles. Then we have the following equality in $\mathcal{H}^\infty_{G,rapid}(\mathfrak{g},\mathcal{V})$:*

$$\mathrm{ch}([\sigma]) \equiv \mathrm{ch}(\mathbb{A}(\sigma)).$$

Proof. The proof is very similar to the proof of Theorem 19. We use the same notations. As $\mathcal{V}$ has a G-invariant Euclidean structure, then $\Sigma(\mathcal{V})$ is a compact G-manifold. We need to prove that if $F(t,\sigma)(X)$ is the equivariant curvature of the superconnection $\mathbb{A}(t\sigma)$, the equivariant form

$$\alpha(t)(X) = \mathrm{Str}(i\sigma e^{F(t,\sigma)(X)})$$

satisfies an estimate $\|\alpha(t)(X)\|_{(z,x)} \leq K e^{-ct^2\|x\|^2/2}$ when $\|x\|$ is sufficiently big and X varies in a compact subset of $\mathfrak{g}$. We use Volterra expansion(7) to compute $e^{F(t,\sigma)(X)}$ with

$$A = -t^2v^2,$$

$$Z = p^*(\mu(X)) + it(p^*\nabla)\cdot v + p^*F.$$

We write $Z = Z_0 + Z_1$ with $Z_0 = p^*(\mu(X))$ and $Z_1 = it(p^*\nabla)\cdot v + p^*F$. Remark now that $Z_0 = p^*(\mu(X))$ is independent of t, x while $Z_1(t,z,x)$ has polynomial growth in t, x. Thus when X varies in a compact subset of $\mathfrak{g}$ we can find a constant z_0 such that $\|Z_0(z,x)\| \leq z_0$ uniformly. Thus using (10) we obtain

$$\|e^{F(t,\sigma)(X)(z,x)}\| \leq e^{-ct^2\|x\|^2} e^{2z_0} I(t,x)$$

where $I(t,x)$ is a polynomial in (t,x). The remaining of the proof is identical to that of Theorem 19.

Consider $s \in G$. Each fixed point set $\mathcal{V}(s)$ is a vector bundle over $B(s)$. Repeating the preceding proof for each s, we obtain

Proposition 30 *If σ is a G-invariant good bundle map on $\mathcal{V}$, the bouquet of Chern characters* $\mathrm{bch}(\mathbb{A}(\sigma))$ *is a bouquet of rapidly decreasing equivariant forms on $\mathcal{V}$. Its class* $[\mathrm{bch}(\mathbb{A}(\sigma))]$ *in $\mathcal{K}_{G,rapid}(\mathcal{V})$ depends only of σ.*

Let $[\sigma] \in K_G(\mathcal{V})$. By definition , the class $\mathrm{bch}([\sigma])$ belongs to $\mathcal{K}_{G,cpt}(\mathcal{V})$. A fortiori, $\mathrm{bch}([\sigma])$ determines a class in $\mathcal{K}_{G,rapid}(\mathcal{V})$.

Theorem 31 *Let σ be a good bundle map of G-equivariant vector bundles. We have*

$$\mathrm{bch}([\sigma]) = \mathrm{bch}(\mathbb{A}(\sigma))$$

in $\mathcal{K}_{G,rapid}(\mathcal{V})$.

2.6 Integration of bouquets

We have defined in [19] [20] a notion of integration which allows us to associate to a bouquet of differential forms (more exactly to a twisted bouquet) on a compact G-manifold M *an invariant function on the group G.* To describe the integral of a bouquet, we need to introduce some other equivariant forms.

Definition 32 *Let $\mathcal{V} \to B$ be a G-equivariant real vector bundle over a G-manifold B. Let us suppose that $\mathcal{V} \to B$ has a G-invariant connection ∇ with equivariant curvature $R(X)$, then $J(B, \mathcal{V}, \nabla) \in \mathcal{A}_G^\infty(\mathfrak{g}, B)$ is defined by: for $X \in \mathfrak{g}$*

$$J(B, \mathcal{V}, \nabla)(X) = \det\left(\frac{e^{R(X)/2} - e^{-R(X)/2}}{R(X)}\right).$$

The equivariant form $J(B, \mathcal{V}, \nabla)$ is a closed form. The cohomology class $J(B, \mathcal{V})$ of $J(B, \mathcal{V}, \nabla)$ in $\mathcal{H}_G^\infty(\mathfrak{g}, B)$ is independent of the choice of the G-invariant connection on $\mathcal{V}$. If $TB \to B$ is the tangent bundle, we denote $J(B, TB)$ simply by $J(B)$. Remark that the $J(B)^{[0]}(0) = 1$ so that , if B is compact, $J(B)(X)$ is invertible for X small.

Definition 33 *Let $\mathcal{V} \to B$ be a G-equivariant Euclidean real vector bundle over B with a G-equivariant connection of curvature $R(X)$. Let $s \in G$. Assume that s acts trivially on B. Define for $X \in \mathfrak{g}(s)$:*

$$D_s(\mathcal{V}, \nabla)(X) = \det(1 - s e^{R(X)}).$$

The $G(s)$-equivariant form $D_s(\mathcal{V}, \nabla)$ is a closed equivariant form on B. The cohomology class $D_s(\mathcal{V})$ of $D_s(\mathcal{V}, \nabla)$ is independent of the choice of the G-equivariant connection ∇.

The notion of integration of bouquets of equivariant differential forms on a G-manifold M requires care in the case where the submanifolds $M(s)$ are not canonically oriented (see[19] [20]). Here we will define integration only in the case of main interest to us : M is the cotangent bundle to a compact manifold B.

Let G be a compact Lie group acting on a compact manifold B. Consider the induced action of G on T^*B. Formula (38) implies that $(T^*B)(s)$ is canonically isomorphic to the cotangent bundle $T^*B(s)$ to the manifold $B(s)$. In particular $T^*B(s)$ is canonically oriented. Consider the form $J(B(s)) \in \mathcal{A}_{G(s)}^\infty(\mathfrak{g}(s), B(s))$. As $B(s)$ is compact, if Y is sufficiently small, $J(B(s))(Y)$ is invertible. Similarly the action of s on $\mathcal{N}(B/B(s)) \to B(s)$ allows us to define the equivariant form $D_s(\mathcal{N}(B/B(s))$ on $B(s)$. As $D_s(\mathcal{N}(B/B(s))^{[0]}(0) = \det(1 - s) > 0$, the form $D_s(\mathcal{N}(B/B(s))(Y)$ is invertible for Y small.

Theorem 34 *Let G be a compact Lie group acting on a compact manifold B. Let $\alpha = (\alpha_s)_{s\in G}$ be a bouquet of rapidly decreasing equivariant forms on T^*B. Then there exists a unique G-invariant C^∞ function $\Theta(\alpha)$ on G such that for all $s \in G$ and all $Y \in \mathfrak{g}(s)$ sufficiently small,*

$$\Theta(\alpha)(se^Y) = \int_{(T^*B)(s)} (2i\pi)^{-\dim B(s)} \frac{\alpha_s(Y)}{D_s(\mathcal{N}(B/B(s))(Y)J(B(s))(Y)}.$$

Here $(T^*B)(s) = T^*B(s)$ has its canonical orientation.

Definition 35 *For $\alpha = (\alpha_s)_{s\in G}$ a bouquet of rapidly decreasing equivariant forms on T^*B, we will denote the function $\Theta(\alpha)$ defined in Theorem 34 by $\int_b \alpha$ and we will say that $\int_b \alpha \in C^\infty(G)^G$ is the integral of the bouquet α.*

For each $s \in G$, we will say that the formula of Theorem 34 for $\Theta(\alpha)(se^Y), Y \in \mathfrak{g}(s)$, is the "$s$" part of the formula of $\int_b \alpha$.

The meaning of Theorem 34 is as follows: we can define for each $s \in G$ a function θ^s on a small neighborhood of 0 in $\mathfrak{g}(s)$ by the formula: for Y sufficiently small in $\mathfrak{g}(s)$

$$\theta^s(Y) = \int_{T^*B(s)} (2i\pi)^{-\dim B(s)} \alpha_s(Y) D_s^{-1}(\mathcal{N}(B/B(s))(Y)J^{-1}(B(s))(Y).$$

The theorem asserts that the family of functions $\theta^s, s \in G$ is obtained by restrictions to $s \exp \mathfrak{g}(s)_b$ of a *global* C^∞- function $\Theta(\alpha)$ on G. Clearly if $\Theta(\alpha)$ exists, it is unique.

2.7 Index of invariant elliptic operators

Let G be a compact Lie group acting on a compact manifold B. Let $\mathcal{E}^\pm$ be two G-equivariant vector bundles over B. Let

$$P : \Gamma(B, \mathcal{E}^+) \to \Gamma(B, \mathcal{E}^-)$$

be an elliptic pseudo-differential operator on B. If P commutes with the action of G, the spaces $\operatorname{Ker} P$ and $\operatorname{Coker} P$ are finite dimensional representation spaces for the group G. Let us choose a G-invariant smooth positive density on B and G-invariant Hermitian structure on $\mathcal{E}^\pm$. Let P^* be the adjoint of P. Then $\operatorname{Coker} P$ is isomorphic to $\operatorname{Ker} P^*$. The function

$$\operatorname{index}(P)(g) = \operatorname{Tr}(g, \operatorname{Ker} P) - \operatorname{Tr}(g, \operatorname{Ker} P^*)$$

is an element of $R(G)$. This function is called the equivariant index of G.

Let us give a formula for the index of P in terms of the equivariant cohomology of T^*B. Let $p : T^*B \to B$ be the projection. The principal symbol of P defines a G-invariant bundle map $\sigma(P) : p^*\mathcal{E}^+ \to p^*\mathcal{E}^-$ defined and invertible outside the zero section of T^*B. In particular it defines an element $[\sigma(P)]$ of $K_G(T^*B)$. As explained in Section 1, it is easy to construct a G-invariant good bundle map σ such that $[\sigma] = [\sigma(P)]$ in $K_G(T^*B)$. (see Formula 33).

Let us choose G-invariant connections $\nabla^\pm$ on $\mathcal{E}^\pm \to B$. The bouquet of Chern characters $\operatorname{bch}(\mathbb{A}(\sigma))$ is a bouquet of rapidly decreasing equivariant differential forms on T^*B. Its integral $\int_b \operatorname{bch}(\mathbb{A}(\sigma))$ is a G-invariant function on G.

Theorem 36 *Let $P : \Gamma(B, \mathcal{E}^+) \to \Gamma(B, \mathcal{E}^-)$ be a G-invariant pseudo-differential elliptic operator. Let σ be a good bundle map such that $[\sigma] = [\sigma(P)]$. Then the index of P is the integral of the bouquet $\mathrm{bch}(\mathbb{A}(\sigma))$ of Chern characters of the superconnection $\mathbb{A}(\sigma)$. More explicitly, for every $s \in G$, the "s"-part of the equality*

$$\mathrm{index}(P) = \int_b \mathrm{bch}(\mathbb{A}(\sigma))$$

gives the equality: for all $Y \in \mathfrak{g}(s)$ sufficiently small

$$\mathrm{index}(P)(se^Y) = \int_{T^*B(s)} (2i\pi)^{-\dim B(s)} \frac{\mathrm{ch}_s(\mathbb{A}(\sigma))(Y)}{D_s(\mathcal{N}(B/B(s))(Y) J(B(s))(Y)}.$$

Thus the "s"-part of this formula is a generalisation of the fixed point formula of Atiyah-Segal-Singer. Indeed the index of P near each point $s \in G$ is given by an integral over the fixed submanifold $T^*B(s)$ for the action of s on T^*B. When $Y = 0$, we obtain Atiyah-Segal-Singer formula [6] [8]. Using localisation formulas in equivariant cohomology ([5],[10]) and Theorem 29 on the determination of $\mathrm{ch}([\sigma])$, this theorem follows from Atiyah-Segal-Singer formula by the same argument as the one given in [12].

3 Index of transversally elliptic operators

3.1 Transversally elliptic operators

Let G be a compact Lie group acting on a compact manifold B. Let $\mathcal{E}^\pm$ be two G-equivariant vector bundles over B. Let

$$P : \Gamma(B, \mathcal{E}^+) \to \Gamma(B, \mathcal{E}^-)$$

be a G-invariant pseudodifferential operator with principal symbol σ. By definition, the operator P is transversally elliptic if $\sigma(x, \xi)$ is invertible for all non-zero $\xi \in T_x^*B$ which vanish on the tangent space $T_x(G \cdot x)$ to the G-orbit through x. The choice of a G-invariant smooth positive density on B and of G-invariant Hermitian structures on $\mathcal{E}^\pm$ gives to $\Gamma(B, \mathcal{E}^\pm)$ a pre-Hilbert space structure. Although the space $\mathrm{Ker}\, P$ is not necessarily finite dimensional, the representation T of G inside $\mathrm{Ker}\, P$ is a trace-class representation [2]. Thus we can define

$$\mathrm{Tr}(g, \mathrm{Ker}\, P) = \sum_k (T(g) \cdot e_k, e_k)$$

where e_k is an orthonormal basis of $\mathrm{Ker}\, P$ consisting of C^∞-sections of $\mathcal{E}^+$ which are eigenvectors for the action of the Casimir operator of $\mathfrak{g}$. The infinite sum of the diagonal coefficients of the matrix $T(g)$ with respect to such a basis is convergent in the space of generalized functions on G. Let P^* be the adjoint of P. Define a (generalized) G-invariant function on G by

$$\mathrm{index}(P)(g) = Tr(g, \mathrm{Ker}\, P) - \mathrm{Tr}(g, \mathrm{Ker}\, P^*).$$

It is easy to see [2] that $\mathrm{index}(P)$ depends only of the principal symbol of P and moreover only on the class of σ in the K_G-group of the transverse cotangent bundle. The aim of this section is to state a formula for the index of P. Proofs will be given in a forthcoming article.

3.2 Descent of generalized functions on G

We first recall some definitions and notations. Let M be a manifold. By definition, a test density on M is a smooth density with compact support and the space $C^{-\infty}(M)$ of generalized functions on M is the continuous dual of the space of test densities. The space $C^{\infty}(M)$ is canonically imbedded in $C^{-\infty}(M)$: if Θ is a C^{∞} function on M and μ a test density, the linear form $\mu \to \int_M \Theta(m)d\mu(m)$ is a generalized function on M. We use informally the notation $m \to \Theta(m)$ to denote a generalized function Θ on M and the integral notation $\int_M \Theta(m)d\mu(m)$ to denote the value of Θ on the test density μ, although taking the value of Θ at a point m is meaningless.

If G is a Lie group and e the identity of G, the distribution $(\delta_e, \phi) = \phi(e)$ is well defined. Choose a left invariant Haar measure dg on G. We denote by δ_e/dg the generalized function on G such that

$$\int_G (\delta_e/dg)(g)\phi(g)dg = \phi(e).$$

Let $\mathfrak{g}$ be the Lie algebra of G. The group G acts on $\mathfrak{g}$ by the adjoint action. We denote by $C^{-\infty}(\mathfrak{g})^G$ the space of G-invariant generalized functions on $\mathfrak{g}$. The group G acts by conjugation on $C^{-\infty}(G)$ and we denote by $C^{-\infty}(G)^G$ the space of G-invariant generalized functions on G. Let Θ be a G-invariant generalized function on G. If N is a submanifold of G transverse to the G-orbits, we can define the restriction to N of the generalized function Θ, as Θ is constant on the G-orbits. The restriction $n \to \Theta(n)$ of Θ to N is a generalized function on N.

Let $s \in G$. Let us choose a G-invariant scalar product on $\mathfrak{g}$. For $b > 0$, let $\mathfrak{g}(s)_b$ be the open ball of radius b of $\mathfrak{g}(s)$ and let

$$W(s,b) = \{gs\exp Yg^{-1}, g \in G, Y \in \mathfrak{g}(s)_b\}.$$

There exists $a > 0$ such that for every $b \leq a$, the set $W(s,b)$ is a G-invariant tubular open neighborhood of the orbit $G \cdot s$. If Θ is a G-invariant generalized function on G, the restriction of Θ to the transverse manifold $s\exp\mathfrak{g}(s)_b$ to the G-orbits is well defined. We thus can define a generalized function Θ_s on $\mathfrak{g}(s)_b$ by $\Theta_s(Y) = \Theta(s\exp Y), Y \in \mathfrak{g}(s)_b$. Reciprocally Θ is determined in a neigborhood of $s \in G$ by the generalized function Θ_s on $\mathfrak{g}(s)_b$.

In particular, if P is a transversally elliptic operator we will give formulas for the generalized function $Y \to \text{index}(P)(s\exp Y)$ for Y small in $\mathfrak{g}(s)$. This formula will be reminiscent of the Lefschetz formula for index$(P)(s)$ given by Atiyah-Segal-Singer, when P is elliptic. However it will be crucial to use the *equivariant cohomology* of $M(s)$ in order to obtain a meaningful formula.

3.3 Mean rapidly decreasing equivariant differential forms

Let $\mathcal{V}$ be the total space of a real G-equivariant vector bundle over a compact base B. We denote by $p : \mathcal{V} \to B$ the projection.

If $\alpha \in \mathcal{A}_G^{\infty}(\mathfrak{g}, \mathcal{V})$ is an equivariant form and $\phi(X)dX$ a test density on $\mathfrak{g}$, then

$$(\alpha, \phi dX) = \int_{\mathfrak{g}} \alpha(X)\phi(X)dX$$

is a form on $\mathcal{V}$.

Definition 37 *An equivariant form on $\mathcal{V}$ is rapidly decreasing in $\mathfrak{g}$-mean if for every test density μ on $\mathfrak{g}$, the form (α, μ) is rapidly decreasing on $\mathcal{V}$. Furthermore it is required that the map $\mu \to (\alpha, \mu) \in \mathcal{A}_{rapid}(\mathcal{V})$ is continuous.*

The space of mean rapidly decreasing equivariant forms will be denoted by $\mathcal{A}^{\infty}_{G,mean-rapid}(\mathfrak{g}, \mathcal{V})$. We denote by $\mathcal{H}^{\infty}_{G,mean-rapid}(\mathfrak{g}, \mathcal{V})$ the space $\operatorname{Ker} d_{\mathfrak{g}} / \operatorname{Im} d_{\mathfrak{g}}$ for $d_{\mathfrak{g}} : \mathcal{A}^{\infty}_{G,mean-rapid}(\mathfrak{g}, \mathcal{V}) \to \mathcal{A}^{\infty}_{G,mean-rapid}(\mathfrak{g}, \mathcal{V})$.

If α is a mean rapidly decreasing equivariant form and if $\mathcal{V}$ is G-oriented, the G-invariant generalized function $\int_{\mathcal{V}} \alpha \in C^{-\infty}(\mathfrak{g})$ is defined : for every test function ϕ on $\mathfrak{g}$,

$$\int_{\mathfrak{g}} (\int_{\mathcal{V}} \alpha)(X)\phi(X)dX = \int_{\mathcal{V}} (\int_{\mathfrak{g}} \alpha(X)\phi(X)dX).$$

The map $\int_{\mathcal{V}}$ induces a map

$$\int_{\mathcal{V}} : \mathcal{H}^{\infty}_{G,mean-rapid}(\mathfrak{g}, \mathcal{V}) \to C^{-\infty}(\mathfrak{g})^{G}.$$

Of course if $\alpha \in \mathcal{A}^{\infty}_{G,rapid}(\mathfrak{g}, \mathcal{V})$ is a rapidly decreasing equivariant form, then α is a mean rapidly decreasing form. Let us give a more convincing example of mean rapidly decreasing form. Let $p : T^*B \to B$ be the cotangent bundle to a compact manifold B. Let ω be the canonical 1-form of T^*B: on a vector field X on T^*B, $(\omega, X)_{x,\xi} = (\xi, p_* X)$.

Lemma 38 *If the manifold B is homogeneous, the form $e^{id_{\mathfrak{g}}\omega}$ is a mean rapidly decreasing equivariant form on T^*B.*

Proof. Let us give the proof for the case where $B = G$. Proof in the homogeneous case is similar. We associate to $f \in \mathfrak{g}^*$ the 1-form θ^f on G which is invariant by left translation and coincide with f at the identity of G. Thus $T^*G = G \times \mathfrak{g}^*$. Fix a basis $\{f^1, \dots f^r\}$ of $\mathfrak{g}^*$ and write $f = \sum_i y_i f^i$ for $f \in \mathfrak{g}^*$. Let us denote θ^{f^j} by θ^j. Let us compute $(d_{\mathfrak{g}}\omega)(X) = -(\omega, X_{T^*B}) + d\omega$ in these coordinates. The canonical 1-form ω is given by

$$\omega = \sum_j y_j \theta^j.$$

Thus

$$d\omega = \sum_j dy_j \theta^j + y_j d\theta^j.$$

At the point $(g, f) \in G \times \mathfrak{g}^*$

$$\omega(X_{T^*B})(g, f) = -(f, g^{-1}X).$$

If ϕ is a test function on $\mathfrak{g}$, the differential form $\int_{\mathfrak{g}} e^{i(d_{\mathfrak{g}}\omega)(X)}\phi(X)dX$ on T^*B is equal in (g, f) to

$$e^{id\omega} \int_{\mathfrak{g}} e^{i(f, g^{-1}X)}\phi(X)dX = e^{id\omega} \int_{\mathfrak{g}} e^{i(f,X)}\phi(gX)dX.$$

This is a rapidly decreasing form on $G \times \mathfrak{g}^*$ as Fourier transforms of test functions on $\mathfrak{g}$ are rapidly decreasing on $\mathfrak{g}^*$ while $e^{id\omega}$ is of at most polynomial growth on $\mathfrak{g}^*$.

Remark. The form $e^{id_{\mathfrak{g}}\omega}$ is obviously closed. Let dX be a Lebesgue measure on $\mathfrak{g}$. Let δ_0/dX be the corresponding δ function at $0 \in \mathfrak{g}$ and let dg be the Haar measure on G with tangent measure dX at e. Then

$$(2i\pi)^{-\dim G}\int_{T^*G} e^{id_{\mathfrak{g}}\omega} = \mathrm{vol}(G, dg)\frac{\delta_0}{dX}.$$

Remark that $e^{id_{\mathfrak{g}}\omega} = 1 + d_{\mathfrak{g}}(\omega\frac{e^{id_{\mathfrak{g}}\omega}-1}{d_{\mathfrak{g}}\omega})$, so that $e^{id_{\mathfrak{g}}\omega} - 1$ is $d_{\mathfrak{g}}$-exact in the space $\mathcal{A}_G^\infty(\mathfrak{g}, T^*G)$. However the integral $\int_{T^*G} e^{id_{\mathfrak{g}}\omega}$ is not zero. In particular $e^{id_{\mathfrak{g}}\omega} - 1$ is *not exact* in $\mathcal{A}^\infty_{G,mean-rapid}(\mathfrak{g}, T^*G)$.

3.4 Transversally good bundle maps

Let G be a compact Lie group acting on a compact manifold B. Let $p : T^*B \to B$ be the cotangent bundle to B. We denote a point of T^*B as (z, ξ) with $z \in B$ and $\xi \in T^*_z B$. Consider the closed subset $T^*_G B$ of T^*B:

$$T^*_G B = \{(z, \xi); (\xi, (X_B)_z) = 0 \quad \text{for all } X \in \mathfrak{g}\}.$$

By definition a neighborhood of $(T^*_G B)_{z_0}$ in T^*B is said to be *conic* if it is stable by the action of $\mathbf{R}^+$ on T^*B.

Let $\mathcal{E}^\pm$ be two G-equivariant vector bundles on B. Let $\sigma : p^*\mathcal{E}^+ \to p^*\mathcal{E}^-$ be a morphism of G-equivariant vector bundles. If the support of σ intersects $T^*_G B$ in a compact set, then the restriction of σ to $T^*_G B$ determines an element

$$[p^*\mathcal{E}^+|_{T^*_G B}, \sigma, p^*\mathcal{E}^-|_{T^*_G B}]$$

of $K_G(T^*_G B)$. We will denote it simply by $[\sigma]$. We need to define the set of *good representatives* of elements in $K_G(T^*_G B)$. Consider the superbundle $\mathcal{E} = \mathcal{E}^+ \oplus \mathcal{E}^-$. As in section 2.5, choose G-invariant Hermitian structures on $\mathcal{E}^\pm$ and associate to σ the odd endomorphism $v = v(\sigma)$ of $p^*\mathcal{E}$ defined by

$$v = \begin{pmatrix} 0 & \sigma^* \\ \sigma & 0 \end{pmatrix}. \tag{43}$$

Definition 39 *The bundle map $\sigma : p^*\mathcal{E}^+ \to p^*\mathcal{E}^-$ is called* transversally good *if*

1. *The bundle map $\sigma : p^*\mathcal{E}^+ \to p^*\mathcal{E}^-$ is a smooth G-bundle homomorphism.*
2. *The bundle map σ and all its derivatives have at most polynomial growth in the fiber direction.*
3. *For every $z_0 \in B$, there exist $r > 0$ and $c > 0$ such that*

$$v(\sigma)(z, \xi)^2 \geq c\|\xi\|^2 I_{\mathcal{E}_z}$$

for all (z, ξ) in a conic neighborhood of $(T^*_G B)_{z_0}$ such that $\|\xi\| \geq r$.

The next lemma can be proven similarly to Lemma 17.

Lemma 40 *(i) Every element of $K_G(T^*_G B)$ has a transversally good representative.*

(ii) Let $\sigma_0, \sigma_1 : p^*\mathcal{E}^+ \to p^*\mathcal{E}^-$ be transversally good bundle maps and assume that there exists an homotopy of G-bundle maps with compact support $\sigma_t : p^*\mathcal{E}^+|_{T^*_G B} \to p^*\mathcal{E}^-|_{T^*_G B}$. Then there exists an homotopy $\tau_t : p^*\mathcal{E}^+ \to p^*\mathcal{E}^-$ between σ_0 and σ_1 such that each τ_t is a transversally good bundle map and furthermore such that $\frac{d}{dt}\tau_t$ and all its derivatives have at most polynomial growth in the fiber direction.

Let $\nabla = \nabla^+ \oplus \nabla^-$ be a G-invariant connection on $\mathcal{E} = \mathcal{E}^+ \oplus \mathcal{E}^-$. Let ω be the canonical 1-form on T^*B. We will modify the superconnection $\mathbb{A}(\sigma)$ on $p^*\mathcal{E}$ by adding the canonical 1-form ω. Thus we denote by $\mathbb{A}^\omega(\sigma)$ the superconnection on $p^*\mathcal{E}$ defined by

$$\mathbb{A}^\omega(\sigma) = iv(\sigma) + p^*\nabla + i\omega.$$

(As before, the choice of ∇ remains implicit). Consider the equivariant Chern character $\mathrm{ch}(\mathbb{A}^\omega(\sigma))$ of the bundle $p^*\mathcal{E}$ with superconnection $\mathbb{A}^\omega(\sigma)$.

Proposition 41 *Let σ be a transversally good bundle map. Then the Chern character $\mathrm{ch}(\mathbb{A}^\omega(\sigma))$ is a mean rapidly decreasing equivariant differential form on T^*B. Furthermore its cohomology class in $\mathcal{H}^\infty_{G,mean-rapid}(\mathfrak{g}, T^*B)$ depends only of the element $[\sigma]$ of $K_G(T^*_G B)$.*

Remark 42 *Similarly, for each $s \in G$ the form $\mathrm{ch}_s(\mathbb{A}^\omega(\sigma))$ is rapidly decreasing on $T^*B(s)$ in $\mathfrak{g}(s)$-mean.*

3.5 Indices of transversally elliptic operators

Let G be a compact group acting on a compact manifold B. Let us first show that there are interesting examples of transversally elliptic operators.

If $B = G/H$ is an homogeneous space, then the operator $0 : C^\infty(B) \to 0$ is transversally elliptic. The representation of G in $\operatorname{Ker} P$ is the natural action of G in $L^2(G/H)$ while $\operatorname{Coker} P = 0$. This representation is trace class. If $B = G$, the trace of the representation of G in $L^2(G)$ is equal to the generalized function $(\operatorname{vol} G)\delta_e$.

Let us give another example: let $B = S^3$ be the 3-sphere embedded in $\mathbf{C}^2$: $B = \{(z_1, z_2) \in \mathbf{C}^2; |z_1|^2 + |z_2|^2 = 1\}$. The induced Cauchy-Riemann operator $P = \overline{\partial}_b : C^\infty(B) \to C^\infty(B)$, given by

$$(\overline{\partial}_b \cdot \phi)(z_1, z_2) = z_1\partial_{\overline{z}_2} - z_2\partial_{\overline{z}_1}$$

is S^1-invariant for the action of $S^1 = \{e^{i\theta}\}$ given on the source $C^\infty(B)$ by

$$(T^+(e^{i\theta})\phi)(z_1, z_2) = e^{i\theta}\phi(e^{i\theta}z_1, e^{i\theta}z_2)$$

and on the target $C^\infty(B)$ by

$$(T^-(e^{i\theta})\phi)(z_1, z_2) = e^{-i\theta}\phi(e^{i\theta}z_1, e^{i\theta}z_2).$$

Furthermore $\overline{\partial}_b$ is S^1-transversally elliptic. The kernel $\operatorname{Ker} P$ consists of the restrictions to B of holomorphic functions on $\mathbf{C}^2$. Thus under the action of S^1, $\operatorname{Ker} P$ breaks up as

$$\operatorname{Ker} P = \oplus V_n$$

where V_n is the space of restrictions to B of homogeneous polynomials on $\mathbf{C}^2$ of degree $n-1$:

$$V_n = \mathbf{C}z_1^{n-1} \oplus \mathbf{C}z_1^{n-2}z_2 \oplus \ldots \oplus \mathbf{C}z_2^{n-1}.$$

On the n-dimensional space V_n, the representation $T^+(e^{i\theta})$ acts by $e^{in\theta}$ so that

$$\operatorname{Tr}(T^+(e^{i\theta}), \operatorname{Ker} P) = \sum_{n=1}^{\infty} ne^{in\theta}.$$

The adjoint of P is $P^* = -\overline{z_1}\partial_{z_2} + \overline{z_2}\partial_{z_1}$. The kernel $\operatorname{Ker} P^*$ breaks up under the action of S^1 as

$$\operatorname{Ker} P^* = \oplus_{n=1}^{\infty} \overline{V_n}$$

so that

$$\operatorname{Tr}(T^-(e^{i\theta}), \operatorname{Ker} P^*) = \oplus_{n=1}^{\infty} n e^{-in\theta}.$$

Finally the difference of traces gives

$$\operatorname{index}(P)(e^{i\theta}) = \sum_{n\in\mathbb{Z}} n e^{in\theta}.$$

Let δ_1 be the δ-function at the identity 1 of S^1. Thus

$$\operatorname{index}(P)(e^{i\theta}) = (-2i\pi)\frac{\partial}{\partial\theta}\delta_1(e^{i\theta}).$$

Remark here that the action of an element $g = e^{i\theta}$ on S^3 is free , if $g \neq 1$. According to the general principle of fixed points formulas (see [3]), the generalized function $\operatorname{index}(P)$ is supported on the point 1, hence it is a derivative of the δ function at 1. Further examples may be given of transversally elliptic operators P with respect to free actions where the index of P is a derivative of high order of the delta function at 1.

The principal symbol of P determines an element $[\sigma(P)]$ of $K_G(T^*_G B)$. As in the case of elliptic operators (formula 33), it is easy to construct a transversally good representative σ of $[\sigma(P)]$.

Theorem 43 *Let $P : \Gamma(B, \mathcal{E}^+) \to \Gamma(B, \mathcal{E}^-)$ be a transversally elliptic operator. Let σ be a transversally good representative of $[\sigma(P)]$. Then the index of P is given by the following formula: for every $s \in G$, and all $Y \in \mathfrak{g}(s)$ sufficiently small,*

$$\operatorname{index}(P)(se^Y) = \int_{T^*B(s)} (2i\pi)^{-\dim B(s)} \frac{\operatorname{ch}_s(\mathbb{A}^\omega(\sigma))(Y)}{D_s(\mathcal{N}(B/B(s))(Y) J(B(s))(Y)}.$$

Theorem 43 states an equality of *generalized* functions. More precisely, for each $s \in G$ there is a neighborhood U_s of 0 in $\mathfrak{g}(s)$ such that the formula:

$$(\theta^s, \phi dY) =$$

$$\int_{T^*B(s)} (2i\pi)^{-\dim B(s)} \left(\int_{\mathfrak{g}(s)} \frac{\operatorname{ch}_s(\mathbb{A}^\omega(\sigma))(Y)}{D_s(\mathcal{N}(B/B(s)))(Y) J(B(s))(Y)} \phi(Y) dY \right)$$

defines a generalized function θ^s on U_s.

The right hand side is well defined as, for ϕ with sufficiently small support, $Y \mapsto D_s^{-1}(\mathcal{N}(B/B(s))(Y) J^{-1}(B(s))(Y)\phi(Y)$ is a smooth map from $\mathfrak{g}(s)$ to the space $\mathcal{A}(B(s))$, so that by proposition 41 and remark 42

$$\int_{\mathfrak{g}(s)} \frac{\operatorname{ch}_s(\mathbb{A}^\omega(\sigma))(Y)}{D_s(\mathcal{N}(B/B(s))(Y) J(B(s))(Y)} \phi(Y) dY$$

is rapidly decreasing on the fibers of $T^*B(s)$. The theorem asserts that the family of generalized functions $\theta^s, s \in G$, is obtained by restriction to $s \exp U_s$ of the generalized function $\operatorname{index} P$ on G.

When $s = e$ the identity of G, this formula was stated in [23]. This theorem can be thought of as a fixed point formula. Indeed the index of P near each point $s \in G$ is given by an integral over the fixed point submanifold $T^*B(s)$ for the action of s on T^*B. However it is crucial to introduce the equivariant cohomology of $T^*B(s)$ as we need to know the generalized function $Y \to \text{index}(P)(s \exp Y)$ on a neighborhood of 0 in $\mathfrak{g}(s)$ to determine index(P) in a neighborhood of s.

We have not explicitly mentioned the choice of connections ∇ on the bundles $\mathcal{E}^{\pm} \to B$ as the integral in theorem 43 is obviously independent of the choice of ∇. In the case where σ is a good bundle map (good in all directions), it is easy to see that for each $t \in \mathbf{R}$, the Chern character $\text{ch}(\mathbb{A}^{t\omega}(\sigma))$ is a rapidly decreasing form on T^*B. Furthermore the transgression formula 15 shows that the class of $\text{ch}(\mathbb{A}^{t\omega}(\sigma)) \in \mathcal{H}^{\infty}_{G,rapid}(\mathfrak{g}, T^*B)$ is independent of $t \in \mathbf{R}$. Taking $t = 0$, we obtain Theorem 36 for elliptic operators. Thus Theorem 43 is a generalization of Theorem 36 for the equivariant index of an elliptic operator. In the case where P is only transversally elliptic, it is crucial to consider the superconnection $\mathbb{A}^{\omega}(\sigma)$ instead of the superconnection $\mathbb{A}(\sigma)$. Consider for example the case where B is homogeneous, $\mathcal{E}^+ = B \times \mathbf{C}$ is the trivial bundle and $\mathcal{E}^- = 0$. Thus $P = 0$ and $\sigma = 0$. The connection on T^*B to be considered is the connection $\nabla = d + i\omega$. Theorem 43 above gives indeed (see [14]) the character of the regular representation of G in $L^2(B)$. In this case, Theorem 43 is in agreement with the general conjecture of Duflo-Vergne on *characters of quantized representations* $Q(M, \mathcal{E}, \mathbb{A})$ attached to a vector bundle $\mathcal{E} \to M$ with a connection $\mathbb{A}$ (see [24]). Theorem 43 suggests that, more generally, the objects that one should quantize are the G-invariant superbundles with superconnections.

The proof of Theorem 43 will be given in a forthcoming article, along the following line: it is first possible to show that the map

$$i : K_G(T^*_G B) \to C^{-\infty}(G)^G$$

given on good representatives σ by the right-hand side of the formula in Theorem 43 satisfies the axioms (excision, multiplicativity, etc) of the analytic index. Then Theorem 43 is checked for the set of generators of $K_T(T^*_T S_n)$ given by Atiyah in [2] for the action of a torus on the n-dimensional sphere S_n.

References

[1] M.F.ATIYAH. Collected works. *Clarendon Press, Oxford,* 1988.

[2] M.F.ATIYAH. Elliptic operators and compact groups. *Lecture notes in Mathematics 401*, Springer-Verlag, Berlin-Heidelberg-New-York. 1974

[3] M. F. ATIYAH AND R. BOTT. A Lefschetz fixed-point formula for elliptic complexes: I. *Ann. of Math.*, **86** (1967), 374–407.

[4] M. F. ATIYAH AND R. BOTT. A Lefschetz fixed-point formula for elliptic complexes: II. *Ann. of Math.*, **88** (1968), 451–491.

[5] M. F. ATIYAH AND R. BOTT. The moment map and equivariant cohomology. *Topology,* **23** (1984), 1–28.

[6] M. F. ATIYAH AND G. B. SEGAL. The index of elliptic operators II. *Ann. Math.*, **87** (1968), 531–545.

[7] M. F. ATIYAH AND I. M. SINGER. The index of elliptic operators. I. *Ann. Math.*, **87** (1968), 484-530.

[8] M. F. ATIYAH AND I. M. SINGER. The index of elliptic operators. III. *Ann. Math.*, **87** (1968), 546–604.

[9] N. BERLINE, E. GETZLER AND M. VERGNE. Heat kernels and Dirac operators. *Springer-Verlag*, Grundlehren der math. Wiss. 298. 1992

[10] N. BERLINE ET M. VERGNE. Classes caractéristiques équivariantes. Formule de localisation en cohomologie équivariante. *C. R. Acad. Sci. Paris,* **295** (1982), 539–541.

[11] N. BERLINE ET M. VERGNE. Zéros d'un champ de vecteurs et classes caractéristiques équivariantes. *Duke Math. Journal,* **50** (1983), 539-549.

[12] N. BERLINE AND M. VERGNE. The equivariant index and Kirillov character formula. *Amer. J. of Math,* **107** (1985), 1159–1190.

[13] N. BERLINE AND M. VERGNE. Open problems in representations theory of Lie groups. Proceedings of the eighteenth international symposium, division of mathematics, the Taniguchi foundation, 1986

[14] N. BERLINE ET M. VERGNE. Indice équivariant et caractère d'une représentation induite. In "D-Modules and Microlocal Geometry" Walter de Gruyter 1992

[15] J. BLOCK AND E. GETZLER. Equivariant cyclic homology and equivariant differential forms. Annales de l'Ec. Norm. Sup.; to appear

[16] H. CARTAN. Notions d'algèbre différentielle; applications aux groupes de Lie et aux variétés où opère un groupe de Lie. In "Colloque de Topologie". *C. B. R. M., Bruxelles,* (1950), 15-27.

[17] H. CARTAN. La transgression dans un groupe de Lie et dans un espace fibré principal. In "Colloque de Topologie". *C. B. R. M., Bruxelles,* (1950), 57-71.

[18] M. DUFLO ET M. VERGNE. Orbites coadjointes et cohomologie équivariante. In The orbit method in representation theory. *Birkhäuser, Progress in math.*, **82** (1990), 11–60.

[19] M. DUFLO ET M. VERGNE. Cohomologie équivariante et descente. *Preprint DMI,* (1992), 1–121.

[20] M. DUFLO ET M. VERGNE. Cohomologie équivariante et descente I, II. *C. R. Acad. Sci. Paris,* **316** (1993), 971–976 and 1143–1148.

[21] D.QUILLEN. Superconnections and the Chern character. *Topology,* **24** (1985), 37–41.

[22] G. SEGAL. Equivariant K-theory. *Publ.Math.Inst. Hautes Etudes Sci.*, **34** (1968), 129–151.

[23] M. VERGNE. Sur l'indice des opérateurs transversalement elliptiques. *C. R. Acad. Sci. Paris*, **310** (1990), 329–332.

[24] M. VERGNE. Equivariant cohomology and geometric quantization. *Proceedings of the European congress in Mathematics. Paris 1992*, (Preprint DMI 1993), to appear.

C.I.M.E. Session of D-Modules and Representation Theory

List of Participants

E. ALDROVANDI, SISSA, Via Beirut 2/4, 34014 Trieste

F. BALDASSARRI, Dip. di Mat. Univ., Via Belzoni 7, 35131 Padova

A. BARAN, Acad. of Sci., Bucharest

C. BARTOCCI, Dip. Mat. Univ., Via L.B. Alberti 4, 16132 Genova

J.E. BJORK, Dept. Math. Univ., Stockolm

G. BRATTI, Dip. Mat. Univ., Via Belzoni 7, 35131 Padova

M. CANDILERA, Dip. Mat. Univ., Via Belzoni 7, 35131 Padova

C. CHOU, Dip. Mat. Univ., Via Belzoni 7, 35131 Padova

R. CONTI, Dip. Mat. Univ., Viale Morgagni 67/A, Firenze

M. COSTANTINI, Dip. Mat. Univ., Via Belzoni 7, 35131 Padova

V. CRISTANTE, Dip. Mat. Univ., Via Belzoni 7, 35131 Padova

S. CHEMLA, Dept. of Math. Univ. of Utrecht, Budapestlaan 6, Box 80.010, 3508 Utrecht

P. DOMINICI, Dip. Mat. Univ., Via Buonarroti 2, 56100 Pisa

B. ENRIQUEZ, Ecole Polytechnique, 91128 Palaiseau

D. FRANCO, SISSA, Via Beirut 2/4, 34014 Trieste

G. GEROTTO, Dip. Mat. Univ., Via Belzoni 7, 35131 Padova

R. GIACHETTI, Dip. Mat. Univ., P.za di Porta S. Donato 5, 40127 Bologna

M. GORESKY, Northeastern Univ. of Boston, MA 02115-5096

G. GUSSI, Acad. of Sci., Bucharest

A. KOSYAK, Inst. of Math., Acad. of Sci. of Ukraine, Repin str.3, Kiev, Ukraine

V. LYUBASHENKO, Scuola Normale Superiore, P.za dei Cavalieri 7, 56126 Pisa

R. MAC PHERSON, M.I.T., Cambridge, MA 02130-4304

C. MARASTONI, Dip. Mat. Univ., Via Belzoni 7, 35131 Padova

G. MARINESCU, Univ. Paris VII, UER Math., 2 pl. Jussieu, 75251 Paris

P. PAPI, Dip. Mat. Univ., Via Buonarroti 2, 56100 Pisa

C. REINA, SISSA, Via Beirut 2/4, 34014 Trieste

A. SCALARI, Dip. Mat. Univ., Via Belzoni 7, 35131 Padova

J.P. SCHNEIDERS, Univ. Paris XIII, CSP Math., 93430 Villetaneuse

S. SEIFARTH, IAAS, Hausvogteiplatz 5-7, 1086 Berlin

A. TIRABOSCHI, ICTP, Box 586, 34100 Trieste

Y.L.L. TONG, Dept. of Math., Purdue Univ., W. Lafayette, IN 47907

F. TONIN, Dip. Mat. Univ., Via C. Alberto 10, 10123 Torino

N. TOSE, Dept. of Math., Univ. of Sapporo

E. VASSEROT, E.N.S., 45 r. d'Ulm, 75005 Paris

FONDAZIONE C.I.M.E.
CENTRO INTERNAZIONALE MATEMATICO ESTIVO
INTERNATIONAL MATHEMATICAL SUMMER CENTER

"Integrable Systems and Quantum Groups"

is the subject of the First 1993 C.I.M.E. Session.

The Session, sponsored by the Consiglio Nazionale delle Ricerche and by the Ministero dell'Università e della Ricerca Scientifica e Tecnologica, will take place under the scientific direction of Professors Mauro FRANCAVIGLIA (Università di Torino), Silvio GRECO (Politecnico di Torino), Franco MAGRI (Università di Milano) at Villa "La Querceta", Montecatini Terme (Pistoia), **from June 14 to June 22, 1993.**

Courses

a) **Spectral covers, algebraically completely integrable Hamiltonian systems, and moduli of bundles.** (6 lectures in English)
Prof. Ron DONAGI (University of Pennsylvania)

Outline

Spectral covers allow a uniform treatment of a wide variety of algebraically completely integrable Hamiltonian systems, ranging from classical systems such as Jacobi's geodesic flow on an ellipsoid, to recent ones such as Hitchin's commuting flows on the cotangent bundle to the moduli space of stable vector bundles on a curve [H1], and Treibich-Verdier's theory of elliptic solitons [TV]. Our goal is to present an outline of this theory, together with some of the important special cases and applications. Topics to be discussed include:

. Construction of spectral covers, isotypic decomposition of their Picard varieties into generalized Pryms, the distinguished Prym [D2] and its modular interpretation via principal bundles with twisted endomorphisms (generalized 'Higgs bundles'), Kanev's Prym-Tyurin varieties [K] and the n-gonal constructions in Prym theory [D1], the structure of nilpotent cones [L] and its relation with fibers of the Springer resolution.

. Existence of symplectic and Poisson structures, considered both from the modular point of view (following [Ma], [Mu], [T]) and via their infinitesimal cubic invariant (compare [BG]).

. Existence of Lax structures linearizing a given system via spectral covers, and Griffiths' cohomological criterion for linearization [G].

. Examples and applications include:
- Jacobi's system and its generalizations by Beauville and by Adams, Harnad, Hurtubise and Previato.
- Elliptic (and abelian) solitons.
- Hitchin's system for an arbitrary reductive group, with its various applications to the structure of moduli spaces [H1] and to the projectively flat connection in Conformal Field Theory [H2].
- Some non-linear variants of Higgs bundles, living on Mukai spaces.

References

[BG] R. Bryant and P.A. Griffiths, Some observations on the infinitesimal period relations for regular threefolds with trivial canonical bundle, in: Arithmetic and Geometry II, Birkhäuser (1983), 77-102.
[D1] R. Donagi, The tetragonal construction, Bull. AMS 4 (1981), 181-185.
[D2] R. Donagi, Spectral covers, preprint, 1983.
[G] P.A. Griffiths, Linearizing flows and a cohomological interpretation of Lax equations, Amer. J. Math. 107 (1985), 1445-1484.
[H1] N. Hitchin, Stable bundles and integrable systems, Duke Math. J. 54 (1987), 91-114.
[H2] N. Hitchin, Flat connections and geometric quantization, Comm. Math. Phys. 131 (1990), 347-380.

[K] V. Kanev, Spectral curves, simple Lie algebras, and Prym-Tyurin varieties, Proc. Symp. Pure Math. 49 (1989, 627-645.
[L] G. Laumon, Un analogue global du cone nilpotent, Duke Math. J. 57 (1988), 647-671.
[Ma] E. Markman, Spectral curves and integrable systems, UPenn dissertation, 1992.
[Mu] S. Mukai Symplectic structure of the moduli space of sheaves on an abelian or K3 surface, Inv. Math. 77 (1984), 101-116.
[TV] A. Treibich and J.L. Verdier, Solitons elliptiques, The Grothendieck Festschrift, vol. 3, Birkhäuser (1990), 437-480.
[T] A. Tyurin, Symplectic structure on the varieties of moduli of vector bundles on an algebraic surface with p,g>0, Math. USSR Izv. 33 (1989), 139-117.

b) **Geometry of two-dimensional topological field theories.** (6 lectures in English).
Prof. Boris DUBROVIN (Moscow State University and SISSA, Trieste)

Lecture plan

1) Topological symmetric lagrangians and their quantization. Atiyah's axioms of a topological field theory (TFT). Intersection theory on moduli spaces as example of TFT. Topological conformal field theories (TCFT) as twisted N=2 susy theories. Topological deformations of a TCFT.

2) Equations of associativity of the primary chiral algebra as defining relations of a 2D TFT. Differential geometry of the small phase space of a TFT. Classification of massive TCFT by isomonodromy deformation method.

3) Integrable hierarchies associated with arbitrary 2D TFT, their hamiltonian formalism, solutions, and tau-functions. Coupling to topological gravity.

4) Ground state metric as a hermitian metric on the small phase space of a 2D TFT. Calculation of the ground state metric of a massive TCFT by isomonodromy deformations method. Relation to the theory of harmonic maps.

References

1. E. Witten, Surv. Diff. Geom. 1 (1991), 243.
2. R. Dijkgraaf, Intersection theory, integrable hierarchies, and topological field theory. Preprint IASSNS-HEP-91/91, to appear in the Proceedings of the Cargese Summer School on New Symmetry Principles in Quantum Field Theory (1991).
3. B. Dubrovin, Nucl. Phys. B379 (1992), 627.
4. B. Dubrovin, Integrable systems and classification of two-dimensional topological field theories, Preprint SISSA 162/92/FM, Semptember 1992, to appear in the J.-L. Verdier memorial volume, Integrable systems, 1992.
5. B. Dubrovin, Geometry and integrability of topological-antitopological fusion, Preprint INFN-8/92, April 1992, to appear in Comm. Math. Phys.

c) **Integrals of motion as cohomology classes.**. (6 lectures in English).
Prof.Edward FRENKEL (Harvard University)

Outline

Integrals of motion of Toda field theory can be interpreted as cohomology classes. For the classical theory they are cohomology classes of the nilpotent subalgebra of the corresponding finite-dimensional or affine Kac-Moody algebra. For the quantum theory they are cohomology classes of the quantized universal enveloping algebra of the nilpotent subalgebra. This definition makes possible to prove the existence of "big" algebras of integrals of motion in these theories, associated to finite-dimensional Lie algebras, these algebras are nothing but the W-algebras. For the affine Toda field theories, these algebras constitute infinite-dimensional abelian subalgebras of the W-algebras, and they are algebras of integrals of motion of certain deformations of conformal field theories.

References

- B. Feigin, E. Frenkel, Phys. Lett. B 276 (1992), 79-86.
- B. Feigin, E. Frenkel, Int. J. Mod. Phys. 7 (1992), Supplement 1A, 197-215.

d) **Integrable equations and moduli of curves and vector bundles.** (6 lectures in English)
Prof.Emma PREVIATO (Boston University)

Description

The theory of integrable systems/integrable equations of KdV type brought about profound interactions between physics and algebraic geometry over the past twenty years. This course will be an illustration of roughly three areas in the field and of open directions branching out of them: area one, the linearization of certain Hamiltonian flows over Jacobian varieties and generalizations to moduli spaces of vector bundles; two, moduli of special (elliptic) solutions; three, projective realizations of moduli spaces of vector bundles.
A list of topics follows.

Prerequisites

Classical Riemann-surface theory and rudiments of algebraic geometry.

Lecture I	:	Burchnall-Cauchy-Krichever map [ADCKP]
Lecture II	:	Generalization to vector bundles [KN]
Lecture II	:	The elliptic case [K], [M]
Lecture IV	:	The hyperelliptic case [VG]
Lecture V	:	The two-theta map [B]
Lecture VI	:	Verlinde formulas and Kummer varieties [vGP]

References

[ADCKP] E. Arbarello, C. De Concini, V.G. Kac and C. Procesi, Moduli spaces of curves and representation theory, Comm. Math. Phys. 117 (1988), 1-36.

[B] A. Beauville, Fibres de rang 2 sur une courbe, fibre determinant et fonctions theta, II, Bull. Soc. Math. France 119 (1991), 259-291.

[vG] B. van Geemen, Schottky-Jung relations and vector bundles hyperelliptic curves, Math. Ann. 281 (1988), 431-449.

[vGP] B. van Geemen and E. Previato, Prym varieties and the Verlinde formula, Math. Ann. (1933).

[K] I.M. Krichever, Elliptic solutions of the Kadomtsev-Petviashvili equation and integrable systems of particles, Functional Anal. Appl. 14 (1980), 282-290.

[KN] I.M. Krichever and S.P. Novikov, Holomorphic fiberings and nonlinear equations. Finite zone solutions of rank 2, Sov. Math. Dokl. 20 (1979), 650-654.

[M] O.I. Mokhov, Commuting differential operators of rank 3, and nonlinear differential equations, Math. USSR Izvestiya, 35 (1990), 629-655.

FONDAZIONE C.I.M.E.
CENTRO INTERNAZIONALE MATEMATICO ESTIVO
INTERNATIONAL MATHEMATICAL SUMMER CENTER

"Algebraic Cycles and Hodge Theories"

is the subject of the Second 1993 C.I.M.E. Session.

The Session, sponsored by the Consiglio Nazionale delle Ricerche and by the Ministero dell'Università e della Ricerca Scientifica e Tecnologica, will take place under the scientific direction of Prof. Fabio BARDELLI (Università di Pisa) at Villa Gualino, Torino, Italy, **from June 21 to June 29, 1993.**

Courses

a) **Infinitesimal methods in Hodge theory.** (8 lectures in English)
Prof. Mark GREEN (University of California, Los Angeles)

Lecture plan

1) The Hodge Theorem. Hodge decomposition and filtrations. The operators L, Λ and H, and the Hodge identities. Principle of two types. Degeneration of the Hodge-De Rham spectral sequence.

2) The Griffiths intermediate Jacobians, The Abel-Jacobi map. Infinitesimal Abel-Jacobi map and the extension class of the normal bundle sequence. Image of cycles algebraically equivalent to zero under the Abel-Jacobi map.

3) Variation of Hodge structure. The Hodge filtration varies analytically. The period map and its derivative. Infinitesimal period relations (Griffiths transversality). Griffiths computation of the infinitesimal period map as a cup product.

4) Hodge theory of hypersurfaces and complete intersections. Derivative of the period map for hypersurfaces. Infinitesimal Torelli for hypersurfaces and complete intersections. Examples of Hodge classes of cycles on hypersurfaces.

5) Mixed Hodge structures. Examples of extension classes.

6) Normal functions. Normal function associated to a primitive Deligne class. Analyticity and infinitesimal relation for normal functions. Infinitesimal invariant of normal function.

7) Koszul cohomology techniques in Hodge theory. Macaulay-Gotzmann theorem. Codimension of the Noether-Lefschetz locus for surfaces. Donagi's generic Torelli theorem for hypersurfaces. Vanishing of the infinitesimal invariant of normal functions for hypersurfaces of high degree.

8) Further applications of Koszul techniques. Nori's connectedness theorem. Abel-Jacobi map for general 3-fold of degree$\geq$ 6. Surjectivity of the general restriction map for rational Deligne cohomology, and the Poincaré-Lefschetz-Griffiths approach to the Hodge conjecture.

Suggested reading

- P. Griffiths and J. Harris, Principles of Algebraic Geometry, Chapters 0-1. This is a good source for the basic facts of Hodge theory, e.g. lecture 1.
- P. Griffiths, Topics in Transcendental Algebraic Geometry. The chapters (by various authors) include some useful surveys as well as more specialized research articles. Chapters I, III, XII, XIII, XIV, XVI, and XVII are probably the most helpful for this course.
- J. Carlson, M. Green, P. Griffiths, J. Harris, "Infinitesimal Variation of Hodge Structure", I-III. Compositio Math. 50 (1083), 109-

324. These contain a lot of information, including of course many interesting topics that won't be covered in these courses. Worth dipping into.
- M. Green, "Koszul cohomology and geometry", in "Lecture on Riemann Surfaces", Proceedings of the ICTP College on Riemann Surfaces, World Scientific 1989. This represents my best effort at an elementary exposition of the Hodge theory of hypersurfaces and Koszul-theoretic techniques. Sections 1,2 and 4 are relevant.
- M. Cornalba and P. Griffiths, "Some transcendental aspects of algebraic geometry" in "Algebraic Geometry - Arcata 1974", Proc. Symp. in Pure Math. 9, AMS (1975), 3-110. This has a relatively painless introduction to mixed Hodge structures. These lectures mostly deal with the differential-geometric aspects of the period map, a beautiful aspect of Hodge theory that we won't cover.
- J. Carlson and C. Peters have a new book on Hodge theory in the works. If it is available in time, it should be an outstanding introduction to many of the topics to be covered.

b) **Algebraic cycles and algebraic aspects of cohomology and K-theory.** (6 lectures in English).
Prof. Jacob MURRE (Universiteit Leiden)

The following subjects will be discussed:

1) Algebraic cycles.
Basic notions. Discussion of the most important equivalence relations. The Chow ring. The Griffiths group. Statement of the principal known facts in codimension greater than one. The definition of higher Chow groups of Bloch. Definition and main properties of Chern classes of vector bundles. The Grothendieck group of vector bundles and sheaves, and its relation to the Chow groups.

2) Deligne-Beilinson cohomology.
Definition and main properties. Examples. Construction of the cycle map; its relation to the classical cycle map and the Abel-Jacobi map.

3) Algebraic cycles and algebraic K-theory.
Introduction to algebraic K-theory. The functors K_0 (see also 1.), K_1 and K_2. The Bloch formula for the Chow groups. Discussion of the regulator map for $K_2(X)$ when X is an algebraic curve.

4) The Hodge Conjecture.
Statement of the (p,p)-conjecture. Survey of the typical known cases. Discussion of some examples. Statement of the generalized Hodge-conjecture as corrected by Grothendieck. Discussion of an example of Bardelli.

5) Some results in codimension 2.
   a. Applications of the Merkurjev-Suslin theorem of algebraic K-theory.
   b. Incidence equivalence and its relation to Abel-Jacobi equivalence.

6) Introduction to motives.
The standard conjectures and something about motives.

References

1. Bloch, S.: Lectures on algebraic cycles. Duke Univ. Math. Ser. IV, 1980.
2. Fulton, W.: Intersection theory. Erg. der Math., 3 Folge, Bd. 2, Springer Verlag, 1984.
3. Esnault, H. and Viehweg, E.: Deligne-Beilinson Cohomology. In: "Beilinson's conjectures and special values of L-function". Perspectives in Math., Vol. 4, Academic Press 1988.
4. Shioda, T.: What is known about the Hodge conjecture? In: Advances Studies in Pure Math., Vol. 1 Kinokuniya Comp. and North Holland, Tokyo 1983.
5. Murre, J.P.: Applications of algebraic K-theory to algebraic geometry. Proc. Conf. Alg. Geom. Sitges 1983, Springer LNM 1124.

c) **Transcendental methods in the study of algebraic cycles.** (8 lectures in English).
Prof. Claire VOISIN (Université de Orsay, Paris)

Outline of the lectures

1) Divisors.
Weil divisors. Cartiers divisors and line bundles; rational and linear equivalence; GAGA principle. The exponential exact sequence and its consequences:

- homological equivalence=algebraic equivalence for divisors
- the Lefschetz theorem on (1,1) classes; Neron-Severi group
- Hodge structure on H^1 and abelian varieties: the Picard variety
- the existence of Poincaré divisor

2) Topology and Hodge theory
Morse theory on affine varieties and the weak Lefschetz theorem. The Hodge index theorem. Consequences: The hard Lefschetz theorem and the Lefschetz decomposition. Applications:
- reduction to the primitive middle dimensional cohomology; degeneracy of Leray spectral sequences; semi-simplicity of the category of polarized Hodge structures.

3) Noether-Lefschetz locus
Deformations of Hodge classes. The Noether-Lefschetz loci; algebraicity of the components; local study (application of transversality of the period map to the codimension, infinitesimal description). Relation with the deformation theory of cycles; ghe semi-regularity property and Bloch-Kodaira theorem.

4) Monodromy
Nodal varieties. Lefschetz degenerations and Lefschetz pencils. Vanishing cycles and cones over them. The Picard-Lefschetz formula and applications of Noether-Lefschetz type. Discussion of the Hodge theory of the vanishing cycles on the central fibre.

5) O-cycles I
O-cycles and holomorphic forms on varieties; Mumford's theorem on the infinite dimensionality of the CH_0 group. Roitman's theorem: $CH^o{}_0$ finite dimensional $\Leftrightarrow$ $CH^o{}_0 \equiv$ Alb. The Bloch's conjecture for surfaces and Bloch-Kas-Lieberman theorem.

6) O-cycles II
The proof of the Bloch conjecture for Godeaux type surfaces; Bloch-Srinivas theorem and consequences of "CH_0 small" on algebraic cycles and Hodge theory of a variety.

7) Griffiths group
One cycles on threefolds; Abel-Jacobi map on cycles algebraically equivalent to zero. Normal functions and their Hodge classes: The theorem of Griffiths. Statement of Clemens theorem and further examples.

8) Application of the NL locus to threefolds
M. Green's criterion for density of the Noether-Lefschetz locus. Applications to one-cycles on threefolds:
- parametrization of certain sub-Hodge structures by algebraic cycles
- infinitesimal proof of Clemens theorem
- generalization of Griffiths theorem to any Calabi-Yau threefold

References

- A. Weil: Variétés Kahleriennes, Actualités scientifiques et industrielles.
- J. Milnor: Morse theory, Annals of Math. Studies, Study 21, Princeton Univ. Press.
- Carlson, Green, Griffiths, Harris: Compositio Math. Vol. 50 (three articles).
- P. Deligne: Théorie de Hodge II, I.H.E.S. Publ. Math. 40, (1971), 5-58.
- P. Griffiths: On the periods of certain rational integrals I, II, Ann. of Math. 90 (1969), 460-541.
- P. Griffiths: Topics in transcendental algebraic geometry, Annals of Math. Studies, Study 106, Princeton Univ. Press.
- H. Clemens: Double solids, Advances in Math. Vol. 47 (1983).

FONDAZIONE C.I.M.E
CENTRO INTERNAZIONALE MATEMATICO ESTIVO
INTERNATIONAL MATHEMATICAL SUMMER CENTER

"Modelling and Analysis of Phase Transition and Hysteresis Phenomena"

is the subject of the Third 1993 C.I.M.E. Session.

The Session, sponsored by the Consiglio Nazionale delle Ricerche and by the Ministero dell'Università e della Ricerca Scientifica e Tecnologica, will take place under the scientific direction of Prof. Augusto VISINTIN (Università di Trento) at Villa "La Querceta", Montecatini Terme (Pistoia), **from July 13 to July 21, 1993.**

Courses

a) **Hysteresis operators.** (6 lectures in English)
Prof. Martin BROKATE (Universität Kaiserslautern)

Course outline

1) Scalar hysteresis operators.
Example of hysteresis models. Hysteresis operators. Continuity properties. Memory properties. Applications.

2) Vector hysteresis operators

3) Hysteresis operators and differential equations.
Ordinary differential equations with hysteresis. Parabolic equations with hysteresis. Hyperbolic equations with hysteresis. Shape memory alloys. Control problems with hysteresis.

References

1. **Books:**
- Brokate, M.: Optimal control of ordinary differential equations with nonlinearities of hysteresis type. Peter Lang Verlag, Frankfurt 1987. (In German; English translation in: Automation and Remote Control, 52 (1991) and 53 (1992)).
- Krasnoselskii, M.A., Pokrovskii, A.V.: Systems with hysteresis. Springer 1969.
- Mayergoyz, I.D.: Mathematical models of hysteresis. Springer 1991.

2. **Survey:**
- Visintin, A.: Mathematical models of hysteresis. In: Topics in nonsmooth mechanics (eds. J. J. Moreau, P.D. Panagiotopoulos, G. Strang), Birkhäuser 1988, 295-326.

3. **Papers:**
- Brokate, M., Visintin, A.: Properties of the Preisach model for hysteresis. J. Reine Angew. Math. 402 (1989), 1-40.
- Krejčí, P.: A Monotonicity method for solving hyperbolic problems with hysteresis. Apl. Mat. 33 (1988), 197-203.
- Krejčí, P.: Hysteresis memory preserving operators. Applications of Math. 36 (1991), 305-326.
- Krejčí, P.: Vector hysteresis models. European J. Appl. Math. 22 (1991), 281-292.
- Visintin, A.: A model for hysteresis of distributed systems. Ann. Mat. Pura Appl. 131 (1982), 203-231.
- Visintin, A.: Rheological models and hysteresis effects. Rend. Sem. Mat. Univ. Padova 77 (1987), 213-243.

b) **Systems of nonlinear PDEs arising from dynamical phase transition.** (6 lectures in English).
Prof. Nobuyuki KENMOCHI (Ghiba University)

Outline of the contents

Systems of nonlinear PDEs are proposed as mathematical models for thermodynamical phase transition processes such as solidification and melting in solid-liquid systems. These are nonlinear parabolic PDEs and variational inequalities with obstacles and the unknowns are the absolute temperature and the order parameter representing the physical situation of the materials. We analyze these models from the following points (1)-(4) of view:

(1) Physical background of the problem
(2) Abstract treatment of the problem
(3) Existence and uniqueness results
(4) Asymptotic stability for the solutions

The basic literature references for the subjects

Nonlinear PDEs:
- D. Gilbarg and N. S. Trudinger, Elliptic Partial Differential Equations of Second Order, Springer-Verlag, Berlin, 1983.
- H. Brézis, Problémes unilatéraux, J. Math. pures appl., 51 (1972), 1-168.

Convex Analysis:
- J. L. Lions, Quelques méthodes de résolution des problémes aux limites non linéaires, Dunod, Gauthier-Villars, Paris, 1969.
- H. Brézis, Opérateurs maximaux monotones et semi-groupes de contractions dans les espaces de Hilbert, North-Holland, Amsterdam, 1973.

c) **Quasiplasticity and Pseudoelasticity in Shape Memory Alloys.** (6 lectures in English).
Prof. Ingo MÜLLER (Technical University Berlin)

Course outline

1. Phenomena.
The phenomena of quasiplasticity and pseudoelasticity in shape memory alloys are described and documented. They are due to a martensitic-austenitic phase transition and to the twinning of the martensitic phase, which is the lowtemperature phase.

2. Model.
A strctural model is introduced wihich is capable of simulating the observed phenomena. The model consists of lattice layers in a potential which has three potential wells, one metastable. Adjacent layers are coherent and their formation requires an extra energy, the coherency energy.

3. Statistical Mechanics.
Statistical Mechanics of the model provides a non-convex free energy and - consequently - a nonmonotone load deformation curve. This is appropriate for pseudoelasticiy. The proper description of quasiplasticity requires a kinetic theory of the model, akin to the theory of activated processes in chemistry.

4. Hysteresis.
Minimization of the free energy under constant deformation leads us to conclude that the observed hysteresis in the pseudo-elastic range is due to the coherency energy. The phase equilibria are unstable and this explains the occurence of internal yield and recovery in pseudoelasticity. A simple mathematical construct for the non-convex free energy permits the description of many observed phenomena inside the hysteresis loop.

5. Thermodynamics.
A systematic exploitation of the first and second law of thermodynamics allows us to predict the thermal and caloric side effects of pseudoelastic deformation.

6. Metastability.
The nature of the metastable states inside the hysteresis loop is as yet not well understood. But there are partial results. They concern observations of the number of interfaces during the phase transition and the role of a "fluctuation temperature" which activates the body to the extent that its entropy can approach its maximum value.

d) **Variational methods in the Stefan problem.** (6 lectures in English)
Prof. José Francisco RODRIGUES (CMAF/Universidade de Lisboa)

Outline of contents

The Stefan problem is one of the simplest possible macroscopic models for phase changes in a pure material when they occur either by heat conduction or diffusion. Its history provides a helpful example of the interplay between free boundary problems and the real world. This course intends to introduce this model problem and to develop an exposition of the variational methods applied to the study of weak solutions for multidimensional problems.

Plan:

1. Introduction to the mathematical-physics models
2. Analysis of the one-phase problem via variational inequalities I
3. Analysis of the two-phase problem via variational inequalities II
4. Study of the enthalpy formulation via Galerkin method
5. Analysis of more complex Stefan problems

Some basic literature

1. G. Duvaut & J. L. Lions, Les inéquations en mécanique et en physique, Dunod, Paris, 1972 (English transl. Springer, Berlin, 1976).
2. A. Friedman, Variational principles and free boundary value problems, Wiley, New York, 1982.
3. D. Kinderlehrer & G. Stampacchia, An introduction to variational inequalities and their application, Academic Press, New York, 1980.
4. J. L. Lions, Sur quelques questions d'analyse, de mécanique et de control optimal, Press Univ. Montréal, 1976.
5. A. M. Mermanov, The Stefan problem, W. De Gruyter, Berlin, 1992.
6. I. Pawlow, Analysis and control of evolution multiphase problems with free boundaries, Polska Akad. Nauk, Warszawa, 1987.
7. J. F. Rodrigues, Obstacle problems in mathematical physics, North-Holland, Amsterdam, 1987.
8. J. F. Rodrigues (Editor), Mathematical models for phase change problems, ISNM n. 88, Birkhäuser, Basel, 1989.
9. E. Zeidler, Nonlinear functional analysis and its application, Vol. II/B, Nonlinear monotone operators, Springer Verlag, New York, 1990.

e) **Numerical aspects of free boundary and hysteresis problems.** (6 lectures in English).
Prof. Claudio VERDI (Università di Pavia)

Summary

1. Time discretization of strongly nonlinear parabolic equations
   - 1.1 Nonlinear methods
   - 1.2 Linear methods
   - 1.3 Applications to problem with hysteresis

2. Full discretization
   - 2.1 Finite element spaces
   - 2.2 Nonlinear schemes
   - 2.3 Linear schemes
   - 2.4 Stability of fully discrete schemes
   - 2.5 Error estimates
   - 2.6 Approximation of free boundaries

3. Adaptive finite element methods for parabolic free boundary problems

Basic references

1. P.G. Ciarlet, The finite element method for elliptic problems, North-Holland, Amsterdam, 1978.
2. V. Thomee, Galerkin Finite Element Methods for Parabolic Problems, Lecture Notes in Mathematics 1054, Sprin ger Verlag, Berlin, 1984.
3. J. M. Ortega and W. C. Rheinboldt, Iterative Solution of Nonlinear Equations in Several Variables, Academic Press, New York, 1970.
4. R. H. Nochetto, Finite element methods for parabolic free boundary problems, in: Advances in Numerical Analysis, Vol. I: Nonlinear Partial Differential Equations and Dynamical Systems, Oxford Academic Press, 1991, 34-95.

5. R. H. Nochetto and C. Verdi, Approximation of degenerate parabolic problems using numerical integration, SIAM J. Numer. Anal., 25 (1988), 784-814.
6. R. H. Nochetto and C. Verdi, An efficient linear scheme to approximate parabolic free boundary problems: error estimates and implementation, Math. Comp., 51 (1988), 27-53.
7. R. H. Nochetto, M. Paolini and C. Verdi, Adaptive finite element method for the two-phase Stefan problem in two space dimension. Part I: Stability and error estimates, Math. Comp., 57 (1991), 73-108); Supplement, Math. Comp. 57 (1991), S1-S11.

LIST OF C.I.M.E. SEMINARS

		Publisher
1954 -	1. Analisi funzionale	C.I.M.E.
	2. Quadratura delle superficie e questioni connesse	"
	3. Equazioni differenziali non lineari	"
1955 -	4. Teorema di Riemann-Roch e questioni connesse	"
	5. Teoria dei numeri	"
	6. Topologia	"
	7. Teorie non linearizzate in elasticità, idrodinamica, aerodinamica	"
	8. Geometria proiettivo-differenziale	"
1956 -	9. Equazioni alle derivate parziali a caratteristiche reali	"
	10. Propagazione delle onde elettromagnetiche	"
	11. Teoria della funzioni di più variabili complesse e delle funzioni automorfe	"
1957 -	12. Geometria aritmetica e algebrica (2 vol.)	"
	13. Integrali singolari e questioni connesse	"
	14. Teoria della turbolenza (2 vol.)	"
1958 -	15. Vedute e problemi attuali in relatività generale	"
	16. Problemi di geometria differenziale in grande	"
	17. Il principio di minimo e le sue applicazioni alle equazioni funzionali	"
1959 -	18. Induzione e statistica	"
	19. Teoria algebrica dei meccanismi automatici (2 vol.)	"
	20. Gruppi, anelli di Lie e teoria della coomologia	"
1960 -	21. Sistemi dinamici e teoremi ergodici	"
	22. Forme differenziali e loro integrali	"
1961 -	23. Geometria del calcolo delle variazioni (2 vol.)	"
	24. Teoria delle distribuzioni	"
	25. Onde superficiali	"
1962 -	26. Topologia differenziale	"
	27. Autovalori e autosoluzioni	"
	28. Magnetofluidodinamica	"

1963 - 29. Equazioni differenziali astratte "
30. Funzioni e varietà complesse "
31. Proprietà di media e teoremi di confronto in Fisica Matematica "

1964 - 32. Relatività generale "
33. Dinamica dei gas rarefatti "
34. Alcune questioni di analisi numerica "
35. Equazioni differenziali non lineari "

1965 - 36. Non-linear continuum theories "
37. Some aspects of ring theory "
38. Mathematical optimization in economics "

1966 - 39. Calculus of variations Ed. Cremonese, Firenze
40. Economia matematica "
41. Classi caratteristiche e questioni connesse "
42. Some aspects of diffusion theory "

1967 - 43. Modern questions of celestial mechanics "
44. Numerical analysis of partial differential equations "
45. Geometry of homogeneous bounded domains "

1968 - 46. Controllability and observability "
47. Pseudo-differential operators "
48. Aspects of mathematical logic "

1969 - 49. Potential theory "
50. Non-linear continuum theories in mechanics and physics and their applications "
51. Questions of algebraic varieties "

1970 - 52. Relativistic fluid dynamics "
53. Theory of group representations and Fourier analysis "
54. Functional equations and inequalities "
55. Problems in non-linear analysis "

1971 - 56. Stereodynamics "
57. Constructive aspects of functional analysis (2 vol.) "
58. Categories and commutative algebra "

1972 -	59.	Non-linear mechanics		"
	60.	Finite geometric structures and their applications		"
	61.	Geometric measure theory and minimal surfaces		"
1973 -	62.	Complex analysis		"
	63.	New variational techniques in mathematical physics		"
	64.	Spectral analysis		"
1974 -	65.	Stability problems		"
	66.	Singularities of analytic spaces		"
	67.	Eigenvalues of non linear problems		"
1975 -	68.	Theoretical computer sciences		"
	69.	Model theory and applications		"
	70.	Differential operators and manifolds		"
1976 -	71.	Statistical Mechanics		Ed Liguori, Napoli
	72.	Hyperbolicity		"
	73.	Differential topology		"
1977 -	74.	Materials with memory		"
	75.	Pseudodifferential operators with applications		"
	76.	Algebraic surfaces		"
1978 -	77.	Stochastic differential equations		"
	78.	Dynamical systems		Ed Liguori, Napoli and Birhäuser Verlag
1979 -	79.	Recursion theory and computational complexity		"
	80.	Mathematics of biology		"
1980 -	81.	Wave propagation		"
	82.	Harmonic analysis and group representations		"
	83.	Matroid theory and its applications		"
1981 -	84.	Kinetic Theories and the Boltzmann Equation	(LNM 1048)	Springer-Verlag
	85.	Algebraic Threefolds	(LNM 947)	"
	86.	Nonlinear Filtering and Stochastic Control	(LNM 972)	"
1982 -	87.	Invariant Theory	(LNM 996)	"
	88.	Thermodynamics and Constitutive Equations	(LN Physics 228)	"
	89.	Fluid Dynamics	(LNM 1047)	"

1983 -	90.	Complete Intersections	(LNM 1092)	"
	91.	Bifurcation Theory and Applications	(LNM 1057)	"
	92.	Numerical Methods in Fluid Dynamics	(LNM 1127)	"
1984 -	93.	Harmonic Mappings and Minimal Immersions	(LNM 1161)	"
	94.	Schrödinger Operators	(LNM 1159)	"
	95.	Buildings and the Geometry of Diagrams	(LNM 1181)	"
1985 -	96.	Probability and Analysis	(LNM 1206)	"
	97.	Some Problems in Nonlinear Diffusion	(LNM 1224)	"
	98.	Theory of Moduli	(LNM 1337)	"
1986 -	99.	Inverse Problems	(LNM 1225)	"
	100.	Mathematical Economics	(LNM 1330)	"
	101.	Combinatorial Optimization	(LNM 1403)	"
1987 -	102.	Relativistic Fluid Dynamics	(LNM 1385)	"
	103.	Topics in Calculus of Variations	(LNM 1365)	"
1988 -	104.	Logic and Computer Science	(LNM 1429)	"
	105.	Global Geometry and Mathematical Physics	(LNM 1451)	"
1989 -	106.	Methods of nonconvex analysis	(LNM 1446)	"
	107.	Microlocal Analysis and Applications	(LNM 1495)	"
1990 -	108.	Geoemtric Topology: Recent Developments	(LNM 1504)	"
	109.	H_∞ Control Theory	(LNM 1496)	"
	110.	Mathematical Modelling of Industrial Processes	(LNM 1521)	"
1991 -	111.	Topological Methods for Ordinary Differential Equations	(LNM 1537)	"
	112.	Arithmetic Algebraic Geometry	(LNM 1553)	"
	113.	Transition to Chaos in Classical and Quantum Mechanics	to appear	"
1992 -	114.	Dirichlet Forms	(LNM 1563)	"
	115.	D-Modules, Representation Theory, and Quantum Groups	(LNM 1565)	"
	116.	Nonequilibrium Problems in Many-Particle Systems	(LNM 1551)	"

1993 -	117. Integrable Systems and Quantum Groups	to appear	"
	118. Algebraic Cycles and Hodge Theories	to appear	"
	119. Modelling and Analysis of Phase Transition and Hysteresis Phenomena	to appear	"

Printing: Weihert-Druck GmbH, Darmstadt
Binding: Buchbinderei Schäffer, Grünstadt

Vol. 1472: T. T. Nielsen, Bose Algebras: The Complex and Real Wave Representations. V, 132 pages. 1991.

Vol. 1473: Y. Hino, S. Murakami, T. Naito, Functional Differential Equations with Infinite Delay. X, 317 pages. 1991.

Vol. 1474: S. Jackowski, B. Oliver, K. Pawałowski (Eds.), Algebraic Topology, Poznań 1989. Proceedings. VIII, 397 pages. 1991.

Vol. 1475: S. Busenberg, M. Martelli (Eds.), Delay Differential Equations and Dynamical Systems. Proceedings, 1990. VIII, 249 pages. 1991.

Vol. 1476: M. Bekkali, Topics in Set Theory. VII, 120 pages. 1991.

Vol. 1477: R. Jajte, Strong Limit Theorems in Noncommutative L_2-Spaces. X, 113 pages. 1991.

Vol. 1478: M.-P. Malliavin (Ed.), Topics in Invariant Theory. Seminar 1989-1990. VI, 272 pages. 1991.

Vol. 1479: S. Bloch, I. Dolgachev, W. Fulton (Eds.), Algebraic Geometry. Proceedings, 1989. VII, 300 pages. 1991.

Vol. 1480: F. Dumortier, R. Roussarie, J. Sotomayor, H. Żołądek, Bifurcations of Planar Vector Fields: Nilpotent Singularities and Abelian Integrals. VIII, 226 pages. 1991.

Vol. 1481: D. Ferus, U. Pinkall, U. Simon, B. Wegner (Eds.), Global Differential Geometry and Global Analysis. Proceedings, 1991. VIII, 283 pages. 1991.

Vol. 1482: J. Chabrowski, The Dirichlet Problem with L^2-Boundary Data for Elliptic Linear Equations. VI, 173 pages. 1991.

Vol. 1483: E. Reithmeier, Periodic Solutions of Nonlinear Dynamical Systems. VI, 171 pages. 1991.

Vol. 1484: H. Delfs, Homology of Locally Semialgebraic Spaces. IX, 136 pages. 1991.

Vol. 1485: J. Azéma, P. A. Meyer, M. Yor (Eds.), Séminaire de Probabilités XXV. VIII, 440 pages. 1991.

Vol. 1486: L. Arnold, H. Crauel, J.-P. Eckmann (Eds.), Lyapunov Exponents. Proceedings, 1990. VIII, 365 pages. 1991.

Vol. 1487: E. Freitag, Singular Modular Forms and Theta Relations. VI, 172 pages. 1991.

Vol. 1488: A. Carboni, M. C. Pedicchio, G. Rosolini (Eds.), Category Theory. Proceedings, 1990. VII, 494 pages. 1991.

Vol. 1489: A. Mielke, Hamiltonian and Lagrangian Flows on Center Manifolds. X, 140 pages. 1991.

Vol. 1490: K. Metsch, Linear Spaces with Few Lines. XIII, 196 pages. 1991.

Vol. 1491: E. Lluis-Puebla, J.-L. Loday, H. Gillet, C. Soulé, V. Snaith, Higher Algebraic K-Theory: an overview. IX, 164 pages. 1992.

Vol. 1492: K. R. Wicks, Fractals and Hyperspaces. VIII, 168 pages. 1991.

Vol. 1493: E. Benoît (Ed.), Dynamic Bifurcations. Proceedings, Luminy 1990. VII, 219 pages. 1991.

Vol. 1494: M.-T. Cheng, X.-W. Zhou, D.-G. Deng (Eds.), Harmonic Analysis. Proceedings, 1988. IX, 226 pages. 1991.

Vol. 1495: J. M. Bony, G. Grubb, L. Hörmander, H. Komatsu, J. Sjöstrand, Microlocal Analysis and Applications. Montecatini Terme, 1989. Editors: L. Cattabriga, L. Rodino. VII, 349 pages. 1991.

Vol. 1496: C. Foias, B. Francis, J. W. Helton, H. Kwakernaak, J. B. Pearson, H_∞-Control Theory. Como, 1990. Editors: E. Mosca, L. Pandolfi. VII, 336 pages. 1991.

Vol. 1497: G. T. Herman, A. K. Louis, F. Natterer (Eds.), Mathematical Methods in Tomography. Proceedings 1990. X, 268 pages. 1991.

Vol. 1498: R. Lang, Spectral Theory of Random Schrödinger Operators. X, 125 pages. 1991.

Vol. 1499: K. Taira, Boundary Value Problems and Markov Processes. IX, 132 pages. 1991.

Vol. 1500: J.-P. Serre, Lie Algebras and Lie Groups. VII, 168 pages. 1992.

Vol. 1501: A. De Masi, E. Presutti, Mathematical Methods for Hydrodynamic Limits. IX, 196 pages. 1991.

Vol. 1502: C. Simpson, Asymptotic Behavior of Monodromy. V, 139 pages. 1991.

Vol. 1503: S. Shokranian, The Selberg-Arthur Trace Formula (Lectures by J. Arthur). VII, 97 pages. 1991.

Vol. 1504: J. Cheeger, M. Gromov, C. Okonek, P. Pansu, Geometric Topology: Recent Developments. Editors: P. de Bartolomeis, F. Tricerri. VII, 197 pages. 1991.

Vol. 1505: K. Kajitani, T. Nishitani, The Hyperbolic Cauchy Problem. VII, 168 pages. 1991.

Vol. 1506: A. Buium, Differential Algebraic Groups of Finite Dimension. XV, 145 pages. 1992.

Vol. 1507: K. Hulek, T. Peternell, M. Schneider, F.-O. Schreyer (Eds.), Complex Algebraic Varieties. Proceedings, 1990. VII, 179 pages. 1992.

Vol. 1508: M. Vuorinen (Ed.), Quasiconformal Space Mappings. A Collection of Surveys 1960-1990. IX, 148 pages. 1992.

Vol. 1509: J. Aguadé, M. Castellet, F. R. Cohen (Eds.), Algebraic Topology - Homotopy and Group Cohomology. Proceedings, 1990. X, 330 pages. 1992.

Vol. 1510: P. P. Kulish (Ed.), Quantum Groups. Proceedings, 1990. XII, 398 pages. 1992.

Vol. 1511: B. S. Yadav, D. Singh (Eds.), Functional Analysis and Operator Theory. Proceedings, 1990. VIII, 223 pages. 1992.

Vol. 1512: L. M. Adleman, M.-D. A. Huang, Primality Testing and Abelian Varieties Over Finite Fields. VII, 142 pages. 1992.

Vol. 1513: L. S. Block, W. A. Coppel, Dynamics in One Dimension. VIII, 249 pages. 1992.

Vol. 1514: U. Krengel, K. Richter, V. Warstat (Eds.), Ergodic Theory and Related Topics III, Proceedings, 1990. VIII, 236 pages. 1992.

Vol. 1515: E. Ballico, F. Catanese, C. Ciliberto (Eds.), Classification of Irregular Varieties. Proceedings, 1990. VII, 149 pages. 1992.

Vol. 1516: R. A. Lorentz, Multivariate Birkhoff Interpolation. IX, 192 pages. 1992.

Vol. 1517: K. Keimel, W. Roth, Ordered Cones and Approximation. VI, 134 pages. 1992.

Vol. 1518: H. Stichtenoth, M. A. Tsfasman (Eds.), Coding Theory and Algebraic Geometry. Proceedings, 1991. VIII, 223 pages. 1992.

Vol. 1519: M. W. Short, The Primitive Soluble Permutation Groups of Degree less than 256. IX, 145 pages. 1992.

Vol. 1520: Yu. G. Borisovich, Yu. E. Gliklikh (Eds.), Global Analysis – Studies and Applications V. VII, 284 pages. 1992.

Vol. 1521: S. Busenberg, B. Forte, H. K. Kuiken, Mathematical Modelling of Industrial Process. Bari, 1990. Editors: V. Capasso, A. Fasano. VII, 162 pages. 1992.

Vol. 1522: J.-M. Delort, F. B. I. Transformation. VII, 101 pages. 1992.

Vol. 1523: W. Xue, Rings with Morita Duality. X, 168 pages. 1992.

Vol. 1524: M. Coste, L. Mahé, M.-F. Roy (Eds.), Real Algebraic Geometry. Proceedings, 1991. VIII, 418 pages. 1992.

Vol. 1525: C. Casacuberta, M. Castellet (Eds.), Mathematical Research Today and Tomorrow. VII, 112 pages. 1992.

Vol. 1526: J. Azéma, P. A. Meyer, M. Yor (Eds.), Séminaire de Probabilités XXVI. X, 633 pages. 1992.

Vol. 1527: M. I. Freidlin, J.-F. Le Gall, Ecole d'Eté de Probabilités de Saint-Flour XX – 1990. Editor: P. L. Hennequin. VIII, 244 pages. 1992.

Vol. 1528: G. Isac, Complementarity Problems. VI, 297 pages. 1992.

Vol. 1529: J. van Neerven, The Adjoint of a Semigroup of Linear Operators. X, 195 pages. 1992.

Vol. 1530: J. G. Heywood, K. Masuda, R. Rautmann, S. A. Solonnikov (Eds.), The Navier-Stokes Equations II – Theory and Numerical Methods. IX, 322 pages. 1992.

Vol. 1531: M. Stoer, Design of Survivable Networks. IV, 206 pages. 1992.

Vol. 1532: J. F. Colombeau, Multiplication of Distributions. X, 184 pages. 1992.

Vol. 1533: P. Jipsen, H. Rose, Varieties of Lattices. X, 162 pages. 1992.

Vol. 1534: C. Greither, Cyclic Galois Extensions of Commutative Rings. X, 145 pages. 1992.

Vol. 1535: A. B. Evans, Orthomorphism Graphs of Groups. VIII, 114 pages. 1992.

Vol. 1536: M. K. Kwong, A. Zettl, Norm Inequalities for Derivatives and Differences. VII, 150 pages. 1992.

Vol. 1537: P. Fitzpatrick, M. Martelli, J. Mawhin, R. Nussbaum, Topological Methods for Ordinary Differential Equations. Montecatini Terme, 1991. Editors: M. Furi, P. Zecca. VII, 218 pages. 1993.

Vol. 1538: P.-A. Meyer, Quantum Probability for Probabilists. X, 287 pages. 1993.

Vol. 1539: M. Coornaert, A. Papadopoulos, Symbolic Dynamics and Hyperbolic Groups. VIII, 138 pages. 1993.

Vol. 1540: H. Komatsu (Ed.), Functional Analysis and Related Topics, 1991. Proceedings. XXI, 413 pages. 1993.

Vol. 1541: D. A. Dawson, B. Maisonneuve, J. Spencer, Ecole d' Eté de Probabilités de Saint-Flour XXI - 1991. Editor: P. L. Hennequin. VIII, 356 pages. 1993.

Vol. 1542: J.Fröhlich, Th.Kerler, Quantum Groups, Quantum Categories and Quantum Field Theory. VII, 431 pages. 1993.

Vol. 1543: A. L. Dontchev, T. Zolezzi, Well-Posed Optimization Problems. XII, 421 pages. 1993.

Vol. 1544: M.Schürmann, White Noise on Bialgebras. VII, 146 pages. 1993.

Vol. 1545: J. Morgan, K. O'Grady, Differential Topology of Complex Surfaces. VIII, 224 pages. 1993.

Vol. 1546: V. V. Kalashnikov, V. M. Zolotarev (Eds.), Stability Problems for Stochastic Models. Proceedings, 1991. VIII, 229 pages. 1993.

Vol. 1547: P. Harmand, D. Werner, W. Werner, M-ideals in Banach Spaces and Banach Algebras. VIII, 387 pages. 1993.

Vol. 1548: T. Urabe, Dynkin Graphs and Quadrilateral Singularities. VI, 233 pages. 1993.

Vol. 1549: G. Vainikko, Multidimensional Weakly Singular Integral Equations. XI, 159 pages. 1993.

Vol. 1550: A. A. Gonchar, E. B. Saff (Eds.), Methods of Approximation Theory in Complex Analysis and Mathematical Physics IV, 222 pages, 1993.

Vol. 1551: L. Arkeryd, P. L. Lions, P.A. Markowich, S.R. S. Varadhan. Nonequilibrium Problems in Many-Particle Systems. Montecatini, 1992. Editors: C. Cercignani, M. Pulvirenti. VII, 158 pages 1993.

Vol. 1552: J. Hilgert, K.-H. Neeb, Lie Semigroups and their Applications. XII, 315 pages. 1993.

Vol. 1553: J.-L- Colliot-Thélène, J. Kato, P. Vojta. Arithmetic Algebraic Geometry. Trento, 1991. Editor: E. Ballico. VII, 223 pages. 1993.

Vol. 1554: A. K. Lenstra, H. W. Lenstra, Jr. (Eds.), The Development of the Number Field Sieve. VIII, 131 pages. 1993.

Vol. 1555: O. Liess, Conical Refraction and Higher Microlocalization. X, 389 pages. 1993.

Vol. 1556: S. B. Kuksin, Nearly Integrable Infinite-Dimensional Hamiltonian Systems. XXVII, 101 pages. 1993.

Vol. 1557: J. Azéma, P. A. Meyer, M. Yor (Eds.), Séminaire de Probabilités XXVII. VI, 327 pages. 1993.

Vol. 1558: T. J. Bridges, J. E. Furter, Singularity Theory and Equivariant Symplectic Maps. VI, 226 pages. 1993.

Vol. 1559: V. G. Sprindžuk, Classical Diophantine Equations. XII, 228 pages. 1993.

Vol. 1560: T. Bartsch, Topological Methods for Variational Problems with Symmetries. X, 152 pages. 1993.

Vol. 1561: I. S. Molchanov, Limit Theorems for Unions of Random Closed Sets. X, 157 pages. 1993.

Vol. 1562: G. Harder, Eisensteinkohomologie und die Konstruktion gemischter Motive. XX, 184 pages. 1993.

Vol. 1563: E. Fabes, M. Fukushima, L. Gross, C. Kenig, M. Röckner, D. W. Stroock, Dirichlet Forms. Varenna, 1992. Editors: G. Dell'Antonio, U. Mosco. VII, 245 pages. 1993.

Vol. 1564: J. Jorgenson, S. Lang, Basic Analysis of Regularized Series and Products. IX, 122 pages. 1993.

Vol. 1565: L. Boutet de Monvel, C. De Concini, C. Procesi, P. Schapira, M. Vergne. D-modules, Representation Theory, and Quantum Groups. Venezia, 1992. Editors: G. Zampieri, A. D'Agnolo. VII, 217 pages. 1993.